PHYSICOCHEMICAL INTERACTIONS OF ENGINEERED NANOPARTICLES AND PLANTS

Nanomaterial-Plant Interactions

PHYSICOCHEMICAL INTERACTIONS OF ENGINEERED NANOPARTICLES AND PLANTS

A Systemic Approach

Edited by

GUADALUPE DE LA ROSA
Universidad de Guanajuato, Guanajuato, Mexico

JOSE R. PERALTA-VIDEA
Department of Chemistry and Biochemistry, The University of Texas at El Paso, United States

ELSEVIER

ACADEMIC PRESS
An imprint of Elsevier

Academic Press is an imprint of Elsevier
125 London Wall, London EC2Y 5AS, United Kingdom
525 B Street, Suite 1650, San Diego, CA 92101, United States
50 Hampshire Street, 5th Floor, Cambridge, MA 02139, United States
The Boulevard, Langford Lane, Kidlington, Oxford OX5 1GB, United Kingdom

Notices

Knowledge and best practice in this field are constantly changing. As new research and experience broaden our understanding, changes in research methods, professional practices, or medical treatment may become necessary.

Practitioners and researchers must always rely on their own experience and knowledge in evaluating and using any information, methods, compounds, or experiments described herein. In using such information or methods they should be mindful of their own safety and the safety of others, including parties for whom they have a professional responsibility.

To the fullest extent of the law, neither the Publisher nor the authors, contributors, or editors, assume any liability for any injury and/or damage to persons or property as a matter of products liability, negligence or otherwise, or from any use or operation of any methods, products, instructions, or ideas contained in the material herein.

ISBN: 978-0-323-90558-9

Publisher: Nikki P. Levy
Acquisitions Editor: Nancy J. Maragioglio
Editorial Project Manager: Catherine Costello
Production Project Manager: Swapna Srinivasan
Cover Designer: Mark Rogers

Typeset by TNQ Technologies

Contents

Section II

Biogeochemical cycle of ENMs before plant exposure explained at the molecular level

Section III

Molecular interaction of ENMs and plants: Reactions and mechanisms

7. Biophysicochemical transformation of ENMs at root level

Alma Hortensia Serafin-Muñoz and Berenice Noriega-Luna

8. Biointeractions of plants—microbes—engineered nanomaterials

Ileana Vera-Reyes, Edgar Vázquez-Núñez, Laura E. Castellano, Diana Ivette Aparicio Bautista, José Humberto Valenzuela Soto and Jessica Denisse Valle-García

9. Chemical transformation and mechanisms of ENMs transport in plants

Guadalupe de la Rosa-Alvarez and Jose R. Peralta-Videa

List of contributors

Rene Loredo-Portales Departamento de Ciencias Ambientales, División de Ciencias de la Vida, Universidad de Guanajuato, Irapuato, Guanajuato, México

Mataz Alcoultabi Department of Mechanical Engineering, University of Texas Rio Grande Valley, TX, United States

Angel Josabad Alonso-Castro Departamento de Farmacia, División de Ciencias Naturales y Exactas, Universidad de Guanajuato, Guanajuato, Guanajuato, México

Diana Ivette Aparicio Bautista Laboratorio de Estructura de Proteínas, Instituto Nacional de Medicina Genómica, INMEGEN, Ciudad de México, México

Jaya Borgatta Corteva Agriscience, Indianapolis, IN, United States

Candy Carranza-Álvarez Facultad de Estudios Profesionales Zona Huasteca, Universidad Autónoma de San Luis Potosí, San Luis Potosí, México

Laura E. Castellano Universidad de Guanajuato, Departamento de Ingenierías Química, Electrónica y Biomédica, División de Ciencias e Ingenierías, Campus León, León, Guanajuato, Mexico

Luis J. Castillo-Pérez Programa Multidisciplinario de Posgrado en Ciencias Ambientales, Universidad Autónoma de San Luis Potosí, San Luis Potosí, San Luis Potosí, México

Edison Castro Department of Chemistry, University of Pittsburgh, Pittsburgh, PA, United States

R.K. Dearth Department of Biology, University of Texas Rio Grande Valley, TX, United States

Guadalupe de la Rosa-Alvarez Department of Chemical, Electrical, and Biomedical Engineering, Universidad de Guanajuato, León, Guanajuato, Mexico

Fabián Fernández-Luqueño Sustainability of Natural Resources and Energy Program, Cinvestav, Saltillo, Coahuila, Mexico

Yulu Ge The University of Texas at El Paso, Department of Chemistry and Biochemistry, El Paso, TX, United States; Nanosystems Engineering Research Center for Nanotechnology-Enabled Water Treatment (NEWT), United States

Tania M. Guardado-Alvarez The Connecticut Agricultural Experiment Station, New Haven, CT, United States

Yuxiong Huang Tsinghua-Berkeley Shenzhen Institute (TBSI), Tsinghua Shenzhen International Graduate School (SIGS), Tsinghua University, Shenzhen, China

Rupesh Kariyat Department of Biology, University of Texas Rio Grande Valley, TX, United States

E. Kotsikorou Department of Chemistry, University of Texas Rio Grande Valley, TX, United States

Mariana Marcos-Hernandez The University of Texas at El Paso, Department of Chemistry and Biochemistry, El Paso, TX, United States; Nanosystems Engineering Research Center for Nanotechnology-Enabled Water Treatment (NEWT), United States

Helia M. Morales Department of Chemistry, University of Texas Rio Grande Valley, TX, United States; Escuela de Ingenierıa y Ciencias, Tecnologico de Monterrey, Monterrey, NL, Mexico

Berenice Noriega-Luna Universidad de Guanajuato, Division de Ingenierias, Campus Guanajuato, Guanajuato, Mexico

J.G. Parsons Department of Chemistry, University of Texas Rio Grande Valley, TX, United States

Jose R. Peralta-Videa Department of Chemistry and Biochemistry, The University of Texas at El Paso, El Paso, TX, United States

Hermes Pérez-Hernandez Universidad Autónoma de Chiapas, Facultad de Ciencias Agrícolas, Huehuetán, Chiapas, México

Alma Hortensia Serafin-Muñoz Universidad de Guanajuato, Division de Ingenierias, Campus Guanajuato, Guanajuato, Mexico

Zoe C. Simon Department of Chemistry, University of Pittsburgh, Pittsburgh, PA, United States

Carlos Tamez The Connecticut Agricultural Experiment Station, New Haven, CT, United States

José Humberto Valenzuela Soto Departamento de Biociencias y Agrotecnología, CONACYT-Centro de Investigación en Química Aplicada, Saltillo, Coah, Mexico

Jessica Denisse Valle-García Programa de Doctorado Transdisciplinario en Desarrollo Científico y Tecnológico para la Sociedad, Cinvestav Unidad Zacatenco, Ciudad de México, Mexico

Edgar Vázquez-Núñez Universidad de Guanajuato, Departamento de Ingenierías Química, Electrónica y Biomédica, División de Ciencias e Ingenierías, Campus León, León, Guanajuato, Mexico

Ileana Vera-Reyes Departamento de Biociencias y Agrotecnología, CONACYT-Centro de Investigación en Química Aplicada, Saltillo, Coah, Mexico

Dino Villagran The University of Texas at El Paso, Department of Chemistry and Biochemistry, El Paso, TX, United States; Nanosystems Engineering Research Center for Nanotechnology-Enabled Water Treatment (NEWT), United States

Jason C. White The Connecticut Agricultural Experiment Station, New Haven, CT, United States

Tao Xu Center for Nanotechnology and Nanotoxicology, Harvard T. H. Chan School of Public Health, Harvard University, Boston, MA, United States

Sheng Yin The University of Texas at El Paso, Department of Chemistry and Biochemistry, El Paso, TX, United States; Nanosystems Engineering Research Center for Nanotechnology-Enabled Water Treatment (NEWT), United States

Yong Zhao W. M. Keck Center for Advanced Microscopy & Microanalysis, University of Delaware, Newark, DE, United States

Fan Zheng Composites and Polymer Engineering Laboratory, South Dakota School of Mines & Technology, Rapid City, SD, United States

Yanjie Zhu Tsinghua-Berkeley Shenzhen Institute (TBSI), Tsinghua Shenzhen International Graduate School (SIGS), Tsinghua University, Shenzhen, China

Nubia Zuverza-Mena The Connecticut Agricultural Experiment Station, New Haven, CT, United States

SECTION I

Introduction

CHAPTER

1

Understanding the interactions of engineered nanomaterials and plants: A systemic perspective

Jose R. Peralta-Videa[1], *and*
Guadalupe de la Rosa-Alvarez[2]

[1] **Department of Chemistry and Biochemistry, The University of Texas at El Paso, El Paso, TX, United States;** [2] **Department of Chemical, Electrical, and Biomedical Engineering, Universidad de Guanajuato, León, Guanajuato, Mexico**

1.1 Introduction

Realistic quantifications of Engineered Nanomaterials (ENMs) in the environment are difficult to achieve. For this reason, several research groups have proposed different models to estimate potential concentrations of ENMs in various media (Gottschalk et al., 2013). This information is important, given the known and unknown toxicological properties of these materials and further potential negative effects on living organisms and the environment.

ENMs reach plant ecosystems through intentional or unintentional pathways. In any case, after release, they may be transformed in the environment by physical, chemical, and biochemical processes. Once in contact with plant systems, biotransformation may still occur, affecting or stimulating plant metabolism. Since plants are the producers of the food chain, it is of paramount importance to understand these mechanisms at the molecular level.

A significant amount of research has determined the effects of ENMs on plants and plant ecosystems. However, relatively few data are available on the molecular origin of such effects. Without this knowledge, future modifications to the structure and composition of ENMs that

Physicochemical Interactions of Engineered Nanoparticles and Plants
https://doi.org/10.1016/B978-0-323-90558-9.00007-3

provide efficiency and specificity will be extremely difficult to achieve. On the other hand, data about molecular interactions will allow researchers and industry to develop new ENMs with proper structure and composition for specific functions. The purpose of this book is to provide a general context of ENMs in the environment, potential interactions and potential transformations, as well as a discussion about known and potential effects of ENMs on plants and plant ecosystems. The importance of an integrative approach is also considered.

1.2 Metallic engineered nanomaterials

1.2.1 Description

According to the American Society for Testing and Materials (ASTM), "nanoparticle (NP) is a sub-classification of ultrafine particle with lengths in two or three dimensions greater than 0.001 μm (1 nm) and smaller than about 0.1 μm (100 nm) and which may or may not exhibit a size-related intensive property" (American Society for Testing and Materials, 2012). Often, it is found that the word nanomaterial is used as a synonym for nanoparticle. However, there are some important differences. According to the European Union, a nanomaterial contains at least 50% of the particles with dimensions between 1 and 100 nm (European Commission Recommendation of October 18, 2011 on the Definition of Nanomaterial, 2011).

Several publications have indicated that the preparation methods of ENMs include top down, bottom up, and metal reduction (Christian et al., 2008; "European Commission Recommendation of October 18, 2011 on the Definition of Nanomaterial," 2011; Peralta-Videa et al., 2016; Stankic et al., 2016). In the top-down methods, NPs are formed by size reduction of large portions of a given material until ultrafine pieces (particulate matter, PM < 0.1 μm) remain. PM < 0.1 μm is also called NP (Stankic et al., 2016). Christian et al. (2008) stated that in top-down methodologies, the size reduction can be achieved by lithographic techniques or using ball mills. In the bottom-up procedures, small units (atoms or molecules) are amassed to obtain larger structures (Peralta-Videa et al., 2016). Fig. 1.1 shows a schematic of the preparation methods, elements used in each method, and the most common types of nanoparticles formed (Peralta-Videa et al., 2016).

Regardless of the method selected for ENMs production, those of the metal(loid)-based type display specific physical and chemical properties that differentiate them from bulk counterparts. Many properties are related to the application. For instance, NPs with magnetic properties are useful in biomedicine (Giner-Casares et al., 2016), as well as to study NP

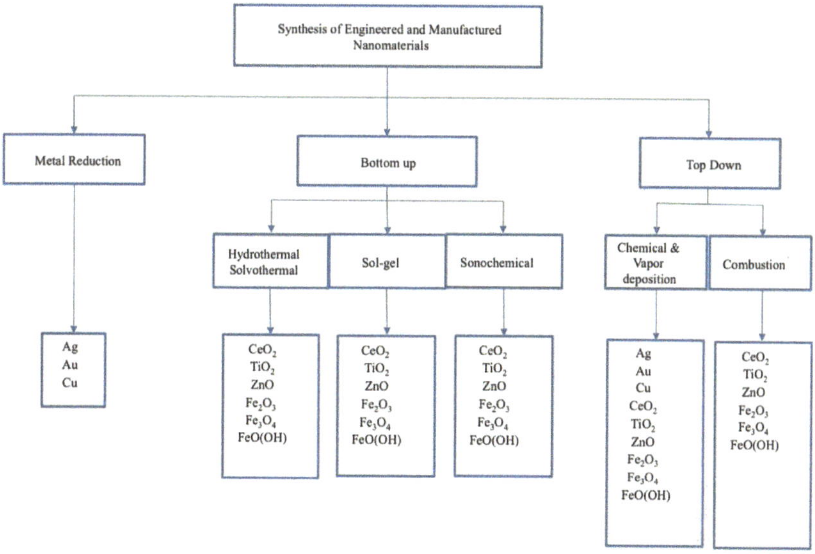

FIGURE 1.1 Traditional methods for the synthesis of engineered metal-based nanoparticles. *Reprinted with permission from Peralta-Videa, J. R., Huang, Y., Parsons, J. G., Zhao, L., Lopez-Moreno, L., Hernandez-Viezcas, J. A., & Gardea-Torresdey, J. L. (2016). Plant-based green synthesis of metallic nanoparticles: Scientific curiosity or a realistic alternative to chemical synthesis?* Nanotechnology for Environmental Engineering, *1(1). Copyright (2016) Springer Nature.*

properties	technique	unit	TiO_2 Evonik 4168063098	CeO_2 Meliorum 121008	ZnO Meliorum 121008
primary size	TEM[a]	nm	27 ± 4	rods: (67 ± 8) × (8 ± 1) (≤10% polyhedra: 8 ± 1 nm)	24 ± 3
particle size in DI water	DLS[a]	nm	194 ± 7	231 ± 16	205 ± 14
phase and structure	XRD[a]		82% anatase and 18% rutile	100% ceria cubic	100% zincite hexagonal
shape/morphology	TEM[a]		semispherical	rods (≤10% Polyhedra)	spheroid
surface area	BET[2]	$m^2\,g^{-1}$	51.5	93.8	42.1
IEP	zetaPALS[a]		6.2	7.5	9.2
EPM in 1 mM KCl	zetaPALS[a]	$10^{-8}\,m^2\,V^{-1}\,s^{-1}$	2.37 ± 0.06	2.19 ± 0.04	1.83 ± 0.11
purity	TGA[a]	wt.%	98.03	95.14	97.27
moisture content	TGA[a]	wt.%	1.97	4.01	1.61

[a] Transmission and scanning electron microscopy (TEM), dynamic light scattering (DLS), X-ray powder diffraction (XRD), isoelectric point (IEP), electrophoretic mobility (EPM), and thermogravimetric analysis (TGA) were done by the UC-CEIN at UCLA. [2] Brunauer–Emmett–Teller analysis (BET) was conducted by Dr. Ponisseril Somasundaran's lab at Columbia University.

FIGURE 1.2 Physicochemical characteristics of metal oxide nanoparticles. *Reprinted with permission from Keller, A. A., Wang, H., Zhou, D., Lenihan, H. S., Cherr, G., Cardinale, B. J., Miller, R., & Zhaoxia, J. I. (2010). Stability and aggregation of metal oxide nanoparticles in natural aqueous matrices.* Environmental Science and Technology, *44(6), 1962–1967. Copyright (2010) American Chemical Society.*

uptake and translocation in plants (Wang et al., 2011). Keller et al. (2010) described the physicochemical properties of three of the most commercially produced metal oxide NPs: TiO_2, ZnO, and CeO_2 (Fig. 1.2). This figure shows the properties, measurement techniques, and units. It is important to point out that several reports about the effects of in-house fabricated nanomaterials on living organisms provide incomplete NPs characterization. On the other hand, researchers often manufacture small quantities of materials whose properties would be challenging to replicate in large-scale production. For this reason, it is difficult to generalize about the effects of specific NPs on particular plant species.

1.2.2 Production

To the best of our knowledge, no recent data on global production of ENMs are available; however, the latest published numbers are extremely useful. Keller et al. (2013) presented the world production as of 2010 (Table 1.1). As data show, SiO_2 ENPs displayed the highest production rate. However, the proportion of SiO_2 that reaches bodies of water was the lowest (2%), followed by CeO_2 (3%), and Cu and Cu oxides (6%). On the other hand, the highest proportion corresponded to TiO_2 NPs, with 18%

TABLE 1.1 Estimated global production and disposal of ENMs as of 2010 assuming maximum production.

ENM	Production	Disposal				Ratio
	Metric tons/ year (estimate)[a]	Landfill	Soil	Air	Water	Water disposal/ Production
SiO_2	99,500	81,200	10,600	1100	2100	2%
TiO_2	88,000	32,600	38,200	1600	15,600	18%
Fe & Fe oxides	42,000	25,700	11,300	600	4300	10%
Al_2O_3	35,000	23,600	8200	500	2600	11%
ZnO	34,000	21,100	8700	600	3700	11%
CeO_2	10,000	8200	1400	100	300	3%
Ag	452	200	150	11	63	14%
Cu & Cu oxides	200	150	36	3	11	6%

[a]*Differences between production and disposal quantities are due to the fact that disposal amounts are the summation of numerous factors.*

Data was extracted From Keller, A. A., McFerran, S., Lazareva, A. & Suh, S. (2013). Global life cycle releases of engineered nanomaterials. Journal of Nanoparticle Research, 15*(6). https://doi.org/10.1007/s11051-013-1692-4.*

reaching aquatic systems. Considering the environmental stability of TiO_2, these ENPs have higher possibilities of impacting living organisms.

The number of nano-enabled products is huge and increases day-by-day. As of November 2020, the Nanodatabase (www.nanodb.dk) reported the existence of 5000 nanoenabled commercial products, 62% (more than 3000) aimed for health and fitness, 17% for home and gardening, and 11% for the automotive industry. The remaining 10% is used in electronics and computers, food and beverages, appliances, and good products (Hansen et al., 2020). The database includes seven pesticide products, but unfortunately the nanomaterial component is not described (https://nanodb.dk/en/search-database/?keyword=pesticide).

1.2.3 Classification

Metal-based ENMs can be classified in several ways including form, composition, dimensions, and state (Hansen et al., 2020; Jeevanandam et al., 2018; Pokropivny & Skorokhod, 2007; Saleh, 2020; Saleh & Gupta, 2016) (Fig. 1.3). The organization based on dimension was proposed by (Pokropivny & Skorokhod, 2007). These researchers presented a large number of examples for each dimension. As expected, some items can be

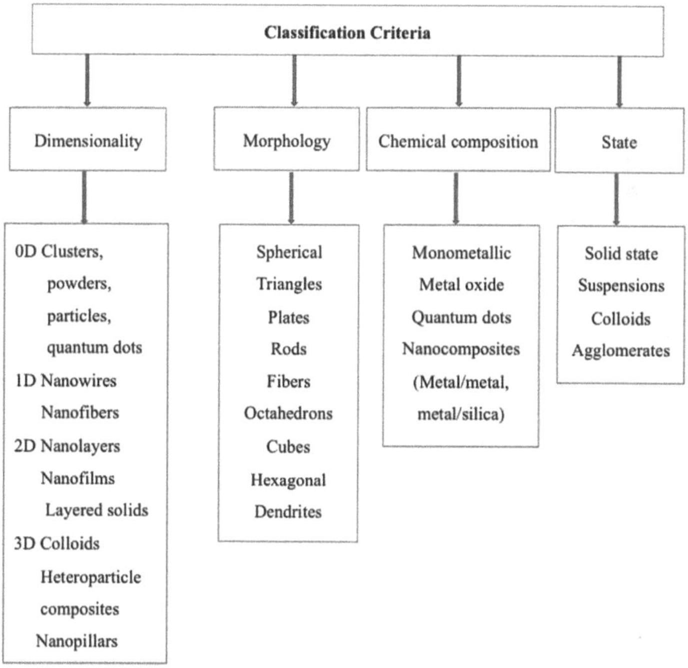

FIGURE 1.3 Classification of the nanomaterials.

found in more than one category. The "chemical composition" category is the most studied for agricultural purposes. A number of experiments have compared the effects of monometallic and bimetallic ENMs on plants. Data have shown that, in the case of copper, metallic ENPs are more toxic than metal oxide ENPs (Hong et al., 2015; Keller et al., 2017). Additionally, the hydroxide form [nano-$Cu(OH)_2$] is more toxic than the metal oxide [nano-CuO] (Keller et al., 2017). Review articles describe the toxicity of many ENMs in plants (Gomez et al., 2021; Keller et al., 2017; Tarrahi et al., 2021); however, data from field applications are not yet available.

1.2.4 Current and future uses of ENMs

The variety of ENM applications increases every day. Several metal(loid)-based ENMs have shown interesting potential applications for agricultural purposes. This has opened the door for a new field of study called "agricultural nanotechnology." The scientific production related to this topic is ever increasing. A search in ScienceDirect.com using "agricultural nanotechnology" as keywords showed 29 papers for 2021 cataloged as "advances in Nano-fertilizers and Nano-agriculture," all of them as book chapters. https://www.sciencedirect.com/search?qs=agricultural%20nanotechnology&years=2021&publicationTitles=779335&offset=25. In addition, several review articles are focused on the application of ENMs in agriculture. For instance, Shafiq et al. (2020) summarized the application of ENMs in the food industry. These researchers described the application of nanosensors, nanoemulsions, nanopesticides, or nanocapsules for the production or preservation of food products. It has been stated that ENMs may contribute to food security by making agriculture more efficient through the smart delivery of nutrients, improvement of plant growth and production through seed priming, as well as the use of nanopesticides and nanoherbicides (White & Gardea-Torresdey, 2018). Ashraf et al. (2021) published a review about the use of TiO_2, Ag, Fe_2O_3, ZnO, and Cu/CuO NPs at several stages of food production. For instance, Ag NPs have been assayed for bacterial diseases control, Fe_2O_3 and ZnO NPs as fertilizers, and Cu/CuO as pesticides. Carboxylic acid modified mesoporous silica NPs have been used to deliver conventional herbicides such as 2,4-D (2,4-dichlorophenoxyacetic acid) and picloram (Prado et al., 2011). Adisa et al. (2019) and Duhan et al. (2017) published comprehensive reviews about the current state and future perspectives of the use of ENMs in agriculture. On the other hand, Gomez et al. (2021) also described ENMs with potential use in agriculture. However, these authors also mentioned that ENMs have the potential to change food quality. For instance,

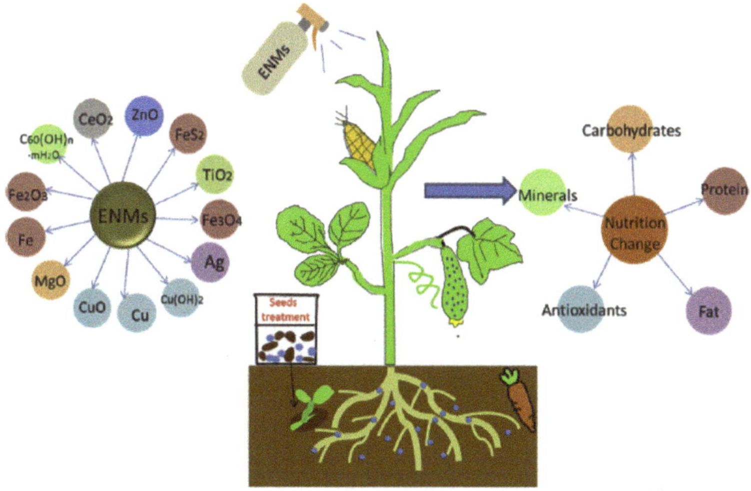

FIGURE 1.4 Materials used as nanofertilizers and observed changes in plants. *Reprinted with permission from Gomez, A., Narayan, M., Zhao, L., Jia, X., Bernal, R. A., Lopez-Moreno, M. L., & Peralta-Videa, J. R. (2021). Effects of nano-enabled agricultural strategies on food quality: Current knowledge and future research needs.* Journal of Hazardous Materials, *401. Copyright (2021) Elsevier.*

changes to the ionome (increase and decrease of essential elements), maturation time (reduced or increased flowering time), sugar, and antioxidants, were reported. Some changes are beneficial, but others are in need of risk assessment (Fig. 1.4). Unfortunately, all the information available in the literature belong to experimental results. To the best of authors' knowledge, no data on the commercial use of ENMs are available.

It is well known that traditional agricultural practices have resulted in contamination of soils and water bodies. Probably the most studied agricultural-derived contaminants are pesticide residues. Worldwide, we are dietarily exposed to several pesticides including carbofuran, captan, diazinon, diquat, malathion, and phorate, among others (Crépet et al., 2021). Nehra et al. (2021) presented an extensive list of pesticides found in fruits, honey, and wheat flour including captan, diazinon, and malathion. Another problem with pesticide residues is the long residence time in soil systems. ENMs may provide a solution to this problem. It has been proven that some ENMs are capable of removing pesticide residues from agricultural soils. For instance, pristine or surface-modified ENMs such as magnetic iron, TiO_2, and ZnO ENMs degraded several pesticides under experimental conditions. Up to 100% of Thiram residues were degraded with ZnO ENPs using a 500 W xenon lamp. N-doped TiO_2 nanosheets were able to degrade 88% of diazinon after 88 min under simulated solar irradiation (Nehra et al., 2021). However, all available information was obtained from experiments at laboratory scale. Unfortunately, no field studies have been reported.

1.2.5 How ENMs enter in contact with plants?

ENMs enter in contact with plants through different routes including direct application, accidental release, or interaction with contaminated soil and/or water. Apart from direct application, probably the highest plant exposure to ENMs results from irrigation practices with reclaimed wastewater. For example, small amounts of silver nanoparticles (AgNPs) persist in the reclaimed water in a water treatment plant of Santa Barbara, CA, US (Cervantes-Avilés et al., 2019). In California, recycled water is frequently used for irrigation (crops and landscape) and also for groundwater recharge (Schulte, 2010). This suggests that many food plants are potentially exposed to AgNPs. One of the most stable ENMs are TiO_2 NPs, which are contained in sunscreen and are released into water bodies during bathing activities. These ENPs are deposited in sediments and eventually resuspended (Cornelis et al., 2021). In the event that TiO_2 impacted water is used for crop irrigation, plants will be accidentally exposed to those ENMs.

Knowledge about the levels of plants exposure to ENMs on a commercial scale is not available. There are several commercial products such as nano Max NPK fertilizer, Nano Ag-Answer, TAG NANO (NKP, PhoS, and Zinc, among others) fertilizers. However, these products include numerous components and differ from traditional fertilizers in various manners (Prasad et al., 2017). The commercial use of ENMs in agriculture is limited by different factors including the lack of knowledge about relevant concentrations, no data on risk assessment, or the uncertainty about specific ENMs that should be applied to particular soils (Gomez et al., 2021). Further missing data include the collateral effects, or the best delivery system for specific products such as nanoherbicides.

1.3 Principles of plant physiology

The objective of this section is to briefly describe some plant physiology principles in order to understand different reactions that occur when plants enter in contact with ENMs. Specifically, a summary of seed germination process and plant growth and development is presented.

1.3.1 Seed germination

Typical seeds of terrestrial plants (angiosperm) contain embryos that develop into plants, a protecting coat called testa, the cotyledon(s) (embryonic leaf), the endosperm (mainly in monocots), the micropyle, and the hilum. Cotyledons and endosperm are sources of food for the embryo prior to germination. The hilum is a scratch in the seed coat that represents the attachment of the ovule to the ovary wall, while the micropyle is a small pore that serves as a path for water entry previous to seed germination. Seeds start the germination process when they are exposed to environmental changes, mainly moisture. The other environmental components for seed germination are oxygen and favorable temperature (Delouche, 1979). During the resumption of the embryonic axis growth, energy and essential molecules are produced by the hydrolysis of carbohydrates, proteins, and lipids, which are present in cotyledons and endosperm. Seed hydration activates proteolytic enzymes that hydrolyze proteins releasing amino acids, amylase that hydrolyzes starch, lipase hydrolyzes lipids, glucanase hydrolyzes glucan (a cover that delays radicle emergency), and phytase hydrolyzes phytin liberating cations, phosphate, and inositol, all of them needed in the growth process of the new plant (Joshi, 2018).

Many papers have described the effects of ENMs on seed germination. In this context, seed priming is the process where seeds are exposed to ENMs before germination. Experiments with CeO_2, MoS_2, FeS_2, and Ag NPs have demonstrated that seed priming improves plant growth and yield (Das et al., 2016; Srivastava et al., 2014). Mahakham et al. (2017) reported that rice seeds nanoprimed with 5 and 10 mg/L Ag NPs increased amylase activity, maintaining starch hydrolysis and upregulating aquaporin genes, which resulted in seedling growth improvement.

1.3.2 Plant growth and development

Growth and development of terrestrial plants depend on three processes: photosynthesis, respiration, and transpiration. Photosynthesis is the production of carbohydrates by plant leaves (and to a lesser extent by green stems) through the use of sunlight, carbon dioxide, and water. The well-known photosynthesis formula is displayed in Eq. (1.1):

$$6\,CO_2(\text{carbon dioxide}) + 6H_2O\ (\text{water}) + \text{sunlight} \rightarrow C_6H_{12}O_6(\text{sugar}) + 6O_2 \tag{1.1}$$

It is important to point out that artificial light (LED) can be used by plants as source of energy to perform photosynthesis, although not as efficient as sunlight.

Photosynthesis takes place in small organelles called chloroplasts, which contain chlorophyll, a pigment that absorbs the light that provides the energy needed to split water molecules. This part of the process releases oxygen atoms to the environment through stomata, while hydrogen atoms take part of the proton gradient in internal membranes of the chloroplast. The other component of photosynthesis is carbon dioxide (CO_2). This gas enters the leaves through the stomata, which are small pores located in the abaxial and adaxial parts of leaves and also in green stems. During photosynthesis, carbon dioxide and water molecules are split, and their respective atoms are reorganized and converted to carbohydrates. During the respiration process, plants transform carbohydrates into CO_2, water, and energy through different oxidation reactions (Eq. 1.2):

$$C_6H_{12}O_6(\text{sugar}) + 6O_2(\text{oxygen}) \rightarrow 6CO_2\ (\text{carbon dioxyde}) + 6H_2O\ (\text{water}) + \text{energy (ATP)} \tag{1.2}$$

Plants also store carbohydrates and are able to form other compounds including oils and proteins. These molecules are moved basipetally from the leaves to the rest of the plant body through the phloem, part of the transport system that is used to move the processed sap.

Plants absorb water from the growth media through the roots. This water is lost in the form of vapor through transpiration, other basic process for plant growth. In addition to water, plants absorb molecules and atoms dissolved/suspended in the aqueous media. However, not all of them are needed for growth. Absorption of water is performed through the xylem, the component of the transport system used to move the raw sap. When the water is loaded with ENMs (accidentally or on purpose), they enter the roots through the xylem and are moved acropetally along with water, nutrients, and other molecules. Herein, the transport can be performed passively through the apoplast (intercellular spaces) or actively via symplast (the cell plasma membrane) (Medina-Velo et al., 2017). The transport system includes the Casparian strip that limits the input of water and nutrients into the symplastic pathway. However, as shown in Fig. 1.5, ENMs (i.e., CeO_2 ENPs) are able to penetrate the transport system. It turns out that the Casparian strip is not well developed in the growing regions of the root, such as the root tip or the emission point of secondary roots. Montes et al. (2017) proposed a model for the uptake and transport of ENMs in plants (Fig. 1.6). Many ENMs or their dissolved ions enter the root at emission points of secondary roots affecting plant growth and development. For instance, CuO ENMs have caused root growth reduction. Margenot et al. (2018) exposed 10-day old seedlings of lettuce (*Lactuca sativa*) and carrot to 0.8–798.9 mg Cu/L in hydroponics and identified a dose-dependent reduction in root length for both species. However, the effect was higher in lettuce plants. It is possible that the varietal effects were due to the higher number of secondary roots in the root system of lettuce plants. It is important to note that, besides the number of secondary roots, there are many factors affecting the uptake of ENMs by plants. Medina-Velo et al. (2017) demonstrated that the uptake of Zn by red kidney beans (*Phaseolus vulgaris*) exposed to ZnO ENMs was significantly affected by the presence of soil organic matter. On the other hand, Majumdar et al.(2015) demonstrated that exposure of red kidney bean to CeO_2 NPs in soil with high organic matter content caused more changes in nutrient quality, compared to plants cultivated in soil with

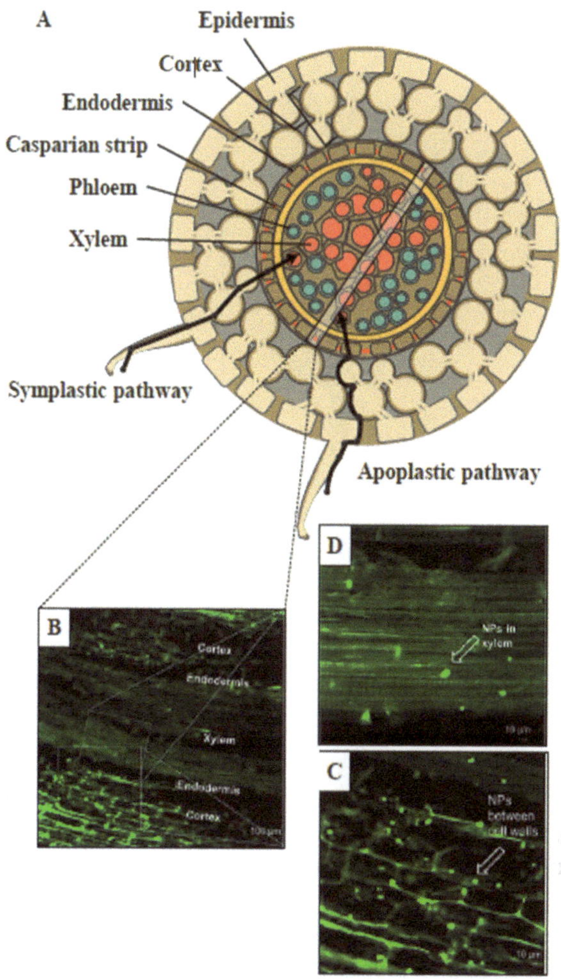

FIGURE 1.5 (A) Diagram of the root structure showing the apoplastic (through the cell walls, nonliving tissues) and symplastic (through the cell plasma membranes, living tissues) pathways for solutes' transport (7); (B) Confocal image of corn root treated with fluorescein isothiocyanate-labeled CeO_2 engineered nanomaterials (ENMs). (C) Aggregates of CeO2 distributed through cell walls in the cortex. (D) CeO_2 ENM aggregates within the xylem. (C–D) Magnifications of the squares in (B) (12) Note: NP, nanoparticle. Numbers 7 and 12 are references cited in Medina-Velo, Peralta-Videa, and Gardea-Torresdey (2017). *Reprinted with permission from Medina-Velo, I. A., Peralta-Videa, J. R., & Gardea-Torresdey, J. L. (2017). Assessing plant uptake and transport mechanisms of engineered nanomaterials from soil.* MRS Bulletin, 42*(5), 379–383. Copyright (2017) Springer Nature.*

lower organic matter content. These findings suggested that the uptake and effects of dissolved/suspended materials depends on the variety and environmental conditions including soil components.

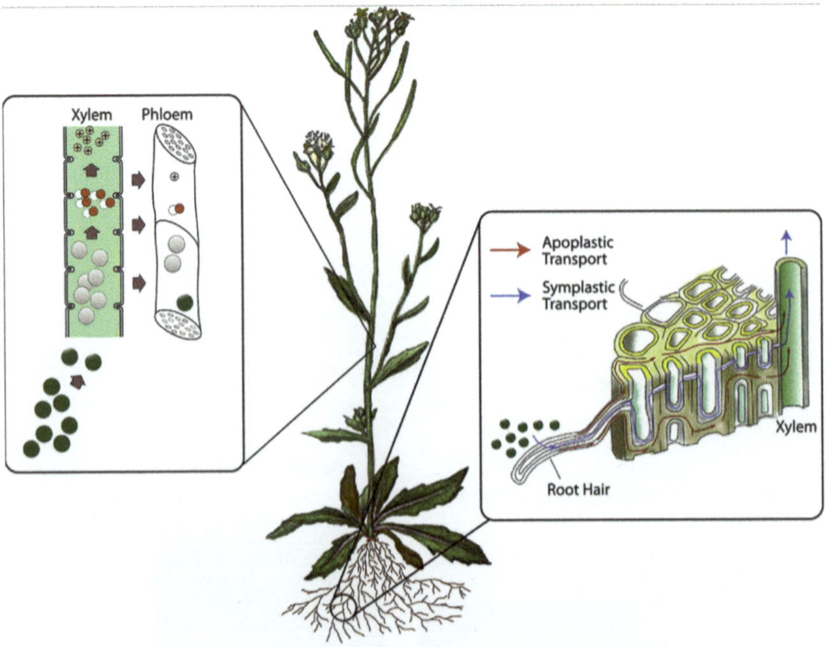

FIGURE 1.6 Model showing pathways for uptake and transport of ENMs in plants. *Reprinted with permission from Montes, A., Bisson, M. A., Gardella, J. A., Jr. & Aga, D. S. (2017). Uptake and transformations of engineered nanomaterials: Critical responses observed in terrestrial plants and the model plant* Arabidopsis thaliana. Science of the Total Environment, *607(608), 1497–1516. Copyright (2017) Elsevier.*

1.4 What does physicochemical interaction mean?

In order to properly understand how the structure and nature of nonliving and living entities are modified by means of their interactions, some fundamental concepts are worth discussing. Physics and chemistry are the basis to explain biochemical and other natural processes where nonliving and living things are involved. We can recall the differences between a physical and a chemical phenomenon (Fig. 1.7). In the first one, the atomic and molecular structure of matter is not modified. However, some changes in properties occur (i.e., density, and solubility, among others). Fig. 1.7A exemplifies the process of solvation where sodium chloride (NaCl, a solid salt at room temperature) is solubilized in water. During solvation, NaCl is separated into ions, which are surrounded by water molecules. After water evaporation, solid NaCl is recovered. A physical phenomenon also occurs without the need of an additional substance. For example, when water is evaporated by means of temperature.

In a chemical phenomenon matter is modified by means of bond breaking and formation (chemical reactions), producing molecules different to the original ones. Fig. 1.7B shows the reaction between molecular hydrogen (H_2) and molecular oxygen (O_2) to produce water. As this figure shows, the original bonds are broken, and new ones are formed. Thus, a physical interaction between ENMs and the surrounding environment involves changes in their physical properties. On the other hand, a chemical interaction changes the nature of the ENMs and those materials interrelating with them. One example is when ZnO NPs are biotransformed after they interact with the plant rhizosphere (De La Rosa et al., 2011).

1.5 Physicochemical interactions of ENMs and plant ecosystems

After ENMs are released, either intentionally or unintentionally, they reach different environmental compartments. Usually, their final

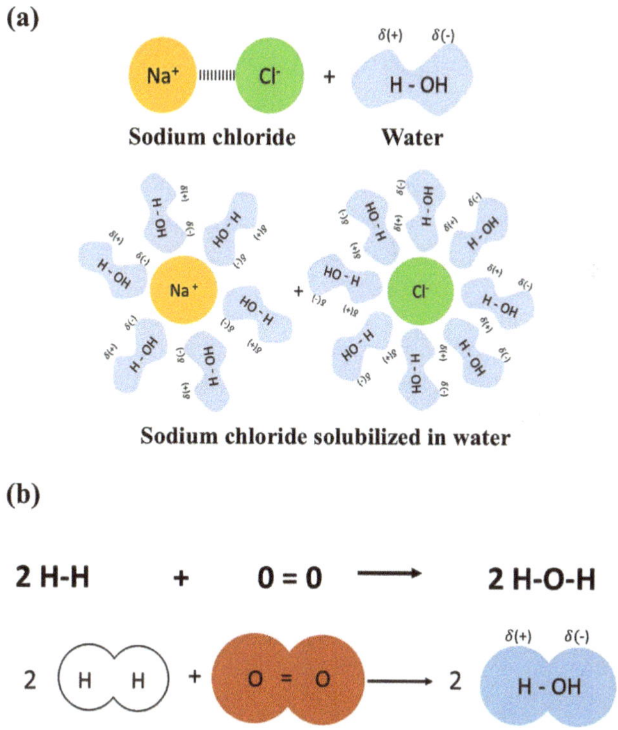

FIGURE 1.7 Schematics of (A) the physical phenomenon of solvation; (B) the chemical reaction between hydrogen and oxygen to produce water.

destinations are soils, sediments, and wastewater treatment plant effluents. Using data from 2010, Keller et al. (2013) evaluated the fate of globally produced ENMs in different environmental compartments. According to their results, most of them (between 63% and 91%) finished their cycle in landfills, while the rest remained in soils, bodies of water, and the atmosphere. In every step of the transport process throughout the environment, NPs interplay in different physical and chemical manners. In the end, we observe the effects of those interactions. Knowledge about those effects on the ecosystems is important for different decision-making purposes. Table 1.2 displays some of the observed effects at the macro, micro, and molecular level.

1.6 Integrative approach toward a beneficial use of ENMs for plants cultivation

It is voiced that by 2050, agricultural production has to increase by about 60% to feed a global population estimated to become 10 billion (White & Gardea-Torresdey, 2018). It is also well known that it will be challenging to achieve such an increase in food needs with current agricultural practices. In recent years, efforts have been directed toward using ENMs in agriculture. Experimental results indicate that some ENMs can be used as nanopesticides and nanofertilizers, with a significant reduction in the inefficiencies of current agricultural practices (White & Gardea-Torresdey, 2018). Researchers reported that in rice farms where a nanofertilizer formed by urea-hydroxyapatite was used, there was a significant increase in grain production with a saving of 50% in urea consumption (Kottegoda et al., 2017). In the case of nanopesticides, Cu-based ENMs had 8% increase in efficiency, compared to the commercial copper hydroxide salt as Kah and Hofmann (2014) have indicated. However, as mentioned by these researchers, there is a lack of information about the mode of action of ENMs, or in the case of precious metals-based ENMs, the application costs at field level would be too high. The lack of knowledge in the reaction mechanisms, mainly of nanopesticides, complicates their application at large scale because no data are available about toxicity to nontarget organisms. Additionally, specific reaction mechanisms of herbicides are missing. Thus, specificity and effectiveness under adverse weather conditions are unknown.

Previous studies have shown the permanence and/or transformation of some ENMs in crop plants (Hernandez-Viezcas et al., 2013; López-Moreno et al., 2010), while other studies have shown the effects on proteins (Majumdar et al., 2015) and genetic material (López-Moreno et al., 2010). However, these are only a few examples of the effects of ENMs in plants. More studies are needed in order to have a more complete picture of the biotransformation products and the omic modifications. Thus, in order to reduce potential negative effects on plants, plant ecosystems, and humans, it is important to consider an integrative approach and this is the aim of the present manuscript.

TABLE 1.2 Some effects of NPs interactions with plant ecosystems at the macro, micro, and molecular level.

Level	NP interaction cause changes in	References
Macro	Soil mechanical properties	Taha and Taha (2012)
	Plant germination	Hoang et al. (2019)
	Plant growth	Margenot et al. (2018)
	Germination time	Younes et al. (2020)
	Plant yield	Elmer and White (2016), Hoang et al. (2019)
	Disease resistance	Elmer and White (2016)
Micro	Cell damage	Ghosh et al. (2016)
	Potential penetration of seed coat by NPs	Younes et al. (2020)
	Impact on soil microbiota	Zhou et al. (2020), McGee et al. (2017)
	Chromosomal aberration	Ghosh et al. (2016), Saha and Dutta Gupta (2017)
Molecular	Modification of plant metabolite production	Mahmoud et al. (2019), Cumplido-Nájera et a. (2019), Hernández-Hernández et al. (2019)
	Potential activation of enzymes	Raliya et al. (2016), Younes et al. (2020)
	Modification of nutrient uptake	Raliya et al. (2016)
	DNA damage	Ghosh et al. (2016)
	ROS production and lipid peroxidation	Ghosh et al. (2016)

1.7 Book summary

This book is divided into three sections. Section I: Introduction, which elaborates on the importance of examining the interaction of ENMs with plants from a systemic perspective. Section II: Biogeochemical cycle of ENMs before plant exposure explained at the molecular level, examines potential transformations of ENMs in different media before they interact with plants. Section III: Reactions and mechanisms related to molecular interaction of ENMs with plants, explains detailed chemical pathways of NPs entrance and transport into plant body.

Chapter 1, "Introduction" discusses why understanding the interactions of plants with ENMs at the molecular level is relevant. What are the implications of the lack of data regarding molecular reactions involved in such interactions? This chapter also provides a general perspective on what we know and what we do not know about plant-ENMs interactions and discusses how a systemic approach is valuable.

Chapter 2 deals with "Source, fate and transport of ENMs in the environment, especially those that may eventually reach plant systems." This chapter provides numbers, predictions, and modeling, among others, regarding the presence of ENMs in air, water, and soil. The questions: Is there any way to know how much NMs would be expected to reach plants? How much may reach edible plants? and Why? are also addressed.

Chapter 3, "Characterization of ENMs in water, soil, and plant media" provides insights about different methodologies and the information that can be gathered when they are applied to the study of ENMs in different media.

Chapter 4, "Biophysicochemical transformations of ENMs in soil" provides or proposes chemical reactions at least at the functional group level. Explains which reactions are favorable or unfavorable to ENMs bioavailability. Discusses the effects of: (a) plant-soil-ENMs interaction; (b) type of soil on ENMs bioavailability and stability; and (c) ENMS on soil fauna and watersheds.

Chapter 5, "Biophysicochemical transformations of ENMs in water" examines what is known about solubility in different aqueous systems, provides information about parameters affecting ENMs solubility and stability, and explains which reactions favor or unfavor ENMs bioavailability and why.

Chapter 6, "Biophysicochemical transformations of ENMs in air," deals with potential physical and chemical reactions of ENMs with atmospheric components (humidity, vapors, gases, colloids, light, etc.). Also, discuss how ENMs present in air interact with plant surface and which plant part is the preferred way of entrance. In addition, provides or proposes chemical reactions at least at the functional group level.

Chapter 7, "Biophysicochemical transformation of ENMs at the root level," discusses bioavailable species of ENMs in soils and chemical reactions explaining their entrance into root system. Proposes chemical reactions at least at the functional group level.

Chapter 8, "Biointeractions of plants-microbes-ENMs" is aimed to discuss the possible effects of beneficial/harmful microorganisms exposed to ENMs on plant physiology and biochemistry.

Chapter 9, "Chemical reactions and mechanisms of ENMs transport in plants," explains why and how ENMs are transported in plants. Includes potential biotransformation of ENMs in the different compartments: xylem, phloem, cell wall, and organelles, among others.

Chapter 10 "Biotransformation in leaves of foliar applied ENMs" explains or hypothesizes what happens when ENMs reach the leaf system. Includes the mechanisms of protection that may allow the leaves to get rid of ENMs.

Chapter 11 "Potential toxicity of ENMs laden plants: are the final products more or less toxic than ENMs per se after they are absorbed into the plant tissues?" deals with ENMs toxicology after they are transformed inside plant systems. This part tries to answer the questions: Do plants reduce ENMs toxicity? And what happens with ENMs that are not transformed? Also proposes detailed chemical reactions and examines potential bioavailability of ENMs when they are laden into the plant tissues.

Chapter 12 "Biodegradation of plants laden with ENMs: Environmental effects" explains biodegradation processes that take place in plants containing ENMs. Also describes the metal species released to soil/air/water matrices and their effects on the environment and living organisms.

References

Adisa, I. O., Pullagurala, V. L. R., Peralta-Videa, J. R., Dimkpa, C. O., Elmer, W. H., Gardea-Torresdey, J. L., & White, J. C. (2019). Recent advances in nano-enabled fertilizers and pesticides: A critical review of mechanisms of action. *Environmental Science: Nano, 6*(7), 2002–2030. https://doi.org/10.1039/c9en00265k

American Society for Testing and Materials (2012). Standard terminology relating to nanotechnology.

Ashraf, S. A., Siddiqui, A. J., Elkhalifa, A. E. O., Khan, M. I., Patel, M., Alreshidi, M., Moin, A., Singh, R., Snoussi, M., & Adnan, M. (2021). Innovations in nanoscience for the sustainable development of food and agriculture with implications on health and environment. *Science of the Total Environment, 768*. https://doi.org/10.1016/j.scitotenv.2021.144990

Cervantes-Avilés, P., Huang, Y., & Keller, A. A. (2019). Incidence and persistence of silver nanoparticles throughout the wastewater treatment process. *Water Research, 156*, 188–198. https://doi.org/10.1016/j.watres.2019.03.031

Christian, P., Von Der Kammer, F., Baalousha, M., & Hofmann, T. (2008). Nanoparticles: Structure, properties, preparation and behaviour in environmental media. *Ecotoxicology, 17*(5), 326–343. https://doi.org/10.1007/s10646-008-0213-1

Cornelis, G., Tuoriniemi, J., Montano, M., Wagner, S., Gallego-Urrea, J. A., Mattsson, K., & Gondikas, A. (2021). 4.Challenges and current approaches toward environmental monitoring of nanomaterials. In *Monitoring environmental contaminants* (pp. 73–108). Elsevier. https://doi.org/10.1016/B978-0-444-64335-3.00003-7

Crépet, A., Luong, T. M., Baines, J., Boon, P. E., Ennis, J., Kennedy, M., Massarelli, I., Miller, D., Nako, S., Reuss, R., Yoon, H. J., & Verger, P. (2021). An international probabilistic risk assessment of acute dietary exposure to pesticide residues in relation to codex maximum residue limits for pesticides in food. *Food Control, 121*. https://doi.org/10.1016/j.foodcont.2020.107563

Cumplido-Nájera, C. F., González-Morales, S., Ortega-Ortíz, H., Cadenas-Pliego, G., Benavides-Mendoza, A., & Juárez-Maldonado, A. (2019). The application of copper nanoparticles and potassium silicate stimulate the tolerance to Clavibacter michiganensis in tomato plants. *Scientia Horticulturae, 245*, 82–89. https://doi.org/10.1016/j.scienta.2018.10.007

Das, C. K., Srivastava, G., Dubey, A., Roy, M., Jain, S., Sethy, N. K., Saxena, M., Harke, S., Sarkar, S., Misra, K., Singh, S. K., Bhargava, K., Philip, D., & Das, M. (2016). Nano-iron pyrite seed dressing: A sustainable intervention to reduce fertilizer consumption in vegetable (beetroot, carrot), spice (fenugreek), fodder (alfalfa), and oilseed (mustard, sesamum) crops. *Nanotechnology for Environmental Engineering, 1*(1). https://doi.org/10.1007/s41204-016-0002-7

De La Rosa, G., Moreno, M. L. L., Viezcas, J. A. H., Montes, M. O., Videa, J. R. P., & Torresdey, J. L. G. (2011). Toxicity and biotransformation of ZnO nanoparticles in the desert plants Prosopis juliflora-velutina, Salsola tragus and Parkinsonia Florida. *International Journal of Nanotechnology, 492*. https://doi.org/10.1504/IJNT.2011.040190

Delouche, J. C. (1979). Seed germination. In *Proceedings of the short course for seedsmen, 1979; 1979-07-Delouche.pdf.* https://ir.library.msstate.edu/bitstream/handle/11668/13984/1979-07-Delouche.pdf?sequence=1&isAllowed=y.

Duhan, J. S., Kumar, R., Kumar, N., Kaur, P., Nehra, K., & Duhan, S. (2017). Nanotechnology: The new perspective in precision agriculture. *Biotechnology Reports, 15*, 11–23. https://doi.org/10.1016/j.btre.2017.03.002

Elmer, W. H., & White, J. C. (2016). The use of metallic oxide nanoparticles to enhance growth of tomatoes and eggplants in disease infested soil or soilless medium. *Environmental Science: Nano, 3*(5), 1072–1079. https://doi.org/10.1039/c6en00146g

European commission recommendation of 18 October 2011 on the definition of nanomaterial. *European Commission, 54*, (2011), 38–40.

Ghosh, M., Jana, A., Sinha, S., Jothiramajayam, M., Nag, A., Chakraborty, A., Mukherjee, A., & Mukherjee, A. (2016). Effects of ZnO nanoparticles in plants: Cytotoxicity, genotoxicity, deregulation of antioxidant defenses, and cell-cycle arrest. *Mutation Research/Genetic Toxicology and Environmental Mutagenesis, 807*, 25–32. https://doi.org/10.1016/j.mrgentox.2016.07.006

Giner-Casares, J. J., Henriksen-Lacey, M., Coronado-Puchau, M., & Liz-Marzán, L. M. (2016). Inorganic nanoparticles for biomedicine: Where materials scientists meet medical research. *Materials Today, 19*(1), 19–28. https://doi.org/10.1016/j.mattod.2015.07.004

Gomez, A., Narayan, M., Zhao, L., Jia, X., Bernal, R. A., Lopez-Moreno, M. L., & Peralta-Videa, J. R. (2021). Effects of nano-enabled agricultural strategies on food quality: Current knowledge and future research needs. *Journal of Hazardous Materials, 401*. https://doi.org/10.1016/j.jhazmat.2020.123385

Gottschalk, F., Sun, T., & Nowack, B. (2013). Environmental concentrations of engineered nanomaterials: Review of modeling and analytical studies. *Environmental Pollution, 181*, 287–300. https://doi.org/10.1016/j.envpol.2013.06.003

Hansen, S. F., Hansen, O. F. H., & Nielsen, M. B. (2020). Advances and challenges towards consumerization of nanomaterials. *Nature Nanotechnology, 15*(12), 964–965. https://doi.org/10.1038/s41565-020-00819-7

Hernández-Hernández, H., Quiterio-Gutiérrez, T., Cadenas-Pliego, G., Ortega-Ortiz, H., Hernández-Fuentes, A. D., De La Fuente, M. C., Valdés-Reyna, J., & Juárez-Maldonado, A. (2019). Impact of selenium and copper nanoparticles on yield, antioxidant system, and fruit quality of tomato plants. *Plants, 8*(10). https://doi.org/10.3390/plants8100355

Hernandez-Viezcas, J. A., Castillo-Michel, H., Andrews, J. C., Cotte, M., Rico, C., Peralta-Videa, J. R., Ge, Y., Priester, J. H., Holden, P. A., & Gardea-Torresdey, J. L. (2013). In situ synchrotron X-ray fluorescence mapping and speciation of CeO 2 and ZnO nanoparticles in soil cultivated soybean (Glycine max). *ACS Nano, 7*(2), 1415–1423. https://doi.org/10.1021/nn305196q

Hoang, S. A., Nguyen, L. Q., Nguyen, N. H., Tran, C. Q., Nguyen, D. V., Vu, Q. N., & Phan, C. M. (2019). Metal nanoparticles as effective promotors for Maize production. *Scientific Reports, 9*(1). https://doi.org/10.1038/s41598-019-50265-2

Hong, J., Rico, C. M., Zhao, L., Adeleye, A. S., Keller, A. A., Peralta-Videa, J. R., & Gardea-Torresdey, J. L. (2015). Toxic effects of copper-based nanoparticles or compounds to lettuce (Lactuca sativa) and alfalfa (Medicago sativa). *Environmental Sciences: Processes and Impacts, 17*(1), 177–185. https://doi.org/10.1039/c4em00551a

Jeevanandam, J., Barhoum, A., Chan, Y. S., Dufresne, A., & Danquah, M. K. (2018). Review on nanoparticles and nanostructured materials: History, sources, toxicity and regulations. *Beilstein Journal of Nanotechnology, 9*(1), 1050–1074. https://doi.org/10.3762/bjnano.9.98

Joshi, R. (2018). Role of enzymes in seed germination. *International Journal of Creative Research Thoughts, 6*, 1481–1485.

Kah, M., & Hofmann, T. (2014). Nanopesticide research: Current trends and future priorities. *Environment International, 63*, 224–235. https://doi.org/10.1016/j.envint.2013.11.015

Keller, A. A., Adeleye, A. S., Conway, J. R., Garner, K. L., Zhao, L., Cherr, G. N., Hong, J., Gardea-Torresdey, J. L., Godwin, H. A., Hanna, S., Ji, Z., Kaweeteerawat, C., Lin, S., Lenihan, H. S., Miller, R. J., Nel, A. E., Peralta-Videa, J. R., Walker, S. L., Taylor, A. A., … Zuverza-Mena, N. (2017). Comparative environmental fate and toxicity of copper nanomaterials. *NanoImpact, 7*, 28–40. https://doi.org/10.1016/j.impact.2017.05.003

Keller, A. A., McFerran, S., Lazareva, A., & Suh, S. (2013). Global life cycle releases of engineered nanomaterials. *Journal of Nanoparticle Research, 15*(6). https://doi.org/10.1007/s11051-013-1692-4

Keller, A. A., Wang, H., Zhou, D., Lenihan, H. S., Cherr, G., Cardinale, B. J., Miller, R., & Zhaoxia, J. I. (2010). Stability and aggregation of metal oxide nanoparticles in natural aqueous matrices. *Environmental Science and Technology, 44*(6), 1962–1967. https://doi.org/10.1021/es902987d

Kottegoda, N., Sandaruwan, C., Priyadarshana, G., Siriwardhana, A., Rathnayake, U. A., Berugoda Arachchige, D. M., Kumarasinghe, A. R., Dahanayake, D., Karunaratne, V., & Amaratunga, G. A. J. (2017). Urea-hydroxyapatite nanohybrids for slow release of nitrogen. *ACS Nano, 11*(2), 1214–1221. https://doi.org/10.1021/acsnano.6b07781

López-Moreno, M. L., De La Rosa, G., Hernández-Viezcas, J. A., Castillo-Michel, H., Botez, C. E., Peralta-Videa, J. R., & Gardea-Torresdey, J. L. (2010). Evidence of the differential biotransformation and genotoxicity of ZnO and CeO2 nanoparticles on soybean (Glycine max) plants. *Environmental Science and Technology, 44*(19), 7315–7320. https://doi.org/10.1021/es903891g

Mahakham, W., Sarmah, A. K., Maensiri, S., & Theerakulpisut, P. (2017). Nanopriming technology for enhancing germination and starch metabolism of aged rice seeds using phytosynthesized silver nanoparticles. *Scientific Reports, 7*(1). https://doi.org/10.1038/s41598-017-08669-5

Mahmoud, A. W. M., Abdelaziz, S. M., El-Mogy, M. M., & Abdeldaym, E. A. (2019). Effect of foliar Zno and FeO nanoparticles application on growth and nutritional quality of red radish and assessment of their accumulation on human health. *Agriculture (Pol'nohospodárstvo), 65*(1), 16–29. https://doi.org/10.2478/agri-2019-0002

Majumdar, S., Almeida, I. C., Arigi, E. A., Choi, H., VerBerkmoes, N. C., Trujillo-Reyes, J., Flores-Margez, J. P., White, J. C., Peralta-Videa, J. R., & Gardea-Torresdey, J. L. (2015).

Environmental effects of nanoceria on seed production of common bean (Phaseolus vulgaris): A proteomic analysis. *Environmental Science and Technology, 49*(22), 13283–13293. https://doi.org/10.1021/acs.est.5b03452

Margenot, A. J., Rippner, D. A., Dumlao, M. R., Nezami, S., Green, P. G., Parikh, S. J., & McElrone, A. J. (2018). Copper oxide nanoparticle effects on root growth and hydraulic conductivity of two vegetable crops. *Plant and Soil, 431*(1–2), 333–345. https://doi.org/10.1007/s11104-018-3741-3

McGee, C. F., Storey, S., Clipson, N., & Doyle, E. (2017). Soil microbial community responses to contamination with silver, aluminium oxide and silicon dioxide nanoparticles. *Ecotoxicology, 26*(3), 449–458. https://doi.org/10.1007/s10646-017-1776-5

Medina-Velo, I. A., Dominguez, O. E., Ochoa, L., Barrios, A. C., Hernández-Viezcas, J. A., White, J. C., Peralta-Videa, J. R., & Gardea-Torresdey, J. L. (2017). Nutritional quality of bean seeds harvested from plants grown in different soils amended with coated and uncoated zinc oxide nanomaterials. *Environmental Science: Nano, 4*(12), 2336–2347. https://doi.org/10.1039/c7en00495h

Medina-Velo, I. A., Peralta-Videa, J. R., & Gardea-Torresdey, J. L. (2017). Assessing plant uptake and transport mechanisms of engineered nanomaterials from soil. *MRS Bulletin, 42*(5), 379–383. https://doi.org/10.1557/mrs.2017.87

Montes, A., Bisson, M. A., Gardella, J. A., Jr., & Aga, D. S. (2017). Uptake and transformations of engineered nanomaterials: Critical responses observed in terrestrial plants and the model plant *Arabidopsis thaliana. Science of the Total Environment, 607*(608), 1497–1516. https://doi.org/10.1016/j.scitotenv.2017.06.190

Nehra, M., Dilbaghi, N., Marrazza, G., Kaushik, A., Sonne, C., Kim, K. H., & Kumar, S. (2021). Emerging nanobiotechnology in agriculture for the management of pesticide residues. *Journal of Hazardous Materials, 401*. https://doi.org/10.1016/j.jhazmat.2020.123369

Peralta-Videa, J. R., Huang, Y., Parsons, J. G., Zhao, L., Lopez-Moreno, L., Hernandez-Viezcas, J. A., & Gardea-Torresdey, J. L. (2016). Plant-based green synthesis of metallic nanoparticles: Scientific curiosity or a realistic alternative to chemical synthesis? *Nanotechnology for Environmental Engineering, 1*(1). https://doi.org/10.1007/s41204-016-0004-5

Pokropivny, V. V., & Skorokhod, V. V. (2007). Classification of nanostructures by dimensionality and concept of surface forms engineering in nanomaterial science. *Materials Science and Engineering C, 27*(5–8), 990–993. https://doi.org/10.1016/j.msec.2006.09.023

Prado, A. G. S., Moura, A. O., & Nunes, A. R. (2011). Nanosized silica modified with carboxylic acid as support for controlled release of Herbicides. *Journal of Agricultural and Food Chemistry, 59*(16), 8847–8852. https://doi.org/10.1021/jf202509g

Prasad, R., Bhattacharyya, A., & Nguyen, Q. D. (2017). Nanotechnology in sustainable agriculture: Recent developments, challenges, and perspectives. *Frontiers in Microbiology, 8*. https://doi.org/10.3389/fmicb.2017.01014

Raliya, R., Tarafdar, J. C., & Biswas, P. (2016). Enhancing the mobilization of native phosphorus in the mung bean rhizosphere using ZnO nanoparticles synthesized by soil fungi. *Journal of Agricultural and Food Chemistry, 64*(16), 3111–3118. https://doi.org/10.1021/acs.jafc.5b05224

Saha, N., & Dutta Gupta, S. (2017). Low-dose toxicity of biogenic silver nanoparticles fabricated by Swertia chirata on root tips and flower buds of Allium cepa. *Journal of Hazardous Materials, 330*, 18–28. https://doi.org/10.1016/j.jhazmat.2017.01.021

Saleh, T. A. (2020). Nanomaterials: Classification, properties, and environmental toxicities. *Environmental Technology & Innovation, 20*, 101067. https://doi.org/10.1016/j.eti.2020.101067

Saleh, T.A., & Gupta, V. K. (2016). (pp. 83–133). Elsevier BV. https://doi.org/10.1016/b978-0-12-804703-3.00004-8.

Schulte, P. (2010). Using recycled water on agriculture: Sea mist farms and sonoma county. In *California farm water success stories* (pp. 55–65).

Shafiq, M., Anjum, S., Hano, C., Anjum, I., & Abbasi, B. H. (2020). An overview of the applications of nanomaterials and nanodevices in the food industry. *Foods, 9*(2). https://doi.org/10.3390/foods9020148

Srivastava, G., Das, C. K., Das, A., Singh, S. K., Roy, M., Kim, H., Sethy, N., Kumar, A., Sharma, R. K., Singh, S. K., Philip, D., & Das, M. (2014). Seed treatment with iron pyrite (FeS 2) nanoparticles increases the production of spinach. *RSC Advances, 4*(102), 58495–58504. https://doi.org/10.1039/c4ra06861k

Stankic, S., Suman, S., Haque, F., & Vidic, J. (2016). Pure and multi metal oxide nanoparticles: Synthesis, antibacterial and cytotoxic properties. *Journal of Nanobiotechnology, 14*, 1–20. https://doi.org/10.1186/s12951-016-0225-6

Taha, M. R., & Taha, O. M. E. (2012). Influence of nano-material on the expansive and shrinkage soil behavior. *Journal of Nanoparticle Research, 14*(10). https://doi.org/10.1007/s11051-012-1190-0

Tarrahi, R., Mahjouri, S., & Khataee, A. (2021). A review on in vivo and in vitro nanotoxicological studies in plants: A headlight for future targets. *Ecotoxicology and Environmental Safety, 208*, 111697. https://doi.org/10.1016/j.ecoenv.2020.111697

Wang, H., Kou, X., Pei, Z., Xiao, J. Q., Shan, X., & Xing, B. (2011). Physiological effects of magnetite (Fe3O4) nanoparticles on perennial ryegrass (Lolium perenne L.) and pumpkin (Cucurbita mixta) plants. *Nanotoxicology, 5*(1), 30–42. https://doi.org/10.3109/17435390.2010.489206

White, J. C., & Gardea-Torresdey, J. (2018). Achieving food security through the very small. *Nature Nanotechnology, 13*(8), 627–629. https://doi.org/10.1038/s41565-018-0223-y

Younes, N. A., Hassan, H. S., Elkady, M. F., Hamed, A. M., & Dawood, M. F. A. (2020). Impact of synthesized metal oxide nanomaterials on seedlings production of three Solanaceae crops. *Heliyon, 6*(1), e03188. https://doi.org/10.1016/j.heliyon.2020.e03188

Zhou, Q., Zhang, X., & Wu, Z. (2020). Impact of TiO 2 and ZnO nanoparticles on soil bacteria and the enantioselective transformation of racemic-metalaxyl in agricultural soil with Lolium perenne: A wild greenhouse cultivation. *Journal of Agricultural and Food Chemistry, 68*(40), 11242–11252. https://doi.org/10.1021/acs.jafc.0c03959

CHAPTER

2

Source, fate and transport of ENMs in the environment, especially those that may eventually reach plant systems

Yanjie Zhu, and Yuxiong Huang

Tsinghua-Berkeley Shenzhen Institute (TBSI), Tsinghua Shenzhen International Graduate School (SIGS), Tsinghua University, Shenzhen, China

2.1 Introduction

The industry in engineered nanomaterials (ENMs) has dramatically increased in recent decades. The market size of global ENMs was estimated at USD 8.5 billion in 2019, which is expected to growth at an annual rate (CAGR) of 13.1% from 2021 to 2027. In addition to consumable products, ENMs have been used as adsorbents and catalysts for environmental remediation (Corsi et al., 2018; Guerra et al., 2018; Wu et al., 2019). Recently, ENMs have been widely applied in agriculture as fertilizers and pesticides, which can help improve plant growth, development and productivity, and overcome biotic and abiotic stress (Abdel-Aziz et al., 2019; Ameen et al., 2021; Chutipaijit et al., 2018). Besides, ENMs also make a substantial contribution to medical applications, as they have been widely used in medical imaging and treatment of some diseases (Maduraiveeran & Jin, 2017; Pellico et al., 2021; Shi et al., 2017; Villanueva-Flores et al., 2020).

On the other hand, ENMs are considered as emerging pollutants due to the potential risks they represent to ecosystems (Huang et al., 2022; Zhu et al., 2019). The large-scale production of ENMs embedded products, as well as the leaching of ENMs from industrial discharge or sewage from households, would inevitably release ENMs into the environment

Physicochemical Interactions of Engineered Nanoparticles and Plants
https://doi.org/10.1016/B978-0-323-90558-9.00006-1

(Bundschuh et al., 2018). Over the past few decades, a significant amount of ENMs released into the environment has accumulated in an underestimated volume (Batley et al., 2013; Eduok et al., 2015). The sewage sludge, wastewater, and wastes of products containing ENMs represent the main flow of ENMs ultimately entering the environment (Kaegi et al., 2021; Tou et al., 2017). The ENMs accumulate on the path from the atmosphere and soil system to the water environment and are being transferred to organisms along with the ecosystem, threatening the terrestrial and the aquatic environments (Abbas et al., 2020; Liu et al., 2022). Currently, very few studies have conducted a quantitative analysis of ENMs concentrations in the environment (Cervantes-Avilés et al., 2019; Keller et al., 2018). Studies have simulated the release path of ENMs in the environment and evaluated the pollutant concentration and diffusion trend based on models (Garner et al., 2017; Sun et al., 2016; Tolaymat et al., 2017). Therefore, it is crucial to study the transmission pathways of ENMs to understand the environmental fate, bioavailability, and ecotoxicity patterns.

As the primary trophic level of ecosystems, plants are the starting point of every complex food chain, and are sensitive to the existence of contaminants in the surrounding environment. Recently, new advances in agricultural production have applied ENMs to assist plant development and crop yield (Du et al., 2019; Kumar et al., 2013; Regier et al., 2015). Plants can efficiently uptake metal ENMs as macronutrients and micronutrients, which can help improve growth, development, and productivity, and overcome abiotic and biotic stress (Khan & Upadhyaya, 2019; Liao et al., 2021; Zhu et al., 2019). On the other hand, the considerable accumulation of ENMs may disturb the soil and plant systems (Ahmed et al., 2021; Rajput et al., 2018), and further result in the accumulation and internalization of ENMs in plants (Avellan et al., 2017; Huang et al., 2017; Ma & Yan, 2018). Exposure to high concentrations of ENMs in the soil can cause severe damage in plant cells, affecting cell machinery, chlorophyll levels, and plant growth (Hernández-Hernández et al., 2019; Lian et al., 2020; Mosa et al., 2018; Regier et al., 2015). This chapter will comprehensively review sources (e.g., anthropogenic activities, agricultural application, industrial emissions), pathways and transport (e.g., aquatic, territory), fate and risks of ENMs to terrestrial plant systems.

2.2 Source of ENMs in the environment

Metallic nanoparticles (MNPs) are released in the environment via three major sources: natural processes, unintentional generation, and human manufactured ENMs. Natural processes such as volcanic eruptions, bush fires, deposition, and rock denudation, among others, could produce natural MNPs (Griffin et al., 2018). In addition, human activities, for example, burning biomass and fossil fuels, may generate MNPs unintentionally (Piccinno et al., 2012). Recently, with the explosive development of nanotechnology and the associated demand and application in various fields, ENMs have been unintentionally and intentionally released into the environment at all stages of products' life cycle (raw materials, production, transportation, usage, and end-of-life management) (Tolaymat et al., 2017).

2.2.1 Anthropogenic activities

Anthropogenic activities unknowingly release a large amount of ENMs into the environment. Metal ENMs are present in cigarette smoke, textile fibers, cleaning products, and cosmetics. Moreover, the mining and use of biomass and fossil fuels will also produce ENMs. The expansion of the transportation system, the consumption of fuels by motor vehicles, airplanes, and, freighters are also important sources of ENMs contamination (Gonet & Maher, 2019). The refining, production, and combustion processes are leading to a significant nuber of nanoparticle (NPs) and secondary NPs, such as titanium (Ti), cadmium (Cd), lead (Pb), and nickel (Ni) NPs that are being released into the environment (Silva et al., 2021).

Landfills also contribute as a source of ENMs. Based on the solid waste management rate, it has been reported that the inventory of silver (Ag), zinc (Zn), and Ti ENMs in landfills in Europe (2020) would be 105, 2077, and 69,000 tons, respectively (Rajkovic et al., 2020). Likewise, a large part of products containing ENMs and household waste would be processed at waste incineration plants (WIPs) (Oischinger et al., 2019). The incineration process can cause changes in the shape, size, structure, and chemical properties of the ENMs. Most of them remain as ashes at the bottom of the incinerator and may eventually enter the landfill after disposal.

2.2.2 Consumable products

According to data released in 2013 by the Consumer Product List of Nanotechnology (CPI), 1814 consumable products were reported to contain ENMs. Ag ENMs are the most commonly used in consumable products, followed by Ti, Zn, and platinum (Pt) ENMs (Vance et al., 2015). Metallic ENMs used in cosmetics and personal care products are released during use, further contaminating the surface water and soil system. Because of their inertness, Ti, Ag, and Zn ENMs are mainly used in sunscreens and toothpaste, as well as in many other products such as white pigments, and food additives colorings, among others. Zn and Ag ENMs are commonly used in antiacne products and antibacterial soaps because of their antibacterial properties (Weir et al., 2012). In addition, Ag ENMs are widely used in food packaging, personal care products, and architectural coatings, among others (Kuenen et al., 2020). In general, most of these ENMs-embedded consumer products would be discarded into the municipal solid waste stream and, eventually, be sent to incinerators or landfills, which would induce the amassment of ENMs in the aquatic and soil environment.

2.2.3 Industrial emissions

Industrial processes are the primary source of ENMs entering the environment. The emissions of ENMs in the first stage of industrial production, formulation, and manufacturing processes were relatively low, as they were expected not to exceed 1% of the total mass of ENMs (Giese et al., 2018). However, in the subsequent use and disposal phases, these products may release high amounts of ENMs into the environment. Large amounts of ENMs are released into the atmosphere by different industrial activities, including point sources (manufacturing units, waste incineration, power plants, wastewater treatment plants, during transportation, and landfills) and nonpoint sources (vehicle emissions and during cleaning or abrasion of nanoproducts) (Peng et al., 2017). It is estimated that approximately 8100 metric tons of ENMs are emitted into the atmosphere each year (Keller & Lazareva, 2014); however, the atmospheric portion of nanoparticles has a relatively short residence period (John et al., 2017). Because of sinking and rainfall, ENMs in the atmospheric medium would return to the land and, eventually, be deposited in soil or water bodies (Giese et al., 2018). In addition, the transformation processes occurring in the atmosphere may affect the interaction and fate of atmospheric ENMs in soil or aquatic systems (Abbas et al., 2020).

The main contributors of ENM to the environment are wastewater and sewage sludge (Loosli et al., 2019). High concentrations of chromium (Cr), copper (Cu), cadmium (Cd), nickel (Ni), zinc (Zn), and lead (Pb) ENMs were observed in wastewater treatment plants (WWTPs) sludge (Huang et al., 2021; Tou et al., 2017). According to a dynamic probabilistic material flow analysis, around 30 tons of Ag ENMs and 3860 tons of Zn ENMs accumulated in wastewater. Thus, wastewater is considered as the largest sink of ENMs (Rajkovic et al., 2020).

2.2.4 Agricultural applications

The application of nanofertilizers has become a promising solution for promoting plant growth and health, increasing yield, and resisting biological stress (Benzon et al., 2015; Elfeky et al., 2013; Liu et al., 2022a, 2022b). The use of ENMs in plants may effectively control biological stresses, including insects, fungi, and weeds (Huang et al., 2019; Liao et al., 2021; Zhang et al., 2021). The residual ENMs from the agricultural applications would remain in the soil system. In addition, many countries are currently using biological waste soil or sludge as a fertilizer supplement in crop planting (Collivignarelli et al., 2019), where ENMs may present a higher concentration (Kuenen et al., 2020). Thus, agricultural applications should also be considered as major sources of ENMs entering the environment; these agricultural residues may cause potential harm to plant growth.

2.2.5 Nanomedicine

Because of their size and shape, ENMs display high efficiency and accuracy as drug delivery systems, sensors, in bioimaging, regenerative medicine, and diagnosis. In fact, the application of nanomaterials in the diagnosis and treatment of diseases has experienced significant growth and expansion (Anselmo & Mitragotri, 2016). In 2016, the global nanomedicine market was valued at 134.4 billion USD (Alexander et al., 2020, pp. 471–491). According to the BCC research forecast, a compound annual growth rate (CAGR) of 14.0% was estimated for this market, which was expected to increase from USD 151.9 billion in 2017 to USD 293.1 billion in 2022 (BCC Research, 2017). The waste and the residues discharged in the nanomedicine field also contribute with an enormous number of ENMs to the environment.

2.2.6 Environmental concentrations of ENMs

Material flow and environmental fate models were used to predict the range or trend of environmental ENMs concentrations (Garner et al., 2017; Rajkovic et al., 2020). As the major source of ENMs into the atmosphere, the emissions of ENMs from diesel to the air were estimated by sampling and measuring exhaust particles emitted during engine operation (Harrison et al., 2018). Taking Beijing, China as an example of a typical megacity, the daily emission of Fe ENMs ranged from 6 to 371 $\mu g/m^3$ (Zhang, Lu, et al., 2020). Likewise, the concentration of Ag ENMs in agricultural soil was estimated between 30 pg/kg and 10 μg/kg (Giese et al., 2018). The concentration range of various ENMs in wastewater systems predicted by dynamic models may span several orders of magnitude. For example, the concentration of Ti ENMs may reach 40,000 μg/L (Sun et al., 2016), while the concentrations of Ce ENMs and Ag ENMs were predicted to be in the levels of ng/L to $\mu g/m^3$ (Giese et al., 2018).

Overall, ENMs have exhibited a trend of an exponentially increasing accumulation in the environment, and the environmental concentrations of ENMs have provided an important reference for their subsequent ecotoxicological assessment.

2.3 Pathway and fate of ENMs in the environment

2.3.1 Atmospheric system

It is estimated that no more than 1.5% of the annual global ENM production was released into the atmosphere (Keller et al., 2013). Despite this, ENMs are primarily released into the atmosphere during production or use as fuels, paint, spraying, and others (Donia & Carbone, 2019; Vance & Marr, 2015). Although the residence period of ENMs in the atmosphere is short, it is critical to understand the transformation of ENMs. ENMs in the atmosphere would be affected by low air pressure, humidity, and temperature, and might be involved in chemical reactions to form new chemical substances, which may, eventually, change ENMs' particle size, morphology, and their chemical properties (Donia & Carbone, 2019). For example, meteorological factors would affect the fate and transport of ENMs in the atmosphere (Keller et al., 2013).

2.3.2 Terrestrial system

The transmission of ENMs in the soil system is mainly caused by landfills, residues of bottom ash from incineration, retention of agricultural ENMs, and sewage sludge (Iavicoli et al., 2017; Kaegi et al., 2021). It is estimated that the global production of ENMs in 2010 exceeded 260,000–309,000 metric tons, of which 63%–91% ended up in landfills, and 8%–28% was released into the soil (Keller et al., 2013). The WIPs would incinerate a large amount of waste embedded ENMs. These ENMs will remain in the bottom ash (P. Vejerano et al., 2014). For example, Ti ENMs, cerium (Ce) ENMs, and Fe ENMs are able to form large aggregates in the bottom ash (Oischinger et al., 2019).

Agglomeration and settlement are the major processes ENMs undergo in the soil system (Donia & Carbone, 2019; Gottschalk & Nowack, 2011). In soils, metallic ENMs would heteroaggregate with organic and inorganic particles, heavy metals, and organisms (Bundschuh et al., 2018). Compared to primary ENMs, the agglomerated ENMs form a larger volume and mass, which reduces their reactivity and bioavailability and increases their sinking properties. Eventually, the ENMs present in the soil may accumulate in the deep layer of the terrestrial system and remain for an extended period of time, resulting in a long-term exposure of the soil ecosystem.

2.3.3 Aquatic system

The ENMs discharged into terrestrial environments, whether consciously or unconsciously, will eventually flow into the aquatic system by rain washing (Abu-Qdais et al., 2021; Goswami et al., 2017; Peng et al., 2017). In addition, numerous ENMs have been discharged into aquatic systems via WWTPs (Collivignarelli et al., 2019; Eduok et al., 2015; Loosli et al., 2019). Part of the metallic ENMs may be dissolved into metallic ions (e.g., Ag^+, Cu^{2+}, and Zn^{2+}) in the aquatic systems (Wang et al., 2017), which may further induce biological toxicity (Rastogi et al., 2017). Furthermore, the ENMs would also form agglomeration with natural minerals and organic colloids in the aquatic media, which would change the ENMs' surface properties and mobility, and further change the consequential environmental fate and risks (Azimzada et al., 2021).

2.3.4 Biosystem

After released into the aquatic and terrestrial systems, ENMs interact with aquatic and terrestrial organisms and enter the food chain, which may result in bioaccumulation and biomagnification (Abdolahpur Monikh et al., 2021). ENMs are able to affect plant seed germination, growth and development, and induce biotic/abiotic stress to various degrees (Nair & Chung, 2014a; Pittol et al., 2017; Vannini et al., 2014; Zhang,Jin, et al., 2020). In addition, these dissolved metallic ions may affect the ion balance in biological cells, and act as a stressor that may induce plants to produce an excessive level of ROS (Qian et al., 2013). For example, it has been reported that Ag^+ ions released by Ag ENMs inhibited ethylene perception in plants and interfered with ethylene biosynthesis, potentially hindering plant growth and development (Syu et al., 2014). Ag^+ ions also lead to a significant increase in the protein oxidation of *Allium cepa*, resulting in the inhibition of antioxidant enzymes (ascorbate peroxidase (APX) and catalase (CAT)) (Cvjetko et al., 2017).

2.4 ENMs-plant interaction

Plants are exposed to ENMs by means of agricultural activities (e.g., nanofertilizers, nanopesticide) or ENMs-contaminated media (e.g., soil, water). Generally, a low concentration of ENMs may help plants resist biotic (diseases, insects) or abiotic stress (salinity, drought, floods, radiation, extremes in temperature, heavy metals, etc.) (Gull et al., 2019). However, high concentrations of metallic ENMs may induce negative effects, including irreversible biological damage. The adverse effects of ENMs on plants are summarized in Table 2.1, and the underlying mechanisms are discussed in the following sections.

2.4.1 Stress induced by dissolved ions

Metal ions released by ENMs dissolution are easily transferred and uptaken by plant systems. For example, it has been reported that Ag ions released from Ag ENMs induced cytotoxicity in the seedling genome and proteomics by changing the gene of a protein (HCF136), which is mainly involved in primary metabolism and cell defense (Vannini et al., 2014). Likewise, Cu ions induced a higher toxicity to wheat than that of Cu ENMs by generating oxidative stress, root growth inhibition, and upregulating the transcription of heavy metal transporters (Zhang et al., 2018).

2.4.2 Bioaccumulation of ENMs at the rhizosphere

In addition to the dissolved metal ions derived from ENMs, particulate ENMs also interact with plants directly. The accumulation and internalization of ENMs in the roots of plants result in an impact on the growth of the root system (Feichtmeier et al., 2015). Typically, the plant root mucus adsorbs positively charged ENMs to prevent translocation into the root tissue, while the negatively charged ENMs are more easily transferred to the apoplast (Avellan et al., 2017). For example, Au ENMs accumulated on the surface of barley plant roots, which were retained by root mucus (Milewska-Hendel et al., 2017). Likewise, the phytotoxicity of Cu ENMs to cucumbers indicated that high levels of Cu ENMs were accumulated in plant root tissues (Mosa et al., 2018). In addition, ENMs may cause mechanical damages to plants, which would further induce oxidative stress to affect gene expression and metabolic pathways (Kohan-Baghkheirati & Geisler-Lee, 2015).

2.4.3 Subcellular effects

ENMs also affect plants at the subcellular level. Specifically, plants pigments (e.g., chlorophyll *a*, chlorophyll *b*) and antioxidant enzymes (e.g., superoxide dismutase (SOD), CAT, and peroxidase oxidase (POX), guaiacol peroxidase (gPOD)) may be altered, causing further damage to cell membranes, proteins, and nucleic acids (Khan et al., 2017). For instance, after exposure to Au ENMs, free radical scavenging activity in plants was significantly improved, resulting in the regulation of

TABLE 2.1 The effects of metal/metallic oxide ENMs on plants.

Metallic NPs	Size	Exposure concentration	Plants	Effects on plants	References
Au Gold	5–20 nm	10–50 mg/L	*Hordeum vulgare* L.	Accumulated root cortex cells	Milewska-Hendel et al. (2017)
	Au^{3+}	9.85–197 mg/L	*Vigna unguiculata* L.	Reduced toxic Au^{3+} by generating Au-NPs	Shabnam et al. (2014)
	10 nm	>3 mg/L	*Hordeum vulgare* L.	Decreased the root growth	Feichtmeier et al. (2015)
	50 nm	100–400 mg/L	*Brassica juncea*	Decreased in overall growth	Bisht et al. (2014)
	24 nm	10, 80 mg/L	*Arabidopsis thaliana*	Improved seed germination rate	Kumar et al. (2013)
				Vegetative growth	
				Free radical scavenging activity	
	60 nm	10–2000 mg/L	*Oryza sativa*	Increases in cell death	Tsi Ndeh et al. (2017)
				Hydrogen peroxide formation	
				Lipid peroxidation in roots and shoots	
Ag Silver	2 nm	125–500 mg/L	*Raphanus sativus*	Reduced the total weight	Zuverza-Mena et al. (2016)
				Reduced the root and shoot lengths	

Continued

TABLE 2.1 The effects of metal/metallic oxide ENMs on plants.—cont'd

Metallic NPs	Size	Exposure concentration	Plants	Effects on plants	References
	5–10 nm	2.7–10.8 mg/L	*Allium cepa*	Inhibited the root growth	Cvjetko et al. (2017)
	8–21 nm	1 mg/L	*Trigonella foenum-graecum* L.	Enhanced of plant growth and diosgenin synthesis	Jasim et al. (2017)
	20 nm	5 mg/L	*Arabidopsis thaliana*	Enhanced the antioxidant activity of genes	Kohan-Baghkheirati & Geisler-Lee (2015)
	10–20 nm	60 mg/L	*Pelargonium zonale*	Decreased petal and leaf abscission	Ghorbanpour & Hatami (2014)
				Increased antioxidative enzymes activities	
	5–15.5 nm	1–10 mg/L	*Triticum aestivum* L.	Induced morphological modifications in root tip cells	Vannini et al. (2014)
	15 nm	0.5–50 mg/L	*Glycine* max L. *cv. Enrei*	Enhanced seedling growth	Mustafa et al. (2015)
				Reduced cytotoxic by-products of glycolysis	

				Increased the abundance of stress-related proteins	
	8–54 nm	0.11–10.8 mg/L	*Arabidopsis*	Exhibited the antimicrobial activity Induced the root growth promotion Induced the ROS accumulation	Syu et al. (2014)
	20 nm	0.2–1 mg/L	*Oryza sativa* L.	Reduced the root elongation, shoot and root fresh weights, total chlorophyll and carotenoids contents	Nair & Chung (2014b)
	2–20 nm	0.2–3 mg/L	*Arabidopsis thaliana*	Inhibits chlorophyll biosynthesis Decreased photosynthetic efficiency Leads to the overproduction of ROS Lead to growth retardation or death of the plant	Qian et al. (2013)

Continued

TABLE 2.1 The effects of metal/metallic oxide ENMs on plants.—cont'd

Metallic NPs	Size	Exposure concentration	Plants	Effects on plants	References
Cu Copper	42 nm	10–250 mg/L	*Solanum lycopersicum*	Decreased the chlorophyll	Hernández-Hernández et al. (2019)
	100–150 nm	0.2–10 mg/L	*Elodea nuttallii*	Increased Cu accumulation from CuO-NPs	Regier et al. (2015)
				Reduced chlorophyll	
				Enhanced peroxidase activity	
	10–30 nm	50–200 mg/L	*Cucumis sativus*	Caused genomic alterations	Mosa et al. (2018)
				Decreased biomass	
				Decreased levels of chlorophyll *a* and chlorophyll *b*	
	50 nm	1–100 mg/L	*Triticum aestivum* L.	Inhibition of oxidative stress and root growth	Zhang et al. (2018)
Ti Titanium	6.5 nm	100–250 mg/L	*Zea mays* L.	Decreased chlorophyll	Lian et al. (2020)
	20–30 nm	50–200 mg/L	*Dracocephalum moldavica*	Increased antioxidant enzyme activity	Gohari et al. (2020)

				Lowered H_2O_2 concentration	
	10–100 nm	20–80 mg/L	*Bipolaris sorokiniana*	Reduced the disease severity	Satti et al. (2021)
				Reduced biotic stress	
				Improved the quality and yield	
	86 nm	10–1500 mg/L	*Raphanus sativus* L.	Provoked transient increase photosynthesis	Tighe-Neira et al. (2020)
				Increased of total phenols concentration	
	90 nm	0.001–10,000 mg/L	*Raphanus sativus* L.	Reduced in root growth	Pittol et al. (2017)
				Affected plant development	
Zn Zinc	68.14 nm	100 mg/L	*Coffea arabica* L.	Increased leaf penetration	Rossi et al. (2019)
				Improved plant performance	
	15–25 nm	10–200 mg/L	*Triticum aestivum* L	Increased grain yield and biomass	Du et al. (2019)
	50 nm	5–500 mg/L	*Glycine* max L. *cv. Enrei*	Reduced the fresh weight and root length	Mustafa et al. (2015)

Continued

TABLE 2.1 The effects of metal/metallic oxide ENMs on plants.—cont'd

Metallic NPs	Size	Exposure concentration	Plants	Effects on plants	References
Fe Iron	5–30 nm	50 mg/L	*Carthamus tinctorius* L.	Reduced impact of drought and improved yield	Davar Zareii et al. (2014)
	10–50 nm	80–200 mg/L	*Arabidopsis thaliana*	Activatived the plasma membrane Hþ-ATPase	Kim et al. (2015)
				Increased chlorophyll content and plant biomass	
	10–50 nm	50–100 mg/L	*Helianthus annuus* L.	Countered drought stress	Martínez-Fernández et al. (2015)
Na Sodium	60 nm	1–100 mg/L	*Pisum sativum* L.	Protect the seedlings against Cr(VI) phytotoxicity	Tripathi et al. (2015)
				Enhanced accumulation of nutrient elements	
				Improved plant growth	
	20–800 nm	10–250 mg/L	*Lycopersicum esculentum Mill.*	Improved plant growth, and	Haghighi et al. (2014)

				chlorophyll and leaf relative water content	
Al Aluminium	5 nm	50–2500 mg/L	*Nigella arvensis* L.	Increased biomass	Chahardoli et al. (2020)
	30-60 nm	50 mg/L	*Glycine max* L. *cv. Enrei*	Increased seedling weight and root length, regulation of energy metabolism and cell death	Mustafa et al. (2015)
Mg Magnesium	47 ± 36 nm	500 mg/L	*Zea mays*	Improved seed germination and growth, enhanced shoot height and root length	Shinde et al. (2020)
	/	20 mg/L	*Raphanus sativus* L.	Increased the plant growth	Hussain et al. (2019)
				Increased total phenolics, flavonoid, and DPPH free radical scavenging activity	
Ni Nickel	5–8 nm	50 mg/L	*Nigella arvensis* L.	Inhibited antioxidant activities	Chahardoli et al. (2020)
Se Selenium	2–20 nm	1–20 mg/L	*Solanum lycopersicum*	Increased the yield decreased the chlorophyll	Hernández-Hernández et al. (2019)
	50–150 nm	10–30 mg/L	*Allium sativum* L.	Enhanced the growth	Ikram et al. (2020)

antioxidant potential (Kumar et al., 2019; Nair & Chung, 2014b; Syu et al., 2014). It has also been found that the exposure of Au ENMs induced lipid peroxidation in rice, causing cell death in roots and shoots (Tsi Ndeh et al., 2017). Likewise, Ag ENMs stimulated the high-level peroxidase activity and lipid peroxidation in the plant, which further induced membrane damage (Jiravova et al., 2016; Kohan-Baghkheirati & Geisler-Lee, 2015). Ag ENMs also significantly affected the lipid and protein structure of plant cells, as well as lignin, pectin, and cellulose (Zuverza-Mena et al., 2016). It has also been reported that Ag ENMs accumulate in plant leaves, and eventually destroy the thylakoid membrane structure, reducing chlorophyll content, and inhibiting plant growth (Qian et al., 2013). Ag ENMs were even found to destroy the xylem and cause irreversible damage to the antioxidant mechanism (Ghorbanpour & Hatami, 2014).

Although the negative effects of ENMs on plants are discussed, when the dosage of ENMs is appropriate, it will lead to a growth increase. For example, it was found that plants treated with 40 mg/L Ti ENMs improved plant performance under salt stress conditions by increasing antioxidant enzymatic activity and promoting enzyme defense systems (Gohari et al., 2020). Similarly, Se and Cu ENMs increased chlorophyll content and improved the enzymatic (SOD, CAT, glutathione peroxidase (GPX) and phenylalanine ammonia-lyase (PAL)) as well as nonenzymatic antioxidant activity (vitamin C and glutathione) in tomato plant leaves (Hernández-Hernández et al., 2019).

2.4.4 Physiological effects

The exposure of plants to ENMs induces physiological changes including modifications in root length and biomass accumulation. Ag ENMs increased petal longevity and decreased petal abscission on *Pelargonium zonale* (Ghorbanpour & Hatami, 2014). Likewise, Ag ENMs significantly affected the growth, nutrient absorption, and macromolecular conformation of radish sprouts, and impaired the absorption of nutrient elements by radish seedlings (Zuverza-Mena et al., 2016). On the other hand, the ENMs treatment help plants to perform better against environmental stress. For example, exposure of wheat to 30 mg/L Se ENMs improved drought resistance and promoted wheat growth (Ikram et al., 2020; Quiterio-Gutiérrez et al., 2019).

2.5 Conclusion

The emergence of nanotechnology has completely changed the development of many fields of human society. The expansion of the nanoindustry and the continuous accumulation of ENMs in the environment pose an obstacle to sustainable development. The development of qualitative and quantitative methods and ENMs modeling provide important approaches to analyze the environmental sources and concentrations of ENMs. The anthropogenic activities, consumable products, industrial emissions, agricultural applications, and biomedical applications were identified as the major release sources of ENMs into the environment. Considering the advancement of analytical methods and modeling, the environmentally realistic concentrations of ENMs ranged from ng/L to mg/L. The environmental transport and fate of ENMs include atmospheric, terrestrial, and aquatic systems as well as biosystems, being WWTPs and WIPs the major intermediate pathways. As the first producer of the ecosystem, plants are exposed to ENMs in environmental media (e.g., water, soil). Abundant studies have revealed the potential ecotoxicological effects of ENMs in plants, from seed germination to plant growth and development, even inducing abiotic stress, causing subcellular and physiological damage.

References

Abbas, Q., Yousaf, B., Amina, Ali, M. U., Munir, M. A. M., El-Naggar, A., Rinklebe, J., & Naushad, M. (2020). Transformation pathways and fate of engineered nanoparticles (ENPs) in distinct interactive environmental compartments: A review. *Environment International, 138*, 105646. https://doi.org/10.1016/j.envint.2020.105646

Abdel-Aziz, H. M. M., Hasaneen, M. N. A., & Omer, A. M. (2019). Impact of engineered nanomaterials either alone or loaded with NPK on growth and productivity of French bean plants: Seed priming vs foliar application. *South African Journal of Botany, 125*, 102–108. https://doi.org/10.1016/j.sajb.2019.07.005

Abdolahpur Monikh, F., Chupani, L., Arenas-Lago, D., Guo, Z., Zhang, P., Darbha, G. K., Valsami-Jones, E., Lynch, I., Vijver, M. G., van Bodegom, P. M., & Peijnenburg, W. J. G. M. (2021). Particle number-based trophic transfer of gold nanomaterials in an aquatic food chain. *Nature Communications, 12*(1), 899. https://doi.org/10.1038/s41467-021-21164-w

Abu-Qdais, H. A., Abu-Dalo, M. A., & Hajeer, Y. Y. (2021). Impacts of nanosilver-based textile products using a life cycle assessment. *Sustainability, 13*(6), 3436. https://doi.org/10.3390/su13063436

Ahmed, B., Rizvi, A., Ali, K., Lee, J., Zaidi, A., Khan, M. S., & Musarrat, J. (2021). Nanoparticles in the soil–plant system: A review. *Environmental Chemistry Letters*. https://doi.org/10.1007/s10311-020-01138-y

Alexander, A., Jotterand, F., & Elrod, J. (2020). *Market considerations for nanomedicines and theranostic nanomedicines* (pp. 471–491).

Ameen, F., Alsamhary, K., Alabdullatif, J. A., & ALNadhari, S. (2021). A review on metal-based nanoparticles and their toxicity to beneficial soil bacteria and fungi. *Ecotoxicology and Environmental Safety, 213*, 112027. https://doi.org/10.1016/j.ecoenv.2021.112027

Anselmo, A. C., & Mitragotri, S. (2016). Nanoparticles in the clinic. *Bioengineering & Translational Medicine, 1*(1), 10–29. https://doi.org/10.1002/btm2.10003

Avellan, A., Schwab, F., Masion, A., Chaurand, P., Borschneck, D., Vidal, V., Rose, J., Santaella, C., & Levard, C. (2017). Nanoparticle uptake in plants: Gold nanomaterial localized in roots of Arabidopsis thaliana by X-ray computed nanotomography and hyperspectral imaging. *Environmental Science & Technology, 51*(15), 8682–8691. https://doi.org/10.1021/acs.est.7b01133

Azimzada, A., Jreije, I., Hadioui, M., Shaw, P., Farner, J. M., & Wilkinson, K. J. (2021). Quantification and characterization of Ti-, Ce-, and Ag-nanoparticles in global surface waters and precipitation. *Environmental Science & Technology, 55*(14), 9836–9844. https://doi.org/10.1021/acs.est.1c00488

Batley, G. E., Kirby, J. K., & McLaughlin, M. J. (2013). Fate and risks of nanomaterials in aquatic and terrestrial environments. *Accounts of Chemical Research, 46*(3), 854–862. https://doi.org/10.1021/ar2003368

BCC Research. (2017). *Nanotechnology in medical applications | BCC research*. Retrieved from https://www.bccresearch.com/market-research/healthcare/nanotechnology-medical-applications-market.html

Benzon, H. R. L., Rubenecia, M. R. U., Ultra, V. U., & Lee, C. (2015). *Nano-fertilizer affects the growth, development, and chemical properties of rice* (Vol. 14).

Bisht, G., Zaidi, M. G. H., & Sandeep, A. (2014). Impact of gold nanoparticles on physiological and bbiochemical characteristics of Brassica juncea. *Journal of Plant Biochemistry & Physiology*. https://doi.org/10.4172/2329-9029.1000133

Bundschuh, M., Filser, J., Lüderwald, S., McKee, M. S., Metreveli, G., Schaumann, G. E., Schulz, R., & Wagner, S. (2018). Nanoparticles in the environment: Where do we come from, where do we go to? *Environmental Sciences Europe, 30*(1), 6. https://doi.org/10.1186/s12302-018-0132-6

Cervantes-Avilés, P., Huang, Y., & Keller, A. A. (2019). Incidence and persistence of silver nanoparticles throughout the wastewater treatment process. *Water Research, 156*, 188–198. https://doi.org/10.1016/j.watres.2019.03.031

Chahardoli, A., Karimi, N., Ma, X., & Qalekhani, F. (2020). Effects of engineered aluminum and nickel oxide nanoparticles on the growth and antioxidant defense systems of Nigella arvensis L. *Scientific Reports, 10*(1), 3847. https://doi.org/10.1038/s41598-020-60841-6

Chutipaijit, S., Whalley, A. J. S., & Sutjaritvorakul, T. (2018). In vitro plant growth promotion by ZnO nanomaterials in indica rice seedlings (Oryza sativa L.). *Materials Today: Proceedings, 5*(7, Part 1), 14944–14949. https://doi.org/10.1016/j.matpr.2018.04.035

Collivignarelli, M. C., Abbà, A., Carnevale Miino, M., & Torretta, V. (2019). What advanced treatments can be used to minimize the production of sewage sludge in WWTPs? *Applied Sciences, 9*(13), 2650. https://doi.org/10.3390/app9132650

Corsi, I., Winther-Nielsen, M., Sethi, R., Punta, C., Della Torre, C., Libralato, G., Lofrano, G., Sabatini, L., Aiello, M., Fiordi, L., Cinuzzi, F., Caneschi, A., Pellegrini, D., & Buttino, I. (2018). Ecofriendly nanotechnologies and nanomaterials for environmental applications: Key issue and consensus recommendations for sustainable and ecosafe nanoremediation. *Ecotoxicology and Environmental Safety, 154*, 237–244. https://doi.org/10.1016/j.ecoenv.2018.02.037

Cvjetko, P., Milošić, A., Domijan, A.-M., Vinković Vrček, I., Tolić, S., Peharec Štefanić, P., Letofsky-Papst, I., Tkalec, M., & Balen, B. (2017). Toxicity of silver ions and differently coated silver nanoparticles in Allium cepa roots. *Ecotoxicology and Environmental Safety, 137*, 18–28. https://doi.org/10.1016/j.ecoenv.2016.11.009

Davar Zareii, F., Roozbahani, A., & Hosnamidi, A. (2014). Evaluation the effect of water stress and foliar application of Fe nanoparticles on yield, yield components and oil percentage of Safflower (Carthamus Tinctorious L.). *International Journal of Advanced Biological and Biomedical Research, 2*(4), 1150–1159.

Donia, D. T., & Carbone, M. (2019). Fate of the nanoparticles in environmental cycles. *International Journal of Environmental Science and Technology, 16*(1), 583–600. https://doi.org/10.1007/s13762-018-1960-z

Du, W., Yang, J., Peng, Q., Liang, X., & Mao, H. (2019). Comparison study of zinc nanoparticles and zinc sulphate on wheat growth: From toxicity and zinc biofortification. *Chemosphere, 227*, 109–116. https://doi.org/10.1016/j.chemosphere.2019.03.168

Eduok, S., Hendry, C., Ferguson, R., Martin, B., Villa, R., Jefferson, B., & Coulon, F. (2015). Insights into the effect of mixed engineered nanoparticles on activated sludge performance. *FEMS Microbiology Ecology, 91*(7). https://doi.org/10.1093/femsec/fiv082

Elfeky, S., Mohammed, M., Khater, M., Osman, Y., & Elsherbini, A. (2013). Effect of magnetite nano-fertilizer on growth and yield of Ocimum basilicum L. *International Journal of Phytomedicines and Related Industries, 3.*

Feichtmeier, N. S., Walther, P., & Leopold, K. (2015). Uptake, effects, and regeneration of barley plants exposed to gold nanoparticles. *Environmental Science and Pollution Research, 22*(11), 8549–8558. https://doi.org/10.1007/s11356-014-4015-0

Garner, K. L., Suh, S., & Keller, A. A. (2017). Assessing the risk of engineered nanomaterials in the environment: Development and application of the nanoFate model. *Environmental Science & Technology, 51*(10), 5541–5551. https://doi.org/10.1021/acs.est.6b05279

Ghorbanpour, M., & Hatami, M. (2014). Spray treatment with silver nanoparticles plus thidiazuron increases anti-oxidant enzyme activities and reduces petal and leaf abscission in four cultivars of geranium (Pelargonium zonale) during storage in the dark. *The Journal of Horticultural Science and Biotechnology, 89*(6), 712–718. https://doi.org/10.1080/14620316.2014.11513142

Giese, B., Klaessig, F., Park, B., Kaegi, R., Steinfeldt, M., Wigger, H., von Gleich, A., & Gottschalk, F. (2018). Risks, release and concentrations of engineered nanomaterial in the environment. *Scientific Reports, 8*(1), 1565. https://doi.org/10.1038/s41598-018-19275-4

Gohari, G., Mohammadi, A., Akbari, A., Panahirad, S., Dadpour, M. R., Fotopoulos, V., & Kimura, S. (2020). Titanium dioxide nanoparticles (TiO_2 NPs) promote growth and ameliorate salinity stress effects on essential oil profile and biochemical attributes of Dracocephalum moldavica. *Scientific Reports, 10*(1), 912. https://doi.org/10.1038/s41598-020-57794-1

Gonet, T., & Maher, B. A. (2019). Airborne, vehicle-derived Fe-bearing nanoparticles in the urban environment: A review. *Environmental Science & Technology, 53*(17), 9970–9991. https://doi.org/10.1021/acs.est.9b01505

Goswami, L., Kim, K.-H., Deep, A., Das, P., Bhattacharya, S. S., Kumar, S., & Adelodun, A. A. (2017). Engineered nano particles: Nature, behavior, and effect on the environment. *Journal of Environmental Management, 196*, 297–315. https://doi.org/10.1016/j.jenvman.2017.01.011

Gottschalk, F., & Nowack, B. (2011). The release of engineered nanomaterials to the environment. *Journal of Environmental Monitoring*, 1145–1155.

Griffin, S., Masood, M. I., Nasim, M. J., Sarfraz, M., Ebokaiwe, A. P., Schäfer, K.-H., Keck, C. M., & Jacob, C. (2018). Natural nanoparticles: A particular matter inspired by nature. *Antioxidants, 7*(1), 3. https://doi.org/10.3390/antiox7010003

Guerra, F. D., Attia, M. F., Whitehead, D. C., & Alexis, F. (2018). Nanotechnology for environmental remediation: Materials and applications. *Molecules, 23*(7), 1760. https://doi.org/10.3390/molecules23071760

Gull, A., Lone, A. A., & Wani, N. U. I. (2019). *Biotic and abiotic stresses in plants*. Retrieved from https://www.intechopen.com/chapters/66714.

Haghighi, M., Abolghasemi, R., & Teixeira da Silva, J. A. (2014). Low and high temperature stress affect the growth characteristics of tomato in hydroponic culture with Se and nano-Se amendment. *Scientia Horticulturae, 178*, 231–240. https://doi.org/10.1016/j.scienta.2014.09.006

Harrison, R. M., Rob MacKenzie, A., Xu, H., Alam, M. S., Nikolova, I., Zhong, J., Singh, A., Zeraati-Rezaei, S., Stark, C., Beddows, D. C. S., Liang, Z., Xu, R., & Cai, X. (2018). Diesel exhaust nanoparticles and their behaviour in the atmosphere. *Proceedings of the Royal*

Society A: Mathematical, Physical and Engineering Sciences, 474(2220), 20180492. https://doi.org/10.1098/rspa.2018.0492

Hernández-Hernández, H., Quiterio-Gutiérrez, T., Cadenas-Pliego, G., Ortega-Ortiz, H., Hernández-Fuentes, A. D., Cabrera de la Fuente, M., Valdés-Reyna, J., & Juárez-Maldonado, A. (2019). Impact of selenium and copper nanoparticles on yield, antioxidant system, and fruit quality of tomato plants. *Plants, 8*(10), 355. https://doi.org/10.3390/plants8100355

Huang, Y., Gao, M., Wang, W., Liu, Z., Qian, W., Chen, C. C., Zhu, X., & Cai, Z. (2022). Effects of manufactured nanomaterials on algae: Implications and applications. *Frontiers of Environmental Science & Engineering, 16*(9), 122.

Huang, Y., Keller, A. A., Cervantes-Avilés, P., & Nelson, J. (2021). Fast multielement quantification of nanoparticles in wastewater and sludge using single-particle ICP-MS. *ACS ES&T Water, 1*(1), 205–213. https://doi.org/10.1021/acsestwater.0c00083

Huang, Y., Zhao, L., & Keller, A. A. (2017). Interactions, transformations, and bioavailability of nano-copper exposed to root exudates. *Environmental Science & Technology, 51*(17), 9774–9783. https://doi.org/10.1021/acs.est.7b02523

Hussain, F., Hadi, F., & Akbar, F. (2019). Magnesium oxide nanoparticles and thidiazuron enhance lead phytoaccumulation and antioxidative response in Raphanus sativus L. *Environmental Science and Pollution Research, 26*(29), 30333–30347. https://doi.org/10.1007/s11356-019-06206-7

Huang, Y., Zhao, L., & Keller, A. A. (2019). Application of metabolomics to discover the implications of nanomaterials for crop plants. In *Nanoscale engineering in agricultural management*.

Jasim, B., Thomas, R., Mathew, J., & Radhakrishnan, E. K. (2017). Plant growth and diosgenin enhancement effect of silver nanoparticles in Fenugreek (*Trigonella foenum-graecum* L.). *Saudi Pharmaceutical Journal, 25*(3), 443–447. https://doi.org/10.1016/j.jsps.2016.09.012

Iavicoli, I., Leso, V., Beezhold, D. H., & Shvedova, A. A. (2017). Nanotechnology in agriculture: Opportunities, toxicological implications, and occupational risks. *Toxicology and Applied Pharmacology, 329*, 96–111. https://doi.org/10.1016/j.taap.2017.05.025

Ikram, M., Raja, N. I., Javed, B., Mashwani, Z.-R., Hussain, M., Hussain, M., Ehsan, M., Rafique, N., Malik, K., Sultana, T., & Akram, A. (2020). Foliar applications of biofabricated selenium nanoparticles to improve the growth of wheat plants under drought stress. *Green Processing and Synthesis, 9*(1), 706–714. https://doi.org/10.1515/gps-2020-0067

Jiravova, J., Tomankova, K. B., Harvanova, M., Malina, L., Malohlava, J., Luhova, L., Panacek, A., Manisova, B., & Kolarova, H. (2016). The effect of silver nanoparticles and silver ions on mammalian and plant cells in vitro. *Food and Chemical Toxicology, 96*, 50–61. https://doi.org/10.1016/j.fct.2016.07.015

John, A. C., Küpper, M., Manders-Groot, A. M. M., Debray, B., Lacome, J.-M., & Kuhlbusch, T. A. J. (2017). Emissions and possible environmental implication of engineered nanomaterials (ENMs) in the atmosphere. *Atmosphere, 8*(5), 84. https://doi.org/10.3390/atmos8050084

Kaegi, R., Gogos, A., Voegelin, A., Hug, S. J., Winkel, L. H. E., Buser, A. M., & Berg, M. (2021). Quantification of individual rare earth elements from industrial sources in sewage sludge. *Water Research X, 11*, 100092. https://doi.org/10.1016/j.wroa.2021.100092

Keller, A. A., Huang, Y., & Nelson, J. (2018). Detection of nanoparticles in edible plant tissues exposed to nano-copper using single-particle ICP-MS. *Journal of Nanoparticle Research, 20*(4), 101. https://doi.org/10.1007/s11051-018-4192-8

Keller, A. A., & Lazareva, A. (2014). Predicted releases of engineered nanomaterials: From global to regional to local. *Environmental Science & Technology Letters, 1*(1), 65–70. https://doi.org/10.1021/ez400106t

Keller, A. A., McFerran, S., Lazareva, A., & Suh, S. (2013). Global life cycle releases of engineered nanomaterials. *Journal of Nanoparticle Research, 15*(6), 1692. https://doi.org/10.1007/s11051-013-1692-4

Khan, M. N., Mobin, M., Abbas, Z. K., AlMutairi, K. A., & Siddiqui, Z. H. (2017). Role of nanomaterials in plants under challenging environments. *Plant Physiology and Biochemistry, 110*, 194–209. https://doi.org/10.1016/j.plaphy.2016.05.038

Khan, Z., & Upadhyaya, H. (2019). Impact of nanoparticles on abiotic stress responses in plants. In *Nanomaterials in plants, algae and microorganisms* (pp. 305–322). Retrieved from https://linkinghub.elsevier.com/retrieve/pii/B9780128114889000159

Kim, J.-H., Oh, Y., Yoon, H., Hwang, I., & Chang, Y.-S. (2015). Iron Nanoparticle-Induced Activation of Plasma Membrane H^+-ATPase Promotes Stomatal Opening in *Arabidopsis thaliana*. *Environmental Science & Technology, 49*(2), 1113–1119. https://doi.org/10.1021/es504375t

Kohan-Baghkheirati, E., & Geisler-Lee, J. (2015). Gene expression, protein function and pathways of Arabidopsis thaliana responding to silver nanoparticles in comparison to silver ions, cold, salt, drought, and heat. *Nanomaterials, 5*(2), 436–467. https://doi.org/10.3390/nano5020436

Kuenen, J., Pomar-Portillo, V., Vilchez, A., Visschedijk, A., Gon, van der, Vázquez-Campos, S., Nowack, B., & Adam, V. (2020). Inventory of country-specific emissions of engineered nanomaterials throughout the life cycle. *Environmental Science: Nano, 7*(12), 3824–3839. https://doi.org/10.1039/D0EN00422G

Kumar, S., Nehra, M., Dilbaghi, N., Marrazza, G., Hassan, A. A., & Kim, K.-H. (2019). Nano-based smart pesticide formulations: Emerging opportunities for agriculture. *Journal of Controlled Release, 294*, 131–153. https://doi.org/10.1016/j.jconrel.2018.12.012

Kumar, V., Guleria, P., Kumar, V., & Yadav, S. K. (2013). Gold nanoparticle exposure induces growth and yield enhancement in Arabidopsis thaliana. *Science of The Total Environment, 461–462*, 462–468. https://doi.org/10.1016/j.scitotenv.2013.05.018

Lian, J., Zhao, L., Wu, J., Xiong, H., Bao, Y., Zeb, A., Tang, J., & Liu, W. (2020). Foliar spray of TiO_2 nanoparticles prevails over root application in reducing Cd accumulation and mitigating Cd-induced phytotoxicity in maize (Zea mays L.). *Chemosphere, 239*, 124794. https://doi.org/10.1016/j.chemosphere.2019.124794

Liao, Y.-Y., Huang, Y., Carvalho, R., Choudhary, M., Da Silva, S., Colee, J., Huerta, A., Vallad, G. E., Freeman, J. H., Jones, J. B., Keller, A., & Paret, M. L. (2021). Magnesium oxide nanomaterial, an alternative for commercial copper bactericides: Field-scale tomato bacterial spot disease management and total and bioavailable metal accumulation in soil. *Environmental Science & Technology, 55*(20), 13561–13570. https://doi.org/10.1021/acs.est.1c00804

Liu, Z., Peng, R., Lv, S., Wang, A., Zhao, L., Dong, S., Yan, D., Keller, A. A., & Huang, Y. (2022). Evidence of indoor dust acting as carrier for metal-based nanoparticles: a study of exposure and oxidative risks. *Environmental Science & Technology Letters, 9*(5), 431–438.

Liu, P., Yang, M., Hermanowicz, S. W., & Huang, Y. (2022). Efficacy-associated cost analysis of copper-based nanopesticides for tomato disease control. *ACS Agricultural Science & Technology*. https://doi.org/10.1021/acsagscitech.2c00098

Liu, P., Zheng, X., Shangguan, S., Zhao, L., Fang, X., Huang, Y., & Hermanowicz, S. W. (2022). Public perceptions and willingness-to-pay for nanopesticides. *Nanomaterials, 12*(8), 1292.

Loosli, F., Wang, J., Rothenberg, S., Bizimis, M., Winkler, C., Borovinskaya, O., Flamigni, L., & Baalousha, M. (2019). Sewage spills are a major source of titanium dioxide engineered (nano)-particle release into the environment. *Environmental Science: Nano, 6*(3), 763–777. https://doi.org/10.1039/C8EN01376D

Ma, X., & Yan, J. (2018). Plant uptake and accumulation of engineered metallic nanoparticles from lab to field conditions. *Current Opinion in Environmental Science & Health, 6*, 16–20. https://doi.org/10.1016/j.coesh.2018.07.008

Maduraiveeran, G., & Jin, W. (2017). Nanomaterials based electrochemical sensor and biosensor platforms for environmental applications. *Trends in Environmental Analytical Chemistry, 13*, 10–23. https://doi.org/10.1016/j.teac.2017.02.001

Martínez-Fernández, D., Vítková, M., Bernal, M. P., & Komárek, M. (2015). Effects of nano-maghemite on trace element accumulation and drought response of Helianthus annuus L. in a contaminated mine soil. *Water, Air, & Soil Pollution, 226*(4), 101. https://doi.org/10.1007/s11270-015-2365-y

Milewska-Hendel, A., Zubko, M., Karcz, J., Stróż, D., & Kurczyńska, E. (2017). Fate of neutral-charged gold nanoparticles in the roots of the Hordeum vulgare L. cultivar Karat. *Scientific Reports, 7*(1), 3014. https://doi.org/10.1038/s41598-017-02965-w

Mosa, K. A., El-Naggar, M., Ramamoorthy, K., Alawadhi, H., Elnaggar, A., Wartanian, S., Ibrahim, E., & Hani, H. (2018). Copper nanoparticles induced genotoxicty, oxidative stress, and changes in superoxide dismutase (SOD) gene expression in cucumber (Cucumis sativus) plants. *Frontiers in Plant Science, 9*, 872. https://doi.org/10.3389/fpls.2018.00872

Mustafa, G., Sakata, K., & Komatsu, S. (2015). Proteomic analysis of flooded soybean root exposed to aluminum oxide nanoparticles. *Journal of Proteomics, 128*, 280–297. https://doi.org/10.1016/j.jprot.2015.08.010

Nair, P. M. G., & Chung, I. M. (2014a). Assessment of silver nanoparticle-induced physiological and molecular changes in Arabidopsis thaliana. *Environmental Science and Pollution Research, 21*(14), 8858–8869. https://doi.org/10.1007/s11356-014-2822-y

Nair, P. M. G., & Chung, I. M. (2014b). Physiological and molecular level effects of silver nanoparticles exposure in rice (Oryza sativa L.) seedlings. *Chemosphere, 112*, 105–113. https://doi.org/10.1016/j.chemosphere.2014.03.056

Oischinger, J., Meiller, M., Daschner, R., Hornung, A., & Warnecke, R. (2019). Fate of nano titanium dioxide during combustion of engineered nanomaterial-containing waste in a municipal solid waste incineration plant. *Waste Management & Research, 37*(10), 1033–1042. https://doi.org/10.1177/0734242X19862603

Pellico, J., Gawne, P.J., & Rosales, de R. T. M. (2021). Radiolabelling of nanomaterials for medical imaging and therapy. *Chemical Society Reviews, 50*(5), 3355–3423. https://doi.org/10.1039/D0CS00384K

Peng, C., Zhang, W., Gao, H., Li, Y., Tong, X., Li, K., Zhu, X., Wang, Y., & Chen, Y. (2017). Behavior and potential impacts of metal-based engineered nanoparticles in aquatic environments. *Nanomaterials, 7*(1), 21. https://doi.org/10.3390/nano7010021

Piccinno, F., Gottschalk, F., Seeger, S., & Nowack, B. (2012). Industrial production quantities and uses of ten engineered nanomaterials in Europe and the world. *Journal of Nanoparticle Research, 14*(9), 1109. https://doi.org/10.1007/s11051-012-1109-9

Pittol, M., Tomacheski, D., Simões, D. N., Ribeiro, V. F., & Santana, R. M. C. (2017). Macroscopic effects of silver nanoparticles and titanium dioxide on edible plant growth. *Environmental Nanotechnology, Monitoring & Management, 8*, 127–133. https://doi.org/10.1016/j.enmm.2017.07.003

Qian, H., Peng, X., Han, X., Ren, J., Sun, L., & Fu, Z. (2013). Comparison of the toxicity of silver nanoparticles and silver ions on the growth of terrestrial plant model Arabidopsis thaliana. *Journal of Environmental Sciences, 25*(9), 1947–1956. https://doi.org/10.1016/S1001-0742(12)60301-5

Quiterio-Gutiérrez, T., Ortega-Ortiz, H., Cadenas-Pliego, G., Hernández-Fuentes, A. D., Sandoval-Rangel, A., Benavides-Mendoza, A., Cabrera-de la Fuente, M., & Juárez-Maldonado, A. (2019). The application of selenium and copper nanoparticles modifies the biochemical responses of tomato plants under stress by Alternaria solani. *International Journal of Molecular Sciences, 20*(8), 1950. https://doi.org/10.3390/ijms20081950

Rajkovic, S., Bornhöft, N. A., van der Weijden, R., Nowack, B., & Adam, V. (2020). Dynamic probabilistic material flow analysis of engineered nanomaterials in European waste treatment systems. *Waste Management, 113,* 118–131. https://doi.org/10.1016/j.wasman.2020.05.032

Rajput, V. D., Minkina, T., Sushkova, S., Tsitsuashvili, V., Mandzhieva, S., Gorovtsov, A., Nevidomskyaya, D., & Gromakova, N. (2018). Effect of nanoparticles on crops and soil microbial communities. *Journal of Soils and Sediments, 18*(6), 2179–2187. https://doi.org/10.1007/s11368-017-1793-2

Rastogi, A., Zivcak, M., Sytar, O., Kalaji, H. M., He, X., Mbarki, S., & Brestic, M. (2017). Impact of metal and metal oxide nanoparticles on plant: A critical review. *Frontiers in Chemistry, 5,* 78. https://doi.org/10.3389/fchem.2017.00078

Regier, N., Cosio, C., von Moos, N., & Slaveykova, V. I. (2015). Effects of copper-oxide nanoparticles, dissolved copper and ultraviolet radiation on copper bioaccumulation, photosynthesis and oxidative stress in the aquatic macrophyte Elodea nuttallii. *Chemosphere, 128,* 56–61. https://doi.org/10.1016/j.chemosphere.2014.12.078

Rossi, L., Fedenia, L. N., Sharifan, H., Ma, X., & Lombardini, L. (2019). Effects of foliar application of zinc sulfate and zinc nanoparticles in coffee (Coffea arabica L.) plants. *Plant Physiology and Biochemistry, 135,* 160–166. https://doi.org/10.1016/j.plaphy.2018.12.005

Satti, S. H., Raja, N. I., Javed, B., Akram, A., Mashwani, Z.-U.-R., Ahmad, M. S., & Ikram, M. (2021). Titanium dioxide nanoparticles elicited agro-morphological and physicochemical modifications in wheat plants to control Bipolaris sorokiniana. *PLoS One, 16*(2), e0246880. https://doi.org/10.1371/journal.pone.0246880

Shabnam, N., Pardha-Saradhi, P., & Sharmila, P. (2014). Phenolics impart Au(3+)-stress tolerance to cowpea by generating nanoparticles. *PLoS One, 9*(1), e85242. https://doi.org/10.1371/journal.pone.0085242

Shi, J., Kantoff, P. W., Wooster, R., & Farokhzad, O. C. (2017). Cancer nanomedicine: Progress, challenges and opportunities. *Nature Reviews Cancer, 17*(1), 20–37. https://doi.org/10.1038/nrc.2016.108

Shinde, S., Paralikar, P., Ingle, A. P., & Rai, M. (2020). Promotion of seed germination and seedling growth of Zea mays by magnesium hydroxide nanoparticles synthesized by the filtrate from Aspergillus niger. *Arabian Journal of Chemistry, 13*(1), 3172–3182. https://doi.org/10.1016/j.arabjc.2018.10.001

Silva, L. F. O., Santosh, M., Schindler, M., Gasparotto, J., Dotto, G. L., Oliveira, M. L. S., & Hochella, M. F., Jr. (2021). Nanoparticles in fossil and mineral fuel sectors and their impact on environment and human health: A review and perspective. *Gondwana Research, 92,* 184–201. https://doi.org/10.1016/j.gr.2020.12.026

Sun, T. Y., Bornhöft, N. A., Hungerbühler, K., & Nowack, B. (2016). Dynamic probabilistic modeling of environmental emissions of engineered nanomaterials. *Environmental Science & Technology, 50*(9), 4701–4711. https://doi.org/10.1021/acs.est.5b05828

Syu, Y., Hung, J.-H., Chen, J.-C., & Chuang, H. (2014). Impacts of size and shape of silver nanoparticles on Arabidopsis plant growth and gene expression. *Plant Physiology and Biochemistry, 83,* 57–64. https://doi.org/10.1016/j.plaphy.2014.07.010

Tighe-Neira, R., Reyes-Díaz, M., Nunes-Nesi, A., Recio, G., Carmona, E., Corgne, A., Rengel, Z., & Inostroza-Blancheteau, C. (2020). Titanium dioxide nanoparticles provoke transient increase in photosynthetic performance and differential response in antioxidant system in Raphanus sativus L. *Scientia Horticulturae, 269,* 109418. https://doi.org/10.1016/j.scienta.2020.109418

Tolaymat, T., El Badawy, A., Genaidy, A., Abdelraheem, W., & Sequeira, R. (2017). Analysis of metallic and metal oxide nanomaterial environmental emissions. *Journal of Cleaner Production, 143,* 401–412. https://doi.org/10.1016/j.jclepro.2016.12.094

Tou, F., Yang, Y., Feng, J., Niu, Z., Pan, H., Qin, Y., Guo, X., Meng, X., Liu, M., & Hochella, M. F. (2017). Environmental risk implications of metals in sludges from waste water treatment plants: The discovery of vast stores of metal-containing nanoparticles. *Environmental Science & Technology, 51*(9), 4831–4840. https://doi.org/10.1021/acs.est.6b05931

Tripathi, D. K., Singh, V. P., Prasad, S. M., Chauhan, D. K., & Dubey, N. K. (2015). Silicon nanoparticles (SiNp) alleviate chromium (VI) phytotoxicity in Pisum sativum (L.) seedlings. *Plant Physiology and Biochemistry, 96*, 189–198. https://doi.org/10.1016/j.plaphy.2015.07.026

Tsi Ndeh, N., Maensiri, S., & Maensiri, D. (2017). The effect of green synthesized gold nanoparticles on rice germination and roots. *Advances in Natural Sciences: Nanoscience and Nanotechnology, 8*(3), 035008. https://doi.org/10.1088/2043-6254/aa724a

Vance, M. E., Kuiken, T., Vejerano, E. P., McGinnis, S. P., Hochella, M. F., Rejeski, D., & Hull, M. S. (2015). Nanotechnology in the real world: Redeveloping the nanomaterial consumer products inventory. *Beilstein Journal of Nanotechnology, 6*, 1769–1780. https://doi.org/10.3762/bjnano.6.181

Vance, M. E., & Marr, L. C. (2015). Exposure to airborne engineered nanoparticles in the indoor environment. *Atmospheric Environment, 106*, 503–509. https://doi.org/10.1016/j.atmosenv.2014.12.056

Vannini, C., Domingo, G., Onelli, E., De Mattia, F., Bruni, I., Marsoni, M., & Bracale, M. (2014). Phytotoxic and genotoxic effects of silver nanoparticles exposure on germinating wheat seedlings. *Journal of Plant Physiology, 171*(13), 1142–1148. https://doi.org/10.1016/j.jplph.2014.05.002

Vejerano, E. P., Leon, E. C., Holder, A. L., & Marr, L. C. (2014). Characterization of particle emissions and fate of nanomaterials during incineration. *Environmental Science: Nano, 1*(2), 133–143. https://doi.org/10.1039/C3EN00080J

Villanueva-Flores, F., Castro-Lugo, A., Ramírez, O. T., & Palomares, L. A. (2020). Understanding cellular interactions with nanomaterials: Towards a rational design of medical nanodevices. *Nanotechnology, 31*(13), 132002. https://doi.org/10.1088/1361-6528/ab5bc8

Wang, S., Liu, Z., Wang, W., & You, H. (2017). Fate and transformation of nanoparticles (NPs) in municipal wastewater treatment systems and effects of NPs on the biological treatment of wastewater: A review. *RSC Advances, 7*(59), 37065–37075. https://doi.org/10.1039/C7RA05690G

Weir, A., Westerhoff, P., Fabricius, L., Hristovski, K., & von Goetz, N. (2012). Titanium dioxide nanoparticles in food and personal care products. *Environmental Science & Technology, 46*(4), 2242–2250. https://doi.org/10.1021/es204168d

Wu, Y., Pang, H., Liu, Y., Wang, X., Yu, S., Fu, D., Chen, J., & Wang, X. (2019). Environmental remediation of heavy metal ions by novel-nanomaterials: A review. *Environmental Pollution, 246*, 608–620. https://doi.org/10.1016/j.envpol.2018.12.076

Zhang, H., Chen, S., Jia, X., Huang, Y., Ji, R., & Zhao, L. (2021). Comparation of the phytotoxicity between chemically and green synthesized silver nanoparticles. *Science of The Total Environment, 752*, 142264. https://doi.org/10.1016/j.scitotenv.2020.142264

Zhang, Q., Lu, D., Wang, D., Yang, X., Zuo, P., Yang, H., Fu, Q., Liu, Q., & Jiang, G. (2020). Separation and tracing of anthropogenic magnetite nanoparticles in the urban atmosphere. *Environmental Science & Technology, 54*(15), 9274–9284. https://doi.org/10.1021/acs.est.0c01841

Zhang, W., Jia, X., Chen, S., Wang, J., Ji, R., & Zhao, L. (2020). Response of soil microbial communities to engineered nanomaterials in presence of maize (Zea mays L.) plants. *Environmental Pollution, 267*, 115608. https://doi.org/10.1016/j.envpol.2020.115608

Zhang, Z., Ke, M., Qu, Q., Peijnenburg, W. J. G. M., Lu, T., Zhang, Q., Ye, Y., Xu, P., Du, B., Sun, L., & Qian, H. (2018). Impact of copper nanoparticles and ionic copper exposure on wheat (Triticum aestivum L.) root morphology and antioxidant response. *Environmental Pollution, 239*, 689–697. https://doi.org/10.1016/j.envpol.2018.04.066

Zhu, Y., Xu, F., Liu, Q., Chen, M., Liu, X., Wang, Y., Sun, Y., & Zhang, L. (2019). Nanomaterials and plants: Positive effects, toxicity and the remediation of metal and metalloid pollution in soil. *Science of The Total Environment, 662*, 414–421. https://doi.org/10.1016/j.scitotenv.2019.01.234

Zuverza-Mena, N., Armendariz, R., Peralta-Videa, J. R., & Gardea-Torresdey, J. L. (2016). Effects of silver nanoparticles on radish sprouts: Root growth reduction and modifications in the nutritional value. *Frontiers in Plant Science, 7*, 90. https://doi.org/10.3389/fpls.2016.00090

CHAPTER

3

Characterization of ENMs in water, soil, and plant media

J.G. Parsons[1], Helia M. Morales[1,2], E. Kotsikorou[1], R.K. Dearth[3], Rupesh Kariyat[3], and Mataz Alcoultabi[4]

[1] Department of Chemistry, University of Texas Rio Grande Valley, TX, United States; [2] Escuela de Ingenieria y Ciencias, Tecnologico de Monterrey, Monterrey, NL, Mexico; [3] Department of Biology, University of Texas Rio Grande Valley, TX, United States; [4] Department of Mechanical Engineering, University of Texas Rio Grande Valley, TX, United States

3.1 Introduction

Engineered Nanomaterials (ENMs) have become an environmental concern due to the increased utilization in consumer products. In 2007, only 54 products contained nanomaterials, which increased to 1015 by 2009 (Vance et al., 2015). As of December 2020, there were 5000 consumer products containing nanomaterials of various compositions (Hansen et al., 2020). Thus, as consumer usage increases, anthropogenic waste is inevitable. Fig. 3.1 depicts the theoretical movement of ENMs through our environment and illustrates the potential for the complete environmental contamination. The hazards of nanomaterials to both environmental and human health are still not completely understood. Thus, there is a need to characterize the impact ENMs have on biotic and abiotic factors that shape our environment.

Investigators focused on the environmental mobility of ENMs, and the accumulation, transformation, and translocation of ENMs in biotic conditions have informed us on the necessity for improved detection methodologies. The development of proper instrumental techniques and methodologies provide the necessary tools to perform these types of studies. Some of these techniques are common-place in the laboratory

Physicochemical Interactions of Engineered Nanoparticles and Plants
https://doi.org/10.1016/B978-0-323-90558-9.00002-4

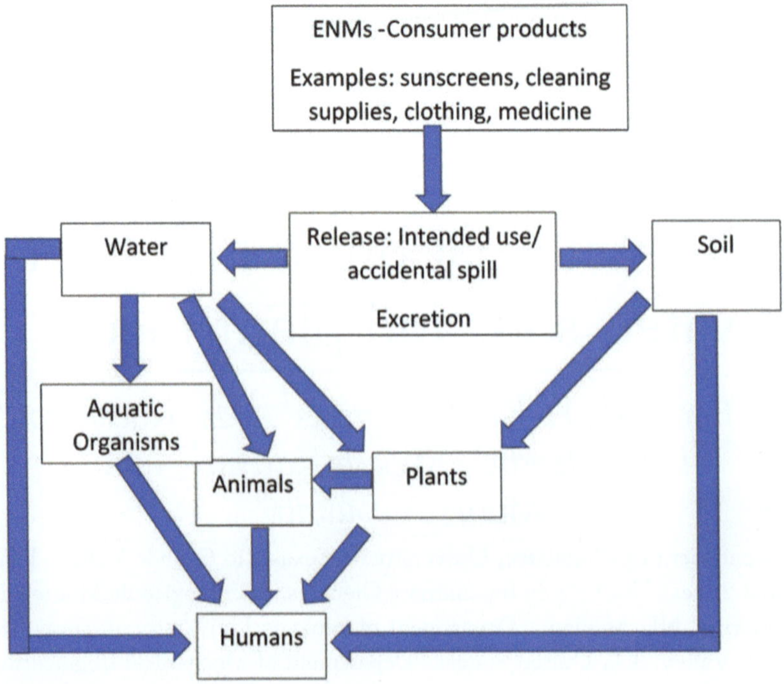

FIGURE 3.1 Flow chart showing theoretical movement and transport of ENMs in the environment.

today and some are still specialized. For example, XRD, TEM, SEM UV-Vis, DLS, and RAMAN are common instrumental techniques used in the detection and characterization of ENMs in environmental-based samples (Alanazi et al., 2019; Dhandapani et al., 2012; Ellis et al., 2016; Farré et al., 2011; Hoque et al., 2012; Kumari et al., 2014; Li et al., 2016; Petosa et al., 2010; Saleh, 2020; Schavkan et al., 2019). More specialized techniques include small angle X-ray scattering (SAXS) and X-ray absorption spectroscopy (XAS) (Farré et al., 2011; Hoque et al., 2012; Saleh, 2020; Schavkan et al., 2019; Zhao et al., 2021). The utilization of SAXS has become a more common practice in the laboratory over the past decade. Also, benchtop XAS systems are available for purchase and laboratory experimentation. Some of the above-mentioned techniques are limited to the media or matrix of the ENMs under investigation, for example, SP-ICP-MS, UV-Vis, and DLS are limited to liquid samples, which implies that the ENMs must be extracted into a liquid medium for analysis while SEM, TEM, and XRD are limited to solid samples, which require sample pretreatment such as drying. RAMAN, XAS, and SAXS are not limited to the matrix since these techniques work in both solid and liquid media.

This chapter will focus on the various methods of characterization of ENMs in the environment. Traditionally, the techniques used for the characterization of nanomaterials in various matrices have included Transmission Electron Microscopy (TEM), Selected Area Electron Diffraction (SAED), Scanning Electron microscopy (SEM), X-ray diffraction (XRD), X-ray Photoemission Spectroscopy (XPS), X-ray Absorption Spectroscopy (XAS), Small Angle X-ray Scatter (SAXS), Wide Angle X-ray Scatter (WAXS), RAMAN, Ultraviolet-Visible spectroscopy (UV-Vis), and Dynamic light scattering (DLS). More recently, techniques such as Single Particle ICP-MS (spICP-MS) are becoming more popular.

3.2 Instrumental Techniques for characterization of ENMs

3.2.1 Ultraviolet-visible spectroscopy (UV-Vis)

Direct and indirect Ultraviolet-Visible spectroscopy (UV-Vis) techniques have been developed for the determination of ENMs in solutions, particularly aqueous media. The direct method for the determination of ENMs in solution has been through plasmon resonance measurement. On the other hand, the indirect methods involve the use of fluorescence dyes to bind to the surface of NPs.

Surface plasmon resonance (SPR) is an electronic process where the electrons at the surface of a metal or conducting NPs are excited by photons of light and are propagated parallel to the surface (Zeng et al., 2017). The SPR is typically expressed as an enhancement of the absorption signal in the UV-VIS; however, it is typically of low intensity. The position of the plasmon peak has been linked to the size of the NPs. Typically, the larger the NPs are, the more red-shifted the position of the plasmon resonance (Lee et al., 2008; Mogensen & Kneipp, 2014; Yeshchenko et al., 2012). Fig. 3.2 shows the plasmon resonance of Au ENMs generated for the synthesis of Au using aqueous alfalfa extracts. The data are provided as an example of the plasmon resonance and to exemplify the observed diffuse absorption.

The insert shows a magnified view of the plasmon resonance for the gold ENMs centered around 535 nm, which is typical of Au ENMs around 50 nm in size. The intensity of the plasmon resonance feature in this case was approximately 0.02 absorption units.

3.2.1.1 Application of UV-Vis in NPs detection

Kim et al. (2019) developed a separation technique to extract gold NPs from biological matrices and used plasmon resonance to determine NPs distribution in tissue extracts of rats. Using UV-Vis, Sikder et al. (2018) investigated the behavior of polyvinyl pyrrolidine (PVP)

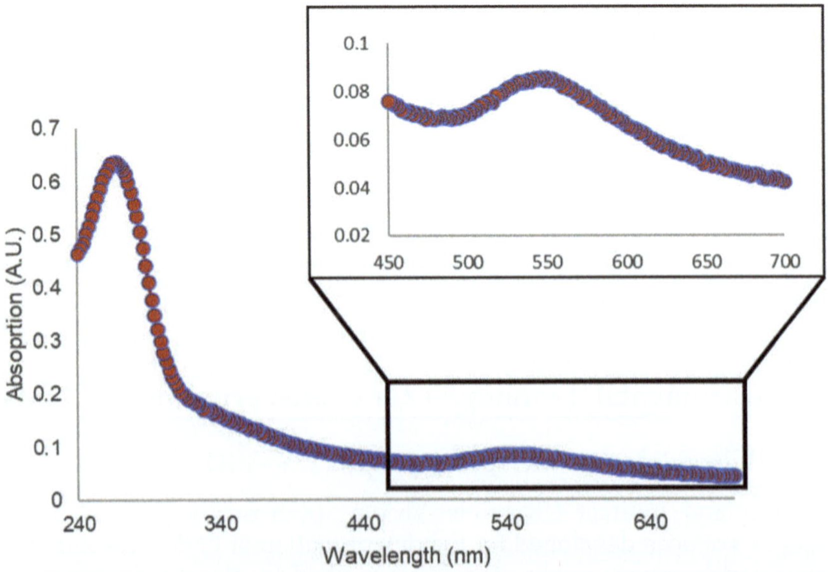

FIGURE 3.2 Example of plasmon resonance for gold NPs in aqueous solution.

stabilized-Ag-NPs in sea water. The authors used the extinction coefficients and maximum absorption wavelength to determine particle size in the UV-Vis spectrum. In that work, Sikder et al. (2018) developed a method to determine the correlation between NPs size, maximum absorption wavelength, and extinction coefficient, which negated the need to measure particle size using other methods. The authors noted that at lower concentrations, NPs size decreased faster than at higher NPs concentrations. The results reported in that study were verified using ICP-MS and AFM. In a closely related study, Jiménez-Lamana and Slaveykova (2016) investigated silver NPs in lake water and the effect of stabilizing agents and dissolved organic matter. Their results indicated that those NPs capped with PVP showed the highest stability. The SP-ICP-MS results reported by Jiménez-Lamana and Slaveykova (2016) were in agreement with the agglomeration observed using UV-Vis surface plasmon resonance. António et al. (2015) used 60 nm silver NPs to study the effects of humic acid and alginate on silver NPs agglomeration using asymmetric field flow fractionation with SP-ICP-MS and UV-Vis. The authors proved the agglomeration of the NPs in simulated seawater. In addition, it was observed that the presence of alginate or humic acids affected the rates of agglomeration. The free silver ENMs and those reacted with alginate showed stability and no agglomeration for over 2 days. Furthermore, the NPs that reacted with humic acids were slightly more stable than those stabilized with alginates. The authors used the Ag surface plasmon resonance at 400–450 nm for the UV-Vis study. In a similar work, Delay et al. (2011) used flow field fractionation with UV-Vis detection to investigate the effect of dissolved natural organic matter (NOM) on the stability of aqueous silver NPs and provide information on chemical structure and composition of ENMs. The authors synthesized their own NPs by the reduction of silver nitrate with sodium borohydride in the presence of different amounts of natural organic materials. The study showed that as NOM increased, NPs stability increased even in the presence of high ionic strength. In addition, the increased stability was caused by the adsorption of NOM to the surface of the NPs.

Zhang et al. (2015) used UV-Vis among other instrumental techniques to determine chloride-induced shape transformation of silver NPs in water. The authors used UV-Vis-NIR extinction spectroscopy and TEM to determine the size of triangular ENMs over time in a chloride solution. The authors concluded that Ag-AgCl NPs were created and hexapode NPs were unstable in aqueous environments. In another study, Li et al. (2019) used UV-Vis to determine the stability of Ag in "real-world" environments. The authors suspended silver NPs both in mineral and tap water and were able to detect the silver NPs without the addition of dyes or electrolytes. The NPs suspended in water displayed plasmon resonances, which were related to their size and shape. Citrate and KCl were added to the solutions to determine the effects on the aging in bottled and tap water. The results showed 30% decrease in the absorbances and a red shift, which indicated the agglomeration of the ENMs. The UV-Vis data proved the stability of NPs in "real-world' environments. Amde et al. (2019) used flow field fractionation with a UV-Vis detector to investigate ZnO ENMs in aqueous solutions. Maximum absorption occurred below 400 nm, which was characteristic of ZnO NPs. The data determined by the UV-Vis analysis were confirmed with other techniques including DLS, TEM, and XRD. The authors tested two ZnO EMNs size ranges (83–97 and 188–200 nm) and the UV-Vis detection on the flow field fractionation system was effective to determine the separation of the NPs in solution. UV-Vis is a rapid method for the detection of ENMs in solution. However, typically a secondary method is necessary for the confirmation of particle size.

3.2.2 Raman spectroscopy

Raman spectroscopy has been applied for the characterization of NPs (Ilie et al., 2017; León et al., 2017; Liu et al., 2017; López-Sánchez et al., 2016; Pandya et al., 2021). Typically, in Raman spectroscopy, the analyte is excited with a laser and the scattering of the light measured. Raman scattering includes stokes, anti-stokes, and Rayleigh scattering, which results in a normal Raman pattern. Raman scattering is proportional to the bond strength in the molecules and provides a spectrum (similar to an IR spectrum) that can be correlated to the bonds present in a sample. With surface-enhanced RAMAN spectroscopy (SERS), the same processes are used as in conventional Raman scattering, but the good contact between the analyte and NPs results in enhanced RAMAN scattering with the same Rayleigh scattering. The two processes are shown in Fig. 3.3.

SERS can be used to determine NPs concentration, the interactions of ENMs with biological materials, organic chemicals, and to follow the transformation of NPs in environmental samples (Bharadwaj et al., 2011; Bharadwaj et al., 2018; Gribanyov et al., 2021; Guo et al., 2016; Lahr & Vikesland, 2014; Quarato et al., 2021; Wei et al., 2019; Yilmaz et al., 2021; Yoshida et al., 2010; Zhang et al., 2016). SERS results from the interaction between the scattered light and the local plasmon resonance at a surface such as that of an NP. The signal enhancement takes place due to the scattering occurring at or near the LSPR and allows the NPs to reemit the radiation from near-by molecules. The NPs act as signal amplifiers causing an additive effect and thus increasing the signal (Bharadwaj et al., 2011; Yoshida et al., 2010).

Traditionally, SERS has been used for the determination of molecules in food, chemicals in the environment, and biological interactions. More recently, SERS has been applied from an environmental perspective for the detection of NPs in plants, algae, fresh water, and sea water samples (Guo et al., 2016; Lahr & Vikesland, 2014; Quarato et al., 2021; Zhang et al., 2016).

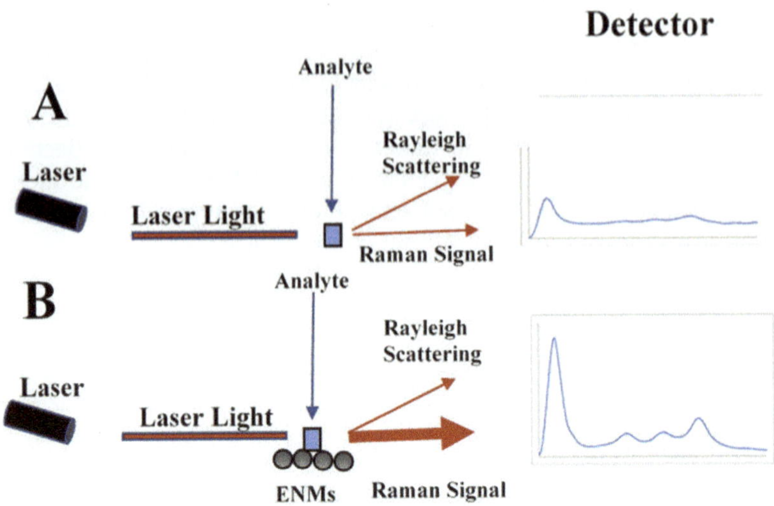

FIGURE 3.3 (A) Conventional Raman spectroscopy for an analyte (B) SERS enhanced RAMAN for the same analyte.

3.2.2.1 *Raman of NPS in environmental matrices*

Recently, Yilmaz et al. (2021) used SERS to observe the synthesis of Ag NPs in live maize seedlings and their subsequent phytotoxicity. The authors prepared chemically stabilized and green synthesized ENMs, and both were functionalized with Ellman's reagent (5,5′-dithiobis-(2-nitrobenzoic acid)) (DTNB). A highly distinctive SERS signal from the roots and shoots of the NPs-treated corn plants was collected, indicating the presence of the nanomaterial on the plant's surface. The study showed the effectiveness of SERS in tracking ENM movement in plants. The authors also noted that the ENMs were located in the epidermis and cortex of the roots and phloem of the shoots. Also, Zhang et al. (2016) were able to map gold NPs in the leaves of spinach plants using SERS. The study was presented as a nondestructive method to detect Au and Ag ENMs in plants. In spinach leaves treated with 15–125 nm Au ENMs, these researchers observed SERS in the spectrum at 1525 cm^{-1}. The Raman mapping of the ENM-treated plants showed aggregation in particular portions of the plants. In a related study, Lahr and Vikesland (2014) used SERS to generate 3D mapping of Au ENMs produced by the green algae Pseudokirchneriella subcapitata exposed to tetrachloroaurate solutions (0.005–0.05 mM gold) at different pHs. The SERS data showed the involvement of glutathione, β-carotene, chlorophyll *a*, hydroxyquinoline, NAD, as well as proteins and peptides. On the other hand, Guo et al. (2016) synthesized different molecules to optimize Ag ENMs SERS signal. Among seven different molecules tested, the best in generating SERS signal was ferric dimethyl-dithiocarbamate (ferbam). In addition, a detection limit of 100 ppb was achieved, and Ag EMNs were distinguished from natural organic matter, dissolved ions, inorganic particles, and biologically relevant molecules. Quarato et al. (2021) studied 15, 100, and 50–80 nm Ag ENMs with a PVP coating. They found out that 15 and 100 nm Ag ENMs were stable in solution, while the 50–80 ENMs were aggregated into large clustered particles. SERS was effective in quantifying both singles and aggregated ENMs in seawater regardless of ENM size.

3.2.3 Dynamic light scattering

Dynamic light scattering (DLS) has been widely used for the determination of NP size in solutions. Typically, DLS determines the hydrated size of an NP, which is different from that determined by other techniques. In DLS, a sample of NPs is irradiated by a laser beam and the light is scattered. This light is detected at specific angles and can be correlated to give an average particle size in the sample as illustrated in Fig. 3.4 where θ is the angle of scattered light from the normal transmitted beam.

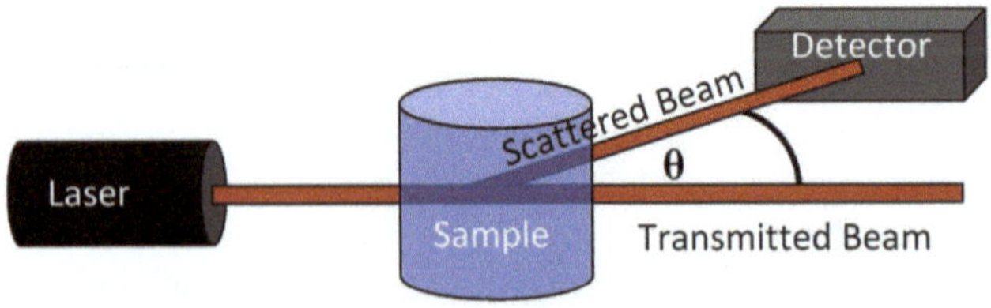

FIGURE 3.4 Diagrammatic representation of the set up for a DLS experiment.

FIGURE 3.5 Diagrammatic representation of a hydration sphere around an ENM in aqueous solution.

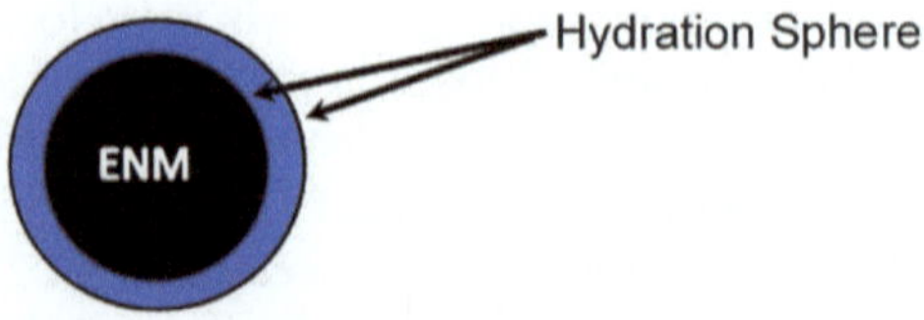

From the recorded spectrum, the size of the NPs and hydrodynamic size can be calculated. The hydrodynamic diameter accounts for the hydration sphere on the surface of the NPs, as exemplified in Fig. 3.5.

In DLS measurements, the particle appears slightly larger than the naked particle (i.e., not in solution), which is observed by other techniques. DLS and UV-Vis are more relevant for the detection of NPs in solution and not for solid phases. To determine NPs in solid samples, extraction/digestion procedures must first be performed.

3.2.3.1 *Applications of DLS in NPs detection*

Tomaszewska et al. (2013) determined the detection limits of DSL spectroscopy for the characterization of polydisperse NPs in solution. The authors noted that DLS was effective for the measurement of narrow particle size distribution in the range of 2–500 nm, and a few large particles could affect the observation of the NPs present in the sample. At a 5% concentration of large particles, the scattering of the smaller particles was completely blocked. Nastulyavichus et al. (2021) used DLS to detect silver ENMs and pathogenic bacteria in food as well as the interactions between the NPs and *Staphylococcus aureus*, *Pseudomonas aeruginosa*, *Listeria monocytogenes*, *Escherichia coli*, and *Salmonella*. The data indicated the leaking of the intracellular content in Gram-positive bacteria (*S. aureus*) and the loss of membrane integrity in the Gram-negative bacteria (*P. aeruginosa*). A study by Baysal et al. (2019) investigated the physicochemical transformation of ZnO and TiO_2 ENMs in sea water and the toxicity on bacteria using DLS after 24 h of exposure. The zeta potential of the ENMS became more negative with longer exposure to seawater. Furthermore, ZnO NP decreased with exposure time while the size of the

TiO_2 ENMs increased with exposure time to seawater. In a simpler study, Keller et al. (2010) used DLS to determine the stability and aggregation of TiO_2, ZnO, and CeO_2 ENMs in seawater, lagoon river water, and ground water. DLS results indicated that the sizes of TiO_2, CeO_2, and ZnO particles were 194, 231, and 205 nm, respectively, which indicated particles were hydrated. The authors noted that the aggregation of the ENMs in seawater was much quicker than in freshwater. In freshwater, the aggregates were greater than 300 nm within the first 2 minutes while in seawater, they (or the aggregates) were greater than 500 nm. In a similar study, Akaighe et al. (2011) investigated humic acid-induced Ag ENMs formation under environmentally relevant conditions. In that study, DLS was used to determine the average particle size under different conditions of temperature, time, and various naturally occurring organic matter. It was observed that neither the soil-derived humic acids nor Suwannee River humic acids generated Ag NMs at room temperature, even when reacted for 25 days. The generation of Ag ENMs was observed within 90 min of reaction at 90°C. The DLS showed that Ag ENMs were polydispersed with respect to particle size and were stable in solution. In a similar study, Gondikas et al. (2020) determined that biological activity in water may play a seasonal role in the stability and instability of ENMs in the environment. Chinnapongse et al. (2011) performed a more comprehensive study on the dispersion of Ag ENMs in freshwater, synthetic seawaters, and simulated estuarine waters. The study used a combination of UV-Vis absorbance spectroscopy, DLS, and AFM to determine the stability of Ag ENMs in the aforementioned environments. Ag EMNs dispersed in seawater with 20 mM Cl^- were unstable while the size approached zero within 10 h of mixing. Lower salinity with NOM showed intermediate stability of the ENMs with time. However, the Ag ENMs dispersed in moderately hard reconstituted water showed signs of agglomeration.

Yin et al. (2015) studied the effects of water chemistry-controlled aggregation and transformation of Ag ENMs in environmental waters. It was determined that both ionic strength and NOM concentration were important parameters affecting Ag ENMs agglomeration. In a separate study, Deonarine and Hsu-Kim (2009) investigated the precipitation of HgS NMs from waters containing naturally occurring organic material. The authors used DLS to monitor the size of the HgS NMs over time. The results showed the presence of NOM in solution stabilized the particles and aggregates with sizes of less than 200 nm were formed. Sulfur containing NOM and organic acids, such as cysteine, also can decrease the growth of HgS NMs. DLS has become an important technique for the characterization of ENMs in aqueous samples with respect to particle agglomeration and particle stability. The only problem noted in the literature for DLS was that larger particles can mask the signal of smaller ENMs in the solution.

3.2.4 Small angle X-ray scattering (SAXS)

The small angle X-ray scattering (SAXS) technique is also based on the scattering of light through a sample. However, SAXS can be used to determine particle sizes in both solution and solid samples. The measurement of sample scattering is similar to that performed by DLS where all the physics and light scattering are the same. SAXS has been used to determine particle size, shape, and pore size (Chen et al., 2012; Harada et al., 2009; Ramallo-López et al., 2007; Schindler et al., 2015; Sciortino et al., 2011; Wang et al., 2012). However, both DLS and SAXS lack the ability to identify ENMs composition, since they provide only structural information. Typically for SAXS, the data collection is recorded from 0.1 to 10 degrees in 2θ. Fig. 3.6 shows a typical SAXS experimental set up.

SAXS has become a more common technique in the laboratory with the development of better X-ray sources and detectors. The application of SAXS for the determination of NP size, shape, and number of scatters are the primary reasons for its use and growth.

3.2.4.1 NPs in water, sediments, and soils

SAXS has not been overly used in environmental studies for ENMs in NOM such as plant, soils, natural waters, and sediments. However, there are a few examples in the literature for these applications, which are outlined below. Tsukimura et al. (2021) used SAXS to quantify the amorphous NPs in clay soils and marine sediments. In that study, it was determined that the size of NPs ranged from 4 to 16 nm in diameter. Similarly, Gentile et al. (2018) used SAXS to determine ferrihydrite ENMs aggregation induced by dissolved organic matter. The study showed that organic compounds extracted from a boreal forest soil induced aggregation of the ferrihydrite ENMs. On the other hand, Vikesland et al. (2016) investigated the aggregation of magnetite ENMs and the subsequent sedimentation. The authors synthesized Fe_3O_4 NPs using two different techniques, one with NaOH as the precipitation base, while the second

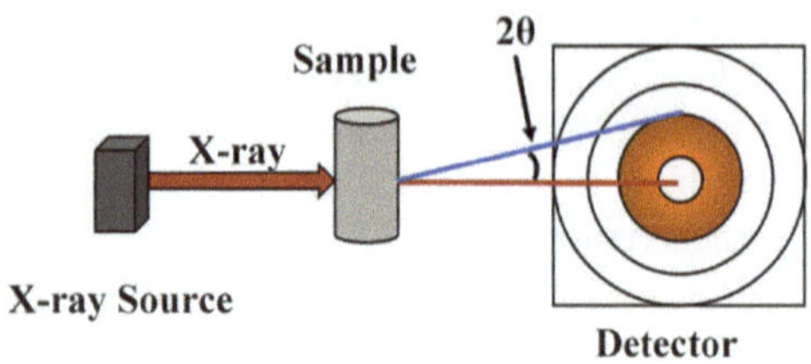

FIGURE 3.6 Diagrammatic representation of a SAXS experiment.

one used TMAOH as the base. From the SAXS analysis, it was found that the particles had a size of 2.1 and 2.2 nm for the TMAOH and NaOH precipitations, respectively. Furthermore, the authors used SAXS to determine the fractal dimension, which showed that the TMAOH synthesized NPs were less densely packed. Legg et al. (2014) investigated the impacts of ionic strength on aggregate structure of 3D NPs as well as the effects on the fate and transport in the environment. The authors combined SAXS and cryogenic TEM to observe the structure of ferrihydrite NP aggregates and how the aggregation changes with ionic strength. The authors noted that in deionized water, the NPs formed stable suspensions with low-density aggregates. In a second column, the authors noted limited deposition on the sand grains. In $NaNO_3$ solutions, the aggregates became denser, higher deposition was observed, and transport was inhibited.

In a different type of study, Kabelitz et al. (2015) followed the in-situ formation of iron oxide NPs using both XANES and SAXS. SAXS was used to provide structural information on the synthesis of NPs in solution. SAXS data showed a NP diameter of 10.8 nm. Also, growth of the NPs in solution resulted in very large-sized precursor akaganeite (FeOCl) crystal, with a diameter of 20 nm (as determined by XRD). The addition of tetraethyl ammonium (TREA) resulted in the formation of small particles that were observed within the first 10 min of reaction. These particles presented a radius of gyration of 1.6 nm. After 14 min of reaction, the NPs were 4.1 nm in size. The addition of HCl to the mixture led to a change in the growth rates of the NPs, which gave a radius of gyration of 3.3 nm corresponding to a 5.6 nm diameter of ENMs. Tobler et al. (2009) used SAXS and DLS to determine the nucleation and growth of silica ENMs in aqueous solution at neutral to slightly basic conditions with silica concentrations of either 640 or 1600 mg/L as well as varied ionic strength from 0.02M to 0.22M. The authors noted that the shape and size dispersion of the NPs did not change under different growing conditions. Similarly, Wang et al. (2010) used a combination of SAXS and UV-Vis studies to investigate the growth of both nanorods and spherical gold NPs in seeded aqueous solutions. The data showed the highest rate of the NPs growth in solution occurred from 0 to 5 min and the rate decreased thereafter. Also, a competitive growth between the spheres, rods, and the size of the ENMs was observed within the first 18 min and the growth of the Au particles stopped after 36 min. Anaraki et al. (2020) used SAXS to determine the effects of ionic strength, NPs concentration, pH, and protein on the stability of silica ENMs in solution. They observed no effect when the diameter of the particles increased to approximately 7 nm after 2.5 h. However, in the presence of protein in the solution, the ENMs started to agglomerate very quickly (under 1 min) while the NPs average diameter increased to approximately 15 nm.

The ability to monitor real-time changes in particle size, agglomeration, and dissolution provides invaluable information on the interactions of ENMs with the environment and potential health effects.

3.2.5 X-ray absorption spectroscopy (XAS)

XAS has become a powerful tool in environmental chemistry and for the detection and determination of NPs in various matrices (Gardea-Torresdey et al., 2002, 2003, 2000; Hernandez-Viezcas et al., 2016; House et al., 2017; Majumdar et al., 2014; Marshall et al., 2007; Parsons et al., 2010; Roberts et al., 2003; Servin et al., 2012; Srabionyan et al., 2014, 2013). XAS is a well-known experimental method which consists of two complementary techniques, Extended X-ray Absorption Fine Structure (EXAFS) and X-ray Absorption Near Edge Structure (XANES). EXAFS gives information on the nearest neighboring atoms, which includes the interatomic distances, the coordination number, and the identity of the nearest neighboring atoms (Parsons et al., 2002). XANES gives similar information, but in addition, this technique provides data on the oxidation state and atomic geometry. Fig. 3.7 shows the relative position of the two spectra in the complete XAS spectrum.

The size of the NPs is determined by performing a calculation on the fitting of the EXAFS region (Borowski, 1997). The data collected by this technique are determined by the following equations:

$$N(\rho) = 1 - \frac{3}{4}\rho + \frac{1}{16}\rho^3 \quad (3.1)$$

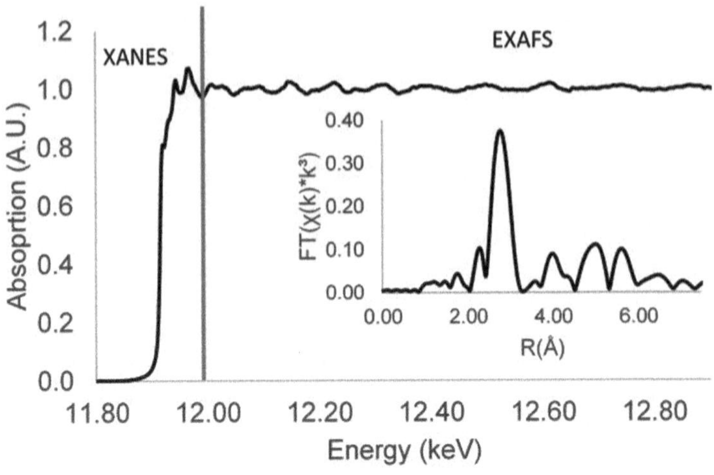

FIGURE 3.7 XAS spectrum showing the separation of the XANES and EXAFS spectrum for a gold foil.

$$N_{\text{ratio}} = \frac{N_{\text{cluster}}}{N_{\text{bulk}}} = 1 - \frac{3}{4}d/R + \frac{1}{16}d^3/R^3 \tag{3.2}$$

$$F(R) = R^3(N_{\text{ratio}} - 1) + \frac{3}{4}\,dR^2 + \frac{1}{16}d^3 \tag{3.3}$$

where N is the average number neighbors, ρ is the reduced cluster size, (Eq. 3.1), R is the nearest neighboring distance, N_{ratio} is the ratio between nearest neighboring atoms in the ENMs and the bulk, d is the particle diameter (Borowski, 1997; Gardea-Torresdey et al., 2000). The above EXAFS analysis for particle size typically works only for small particles, which are 2–3 nm in diameter. For a particle size above 2–3 nm, the first shell EXAFS begins to look like the spectrum of the bulk. In addition, the EXAFS analysis typically only works for single component NPs (Beale & Weckhuysen, 2010; Comaschi et al., 2008; House et al., 2017; Srabionyan et al., 2013, 2014). The types of ENMs investigated using EXAFS for size determination include precious metal catalysts and gold NPs on inactivated tissues of different plants (Beale & Weckhuysen, 2010; Comaschi et al., 2008; House et al., 2017; Srabionyan et al., 2013, 2014).

3.2.5.1 *XANES NPs analysis in soils and plants*

Adediran et al. (2021) used XANES analysis to perform S and P speciation from colloidal particles leached from agricultural soils. It was observed that organic P adsorbed to the surface of Al-containing particles were the most common found P species, approximately 70% of the total (Adediran et al., 2021). However, P adsorbed to Fe-containing particles was maximum 30% of the total P in the soil. In addition, S reduced forms were observed as the major species and indicated reducing redox conditions in the soils. In another study, Settimio et al. (2014) used XANES to investigate the fate and mobility of Ag in soils with respect to aging of the soil. The soils were treated with Ag^+ ions and allowed to age for different time intervals. The mobility of the Ag ions was reduced by 20%–90% within the first 2 weeks of aging and by 6 months, 50%–100% reduction in mobility was observed. In addition, the XANES analysis of the samples showed the reduction of the Ag^+ ions to Ag NPs and a high association of the Ag with S groups in both the soils and Fe-oxyhydroxides. XANES and EXAFS were recently used by Burachevskaya et al. (2021) to investigate the chemical conversion of CuO and CuO NPs in soils and the accumulation in *Hordeum sativum* (spring barley). The authors treated the plants with concentrations from 300 mg/L–10,000 mg/L of either CuO or CuO NPs. The analysis of the soils showed a higher content of Cu^{2+}, which is the mobile form in the soil column. Furthermore, the effects caused by the CuO ENMs were more significant than those observed with the CuO micro-sized particles. The XANES analysis showed that the Cu introduced into the plants was 70% in a coordination environment consistent with CuO. The authors concluded that CuO in the nanoform posed a greater environmental risk than in the micron size. Wang et al. (2016) used XANES to study the dissolution of hydroxyapatite NPs using low molecular weight organic acids. The results showed that under batch conditions, only 30% of the NPs were dissolved. On the other hand, under column (flow) conditions, it was observed that 100% of the NPs were dissolved. These types of ENMs environmental interactions have been further investigated by Sekine et al. (2017), who used XAS to determine the effects of aging of dissolved Cu and Cu-based NPs in five soil types. The authors used 4 days as a short-term and 4 months as the long-term study parameters to investigate dissolved Cu, CuO NPs, and CuS-NPs at 100 mg/kg concentrations. It was observed that low pH soils promoted the dissolution of CuO ENMs while at higher pH, the CuO ENMs were more persistent. These results were similar to those observed for CuS (Sekine et al., 2017). The dissolved Cu ions formed the typical complexes within soils. In addition, the presence of soil organic matter appeared to reduce the rate of dissolution. However, after 4 months, the speciation of Cu was dictated by the soil properties and not the form of Cu added.

Marshall et al. (2007) investigated the accumulation and extraction of gold NPs in *Brassica juncea* grown on soil media. The XANES analysis showed that plants grown on soils containing Au had approximately equal amounts of Au(0) and Au(I). TEM analysis of the samples showed that the gold NPs were approximately 5–50 nm. XAS studies were used by Servin et al. (2017) and Servin et al. (2012) to investigate the exposure of corn, lettuce, soybean, and zucchini to CeO_2 ENMs in a biochar amended soil. The XANES analysis indicated that after 28 days, Ce in the amended soil was still present as CeO_2. Muccifora et al. (2021) studied the accumulation of TiO_2 ENMs in *Pisum sativum* L. (pea) plants using a combination of ICP-OES, XANES, and TEM. The authors treated the peas with biosolids containing one of the following: 800 mg of anatase, 800 mg rutile, a mixture of 50% anatase +50% rutile, or submicron TiO_2. The XANES analysis showed the presence of anatase, rutile, and ilmenite as the main forms of TiO_2 in the plants after treatment, and the preferential uptake of the anatase form into the plants. Wang et al. (2013) demonstrated that *Vigna unguiculata* were less affected by ZnO ENMs than those treated with $ZnCl_2$. In addition, the plants treated with ZnO ENMs caused a rapid dissolution following soil contamination, and no translocation of the ZnO ENMs was observed from the root system to the aerial parts of the plants. In a similar study, Ogunkunle et al. (2019) investigated the uptake and accumulation of Cu in cow pea from soils containing Cu(0) ENMs. The authors observed phytotoxicity caused by the Cu(0) ENMs at concentrations of 250–500 mg/kg. The XANES analysis of the roots showed that Cu was present as Cu^{2+} ions, while in the soil, the Cu consisted of Cu(0) ENMs. Majumdar et al. (2014) also proved the biotransformation of CeO_2 on kidney beans treated with 62.5–500 mg/kg CeO_2 ENMs for 15 days in hydroponics. The XANES analysis showed that most Ce in the plant roots remained as CeO_2 and a small portion was present as Ce^{3+}. In another study, Yang et al. (2020) investigated the accumulation and transport of Ag ENMs in rice plants and the effects of iron-based plaques and dissolved ions. The authors observed that, in the presence of the Fe-plaques, Ag ENMs were oxidized and formed ionic species including AgCl and Ag^+-thiol complexes. Dissolved iron was not very effective at dissolving the Ag ENMs and less silver accumulation was observed. Spielman-Sun et al. (2018) used XANES to follow the change in Cu distribution and speciation in roots of *Triticum aestivum* exposed to CuO, $Cu(OH)_2$, and CuS ENMs. It was found that the more soluble $Cu(OH)_2$ species showed a higher amount of Cu in the root after 1 h of treatment. However, the less soluble CuO and CuS delivered Cu to the roots and leaves for longer periods. One of the earlier papers on NPs in plants was by Gardea-Torresdey et al. (2002), who used XANES to observe the formation of Au(0) NPs within live alfalfa plants. In a similar work, Gardea-Torresdey et al. (2003) used XAS to confirm the presence of Ag NPs in alfalfa plants. XANES analysis showed the Ag in the alfalfa plants was in mixed oxidation state, which was a combination of Ag(I) and Ag(0). On the other hand, Peng et al. (2015) used XANES analysis to investigate CuO accumulation and biotransformation in *Oryza sativa* L. (rice) plants. The authors treated rice plants with 100 ppm of CuO ENMs for 14 days. The data indicated that CuO ENMs were present in the roots and reached the endodermis. The free Cu in the plant was coordinated to cysteine, citrate, and phosphate ligands. Also, the effects of Cu(0) and ZnO ENMs in soils and microbial communities were studied by Collins et al. (2012) over 162 days, concluding that both NPs altered the microbial communities and were mobile within soils at different rates. The XANES analysis showed that after 160 days, no ZnO ENMs remained and a small portion of the Cu(0) ENMs were still present.

XAS has also been used to examine the chemical transformations of NPs in the aquatic environment, which in this chapter includes the water column and sediments (Gogos et al., 2017; Kaegi et al., 2011, 2013; Khaksar et al., 2015; Meier et al., 2016; Thalmann et al., 2016; Tong et al., 2015; Von Der Heyden et al., 2019; Wang, Zhao, et al., 2018). NPs have also been studied using XAS for their ability to transport other chemicals such as phosphorus (Bone et al., 2012; Rick & Arai, 2011). For example, the fate and transport of P in watersheds was investigated by Rick & Arai (2011). The authors focused on the iron species in the soils and the EXAFS showed the presence of Fe-O, Fe-Fe, and Fe-Si interactions and concluded that the iron-based NPs were not as reactive toward P and not as responsible for P transport (Rick & Arai, 2011). Bone et al. (2012)

determined the biotic and abiotic mechanisms for the toxicity and fate of Ag ENMs coated with arabic gum and PVP in the aquatic environment. Data indicated that PVP-Ag ENMs showed high stability with more than 92% of the silver remaining as Ag(0). In addition, 41% of the Arabic gum-stabilized NPs were transformed in the system containing water with plants and sediments. Along the same line of research, Von Der Heyden et al. (2019) determined the speciation of suspended Fe NPs collected from fluvial, marine, and lacustrine surface waters. Using Fe $L_{II,III}$ edge information provided by XAS, the authors identified three groups of Fe-rich nanomaterials in the environment which included Fe(II)-based nanomaterials, Fe(III)-based nanomaterials, and Fe(II)/Fe(III)-(magnetite)-based nanomaterials. The authors also used carbon K-edge XANES to show the presence of carbon coating on the natural nanomaterials in the environment. The authors concluded that the presence of carbon indicated a surfactant coating, which reduced the toxic effects of natural nanomaterials compared to ENMs. Meier et al. (2016) studied the release of Ag ENMs in municipal sewer systems. The authors identified the presence of Ag(0) and Ag_2S ENMs in the sludge before combustion. After combustion at 800°C, the authors observed 80% Ag(0) in the XAS analysis. The presence of the high amount of Ag(0) in the sample was not unexpected due to the high carbon content in these types of wastes. Carbon reacting with metal ions at high temperatures is known to reduce transition metals to their metallic forms (Pindar & Dhawan, 2019). XAS has also been used to determine the kinetics of the sulfidation of CuO NPs in aqueous solution (Gogos et al., 2017). CuO ENMs were reacted with HS^- at concentrations of 26.4–105.6 μm at pH 8. The results showed the conversion of CuO to an amorphous Cu_xS and then into CuS in the covellite form. The authors concluded that the release of CuO in waste waters will rapidly convert to Cu_xS and then crystallize into CuS, which is stable. In another study, Kaegi et al. (2011) investigated the behavior of metallic Ag ENMs in a wastewater treatment plant using XANES. The authors observed free Ag NPs in the effluent only during the initial influx of the particles. XANES analysis showed the majority of the Ag present in the effluent and sludge was Ag_2S. The results of the study suggested that the Ag ENMs were transformed into the Ag_2S within 2 h of release into the effluent. On the other hand, the transformation of ZnO ENMs with phosphate under varying pH conditions was studied by Rathnayake et al. (2014). The XANES data collected by the authors showed that ZNO ENMs reacted to produce a mixture of ZnO and $Zn_3(PO_4)_2$. The reaction at low pH with high phosphate concentration resulted in more $Zn_3(PO_4)_2$ formed. At pH 8, the reaction was significantly slower than the reaction at pH 6, where only 30% $Zn_3(PO_4)_2$ was formed as compared to approximately 70% at pH 6. Wang, Zhao, et al. (2018) studied the colloidal stability of Fe_3O_4 ENMs with varying concentrations of dissolved organic matter. The authors investigated the Fe_3O_4 ENMs stability under different conditions, which included pH, dissolved organic matter, salts, and cationic strength. The authors also performed Ca K-edge XANES studies to investigate the effect of Ca on iron stability. The Ca K-edge data showed that Ca^{2+} formed a complex with a carbonate group, which indicated that the dissolved organic acid was complexed to the Fe-based ENMs and to the Ca^{2+} ion. The authors concluded that in the presence of Ca^{2+}, multilayer adsorption was promoted and the stability of the ENMs was enhanced. Tong et al. (2015) investigated TiO_2 polymorphs in wastewater and sediment samples of a wastewater treatment plant using Ti K-edge XANES studies. The authors collected samples over a 2-month period and the XANES spectra did not show variation in the sample composition from the effluent, activated sludge, or the primary sludge. The XANES data showed the presence of anatase, rutile, and ilmenite in the samples for the most part. However, in the samples collected upstream of the treatment plant, a fourth phase of titanium oxide was present, which was determined to be amorphous in nature.

In summary, XAS has become a common tool for the examination of ENMs in the environment due to the simplicity of the experimental setup and the ability to observe coordination environments. A XAS study does not require pretreatment of the sample such as drying or extraction, which may influence the physical and chemical characteristics of the ENMs under investigation.

3.2.6 X-ray diffraction (XRD)

Powder X-ray diffraction (XRD) is perhaps one of the most common techniques used for the analysis of NPs and has been well documented in the literature (Lu et al., 2015; Luther et al., 2012; Shahwan et al., 2011; Tamez et al., 2016; Wang et al., 2007; Yan et al., 2005). XRD allows for the determination of the crystalline properties of a nanomaterial, which includes crystal lattice and type of material.

For nanomaterials, the determination of the size of a NP is easily obtained using Scherrer's equation.

$$d = \frac{0.9\lambda}{B \cos \theta} \tag{3.12}$$

where d is the crystallite size, 0.9 is a correction factor for using a Gaussian fitting of the diffraction peak, λ is the wavelength of the X-rays used, B is the full width half maximum of the diffraction peak, and cos θ is the diffraction angle.

Scherrer and Debye determined the peak width observed for a diffraction peak in each powder X-ray diffractometer was related to

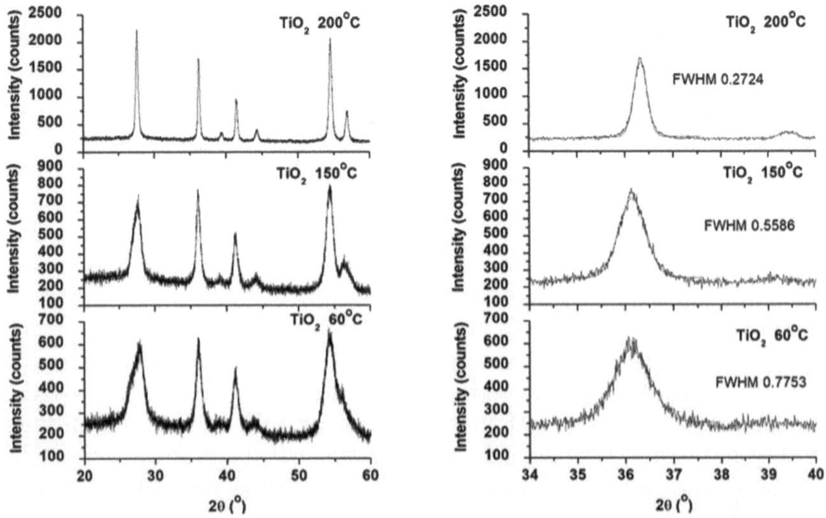

FIGURE 3.8 XRD of TiO_2 NPs synthesized at different temperatures showing change in the diffraction peak width of the 004-diffraction plane located at 36.2 degrees in 2θ.

particle size. An example of peak broadening is shown in Fig. 3.8 for TiO_2 in the rutile phase synthesized at various temperatures.

The full width half maximum (FWHM) becomes smaller as the temperature of the synthesis increases from 60 to 200°C. In addition, the calculated size of the NPs increases as the temperature increases with the average size being approximately 11 nm at 60°C, 15 nm at 150°C, and 30 nm at 200°C according to Scherrer's equation.

There are many examples in the literature for the determination of NPs size and composition using powder XRD (Lu et al., 2015; Luther et al., 2012; Shahwan et al., 2011; Tamez et al., 2016). The use of powder XRD analysis showing the presence of ENMs in soils is also well documented (Cornelis et al., 2011; Gao et al., 2017; Gong et al., 2011; Leonardo et al., 2016; Li et al., 2012; Raju et al., 2012; Silva et al., 2021; Tsao et al., 2013). In addition, XRD has been used extensively to characterize nanomaterials for environmental applications such as adsorption for water treatment.

3.2.6.1 *Examples of application of XRD in environmental studies*

Silva et al. (2021) examined ENMs interactions with toxic elements in fluvial suspended sediments. XRD was used to determine the crystalline phase of the NPs present in the sample (Ti-nanomaterials), which were the anatase and rutile phases. Tsao et al. (2013) compared the separation and identification of soil NPs with laboratory and synchrotron XRD. The authors attempted to use an automatic ultrafiltration device to separate and collect the NPs from soil. The XRD results showed major clay minerals in the <2000 nm size fractions were illite and kaolinite. In addition, minor amounts of goethite and hematite were detected. Illite and kaolinite were observed in the 1–100 nm sized sample using conventional powder XRD. The synchrotron-based XRD also detected goethite and hematite particles. Li et al. (2012) used XRD among other techniques to investigate the extraction and characterization of natural nanomaterials from Chinese soils. The study showed that 12 native NPs were present in the soils: montmorillonite, kaolinite, halloysite, muscovite and a muscovite polytype, hematite, corrensite, recroritem, dickite, calcite, trona, and thermonatrite. The study also indicated that not all soil NPs were extractable/mobile and the stability of the NPs was related to composition, small particle size, small concentration, low ionic strength, and high pH in the extractions. Gao et al. (2017) used XRD to determine the crystalline phase of CuO NPs in Cu ENMs amended soils. The authors determined that CuO present in the soil was tenorite and the CuO extracted was not identified. Cornelis et al. (2011) investigated the solubility and retention of CeO_2 ENMs in soils. The authors used Scherrer's equation to determine the crystallite size, which was of 8 nm for the NPs and 140 nm in diameter for the bulk CeO_2. Burachevskaya et al. (2021) used XRD to analyze soil samples contaminated with CuO in micro and NP forms. The authors did not observe CuO NPs in the soil sample after contamination, which was more than likely due to the diffuse and weak diffraction peaks observed for NPs in general. However, the micron-sized CuO particles were observed in the soils. As mentioned earlier, the presence of the CuO NPs was detected by XANES analysis. X-ray diffraction analysis of the CuO-treated Haplic Chernozem showed that contamination with CuO from the NPs proceeded quicker as compared to the micro-sized particles.

XRD has been applied to investigate the presence of ENMs within live plants and other organisms to a limited extent (Bali & Harris, 2010; Gong et al., 2011; Leonardo et al., 2016; Raju et al., 2012). The limited application for the most part results from a combination of concentration of ENMs within the plants and the diffuse nature of the diffraction patterns of ENMs. Raju et al. (2012) investigated the photosynthesis of intracellular and extracellular gold NPs using *Arachis hypogaea* L. The study showed the presence of gold NPs on the roots of the peanut plants with the 111, 200, 220, and 311 diffraction planes visible in the dried powdered root sample. The authors demonstrated that gold NPs were synthesized extracellularly by the peanut plants. Another study involving algae and NPs was performed by Leonardo et al. (2016) who used XRD to aid in the determination of Ag accumulation in the microalgae *Coccomyxa actinabiotis*. The algae were exposed to 10^{-2} M Ag^+ in aqueous solution for 3 h and then lyophilized. The lyophilized samples were then ground into powder and analyzed using XRD, which confirmed the presence of Ag

NPs. The experiments were performed under both light and dark conditions and confirmed the presence of face-centered cubic silver NPs under both reaction conditions.

Examples of the utilization of XRD as an analytical tool for the investigation of ENMs impact on the environment are limited. However, the examples that do exist show the diversity of how XRD can be applied to environmental nanoscience. XRD is limited in applications due to the need for dry samples. In wet samples, such as ENMs in aqueous samples, the water tends to interfere with the measurements. The direct determination of NPs in an aqueous solution is not possible, which limits some of the applications of the technique.

3.2.7 Scanning (SEM) and transmission electron microscopy (TEM)

Perhaps the second most common technique to observe NPs in different matrices are the electron microscopy techniques, which include scanning electron and transmission electron microscopies. Both techniques use electrons as the source; however, in SEM backscattered electrons are used while TEM uses electrons that are transmitted through the sample.

As mentioned earlier, the major difference in SEM and TEM is the electrons used for the imaging and the placement of the detector in the instrument. The contributions of TEM and SEM in nanoscience have been enormous and instrumental in understanding the transport of ENMs in the environment. Fig. 3.9 shows two TEM and SEM images of gold NPs synthesized by plant extracts (A) and gold NPs synthesized by algae in a growth media (B).

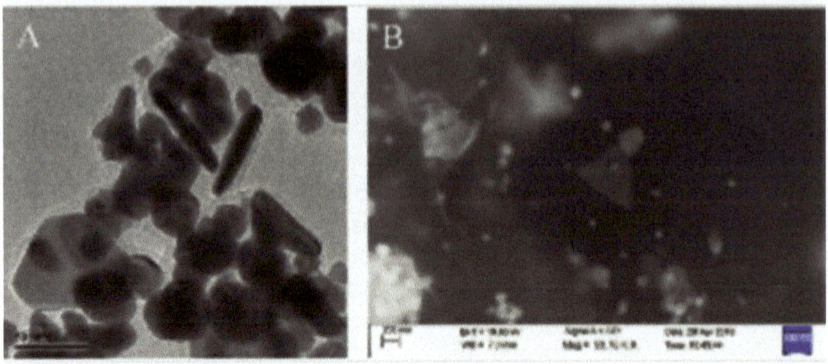

FIGURE 3.9 (A) TEM of gold NPs synthesized from aqueous extract of alfalfa biomass. (B) SEM of gold NPs in growth media with *C. caldarium*.

Both TEM and SEM have also become major analytical techniques for the determination of NPs in environmental samples, which include soils, sediments, and biological samples (Ghosh et al., 2016; Ma et al., 2016; Muccifora et al., 2021; Nath et al., 2018; Schwabe et al., 2013; Vittori Antisari et al., 2013; Wang et al., 2014; Wang, Zhao, et al., 2018; Zhang et al., 2012; Zhang et al., 2017; Zhou et al., 2012). However, the direct determination of ENMs in aqueous samples has not been achieved due to the interactions of the electron beam with water molecules. Yet, with sample pretreatment, the analysis of NPs from aqueous solution is well documented. Below is a discussion of some examples on the applications of TEM and SEM in the imaging of NPs in soils, sediments, and plants.

3.2.7.1 *Environmental applications of TEM and SEM imaging of NPs*

One of the earlier works on the use of TEM to study NPs in plants was performed by Gardea-Torresdey et al. (2002), who showed the formation and growth of Au NPs inside live alfalfa. In that study, alfalfa plants were exposed to potassium tetrachloroaurate in an agar growth media. The plants showed the presence of Au NPs in the roots, stems, and leaves. The NPs showed crystallinity and particle sizes ranged from 5 nm up to 40 nm. This work was followed up by Gardea-Torresdey et al. (2003), who investigated Ag accumulation in alfalfa plants. TEM showed the presence of Ag NPs in the roots and stems of the alfalfa plants grown in agar media. At the root level, 9- nm particles were observed; however, inside the plants, NPs with sizes between 2 and 30 nm were found. In addition, their data indicated that NPs 2–5 nm in size had an icosahedral geometry. Muccifora et al. (2021) studied the uptake and accumulation of TiO_2 ENMs in pea (*Pisum sativum* L.) plants using a combination of ICP-OES, XANES, and TEM. In this case, TEM analysis showed damage in the cells' ultrastructure in all samples treated with TiO_2. On the other hand, Zhang et al. (2012) investigated the biotransformation of CeO_2 NPs in cucumber plants. TEM data showed a large number of CeO_2 NPs adhered to the outside of the root tissues. Also, TEM with energy dispersive spectroscopy (EDS) analysis showed needlelike NPs of $CePO_4$ present in the plants. Nath et al. (2018) traced isotopically labeled Ag, Cu, and Zn NPs inside live plants. The study showed that ENMs were accumulated more significantly in the roots than other portions of the plant. Also, TEM was used to characterize the synthesized NPs for size and behavior in the nutrient solution. The TEM results showed that growth media had little to no effect on the ENMs. Also, Ghosh et al. (2016) applied TEM to study cytotoxicity, genotoxicity, and interference in antioxidant defenses of ZnO ENMs in *Allium cepa, Nicotiana tabacum*, and *Vicia faba*. It was shown that the morphology of the ZnO NPs can be altered inside the plants after accumulation. In addition, TEM of the plants' cells showed damage to the protoplasts, ruptured membranes, damage to the nuclear organization, and vacuolation. On another work, Haverkamp and Marshall (2009) investigated the mechanisms of metal Ag NPs formation in hydroponically grown *Brassica juncea* plants. The TEM results showed the presence of Ag NPs between 2 and 35 nm after treatment of the plants with silver solution. It was determined that the amount of Ag NPs or metal NPs was dependent on the reducing power of the plants and the reductive potential of the metal in question. Corredor et al. (2009) investigated the uptake and transport of carbon-coated magnetic iron NPs in plants. The results showed that the carbon-coated iron magnetic NPs can be absorbed and transported throughout the plant system. Due to this disruption, the authors concluded that these NPs could be used to deliver substances to cells. Ma et al. (2016) examined CeO_2 ENMs and bulk CeO_2 in *Brassica rapa* using TEM as a tool, which showed the average size of the NPs was 19.1 nm. SEM, in conjunction with EDS, was used to determine the location of the CeO_2 ENMs. The bulk samples showed little to no accumulation of CeO_2 on the roots. However, the samples treated with CeO_2 ENMs showed the presence of rodlike NPs on the roots' surface. Zhang et al. (2017) examined the bioavailability of CeO_2 ENMs to *Raphanus sativus* L. from two soils. The authors determined that the fractions of the CeO_2 ENMs were different in the soils, and that soil properties affected ENMs bioavailability and translocation. It was also found that the average size of the CeO_2 ENMs was 19.1 nm before soil treatment and uptake. Wang, Zhao, et al. (2018) observed that the adsorption of Ag-PVP ENMs by natural soils followed the Langmuir model. Also, using TEM, it was determined that the average size of the Ag ENMs was of 10 nm. Moreover, spherical Ag ENMs with diameters of 19–30 nm were observed attached to the soil in areas with high concentrations of iron oxides. Wang et al. (2014) investigated the mobility of water-dispersed ENMs in Ultisol soil. These authors selected TiO_2, C60, single-walled carbon nanotubes, and Ag ENMs. According to their TEM data, ENMs displayed sizes between 4.5 and 25 nm. In addition, they determined that only the Ag ENMs did not form large aggregates and were well dispersed. However, the other ENMs, such as TiO_2, aggregated to form particles of approximately 290 nm. Zhou et al. (2012) reported the effect of clay particles on TiO_2 and Ag ENMs in aqueous environments. Their findings showed that TiO_2 ENMs had an average size of 27 nm and spherical geometry. In addition, the Ag ENMs were 20.8 nm in diameter. The TEM analysis showed large aggregates of montmorillonite with Ag. In the large aggregates, Ag was typically located on the edge of the montmorillonite. In a different type of study, Vittori Antisari et al. (2013) investigated the effects of CeO_2, Fe_3O_4, and SnO_2 ENMs on soil microbial communities and their distribution in soil. The study was performed using SEM with EDS which showed an irregular distribution of the ENMs in the clay fraction. The analysis also proved that in some soils, the ENMs were associated with the organic carbon and microbial biomass-associated carbon. The data indicated that these ENMs can potentially have impacts on microbial communities in soils.

Both TEM and SEM, and variations of these techniques, provide invaluable data in environmental nanoscience for the investigation of ENMs. The ability to determine morphological changes in NPs and the aggregation of ENMs provide powerful insight into the mechanisms of transport, stability, and lifetime of ENMs in the environment.

3.2.8 Single particle ICP-MS (SP-ICP-MS)

In the past decade (2010–2020), SP-ICP-MS has been developed and commercialized. The system development includes changes in the sample introduction system and the faster detection/mass separation portion of the electronics (Bolea-Fernandez et al., 2019; Flores et al., 2021; Gundlach-Graham & Mehrabi, 2020; Montaño et al., 2016). Advances in SP-ICP-MS instrumentation have been a tremendous advantage for NPs research in aqueous-based systems. SP-ICP-MS allows for the direct aspiration of solutions and the distinction of NPs and ions in a sample. The ions in solution are considered to generate continuous signaling events recorded by the detector, and the particles are individual events or small sequential signaling events.

Similar to other techniques discussed in this chapter, each technique has limitations. In the case of SP-ICP-MS it is the need of a solution. Initially, the technique was limited to solutions containing NPs. However, extractions of NPs from biological tissues, soils, and sediments have been achieved, and thus made it possible to use SP-ICP-MS in ENM research.

3.2.8.1 Applications of SP-ICP-MS in the characterization of ENMs in environmental samples

Merrifield et al. (2017) investigated the concentration-dependent transformation of Ag NPs. The study demonstrated that SP-ICP-MS was able to measure particle size and particle number with detection limits of 0.01 and 50 μg/L Au@Ag NPs. The authors determined that the concentration was a key factor in ENMs behavior. Thus, at high ENMs concentrations, aggregation dominated. On the other hand, at lower concentrations, dissolution was the dominant process. Fréchette-Viens et al. (2019) applied SP-ICP-MS to examine and quantify ZnO and Zn containing ENMs in natural waters. The results showed excellent detection limits down to 450 particles per mL of water and quantifiable at 1000 ENMs in rainwater. The study observed samples that were spiked with ZnO ENMs and Zn in naturally occurring colloids. In order to analyze the behavior of TiO_2, Ag, and Au ENMs during drinking water treatment, Donovan et al. (2016b) developed a method to determine ENMs concentrations, size, size distribution, and dissolved metal content in both surface waters and treated drinking water. Removal of the ENMs at 97% was achieved by adding lime softening and alum for coagulation followed by activated carbon treatment. The data also showed that the Ti containing particles and dissolved Ti were present in river water samples, but Au and Ag were not. Long et al. (2016) used SP-ICP-MS to determine average particles size detection limits and stability of Au ENMs in natural waters. The study focused on the discrimination of particle signal and used instrumental dwell time of 3 ms for size detection. Their data showed that stability lasted 6 days and an almost complete degradation at 30 days. Donovan et al. (2016a) simulated drinking water treatment of samples containing ZnO and CeO_2 NPs. SP-ICP-MS results showed particle concentrations of 60% at a minimum and the majority of samples were reduced by 95% after sequential lime, alum, activated carbon, and chlorination treatment. The study concluded that nonstabilized ENMs could be removed effectively by normal water treatment processes. Schwertfeger et al. (2017) developed a method for the extraction of metallic ENMs from soils, using Ag ENMs as a model. From the study, it was determined that tetrasodium pyrophosphate at a ratio of 1:100 soil to reagent was optimum for extraction. This method was 96% effective at removing Ag ENMs. Studies of this type expand the sampling abilities of SP-ICP-MS beyond the liquid-type sample. Bao et al. (2016) used SP-ICP-MS to determine the accumulation and size distributions of Ag ENMs internalized by *Arabidopsis* plants. The authors used an enzymatic digestion (Macerozyme R-10) process on the root and stem samples, which showed no alteration of the Ag ENMS when TEM results were compared. Hayder et al. (2020) applied SP-ICP-MS to evaluate the bioavailability of CeO_2 and CuO ENMs from radish. In this investigation, radishes were grown in hydroponics containing only CeO_2, only CuO, and a mixture of CeO_2 and CuO. In addition, a simulated intestinal fluid was used for digestion of the radishes contaminated with the ENMs. During digestion, CeO_2 were stable and transported to the bloodstream. On the other hand, CuO ENMs were dissolved during the digestion process. Furthermore, in the mixed sample, the size of the CeO_2 decreased while CuO ENMs were again formed. The data indicated both ENMs were bioavailable.

SP-ICP-MS is becoming one of the more widely used tools in the environmental study of ENMs. The ability to distinguish between ions and particles, as well as determine the particle size distribution, provides tremendous amount of information on the stability and dissolution of ENMs in the environment and in living organisms.

3.3 Conclusions

New technology and innovative applications have advanced our ability to detect and characterize ENMs in different environmental matrices. These analytical milestones have benefited from the development of new methodologies applied to different instrumentation. Most current and highly utilized techniques for NPs characterization were discussed in this chapter, which included UV-Vis spectroscopy, DLS, SERS, SAXS, XAS, XANES and EXAFS, XRD, TEM/SEM, and SP-ICP-MS. As mentioned in each section of the chapter, there are limitations for each of these techniques for the characterization of ENMs in environmental matrices-based samples. For example, UV-Vis and DLS do not work in solid samples and require extraction techniques to place the sample in a liquid environment. TEM, SEM, and XRD cannot work well with samples in liquids, thus the samples need to be dried before analysis. SAXS and XAS have size limitations on the particles that can be analyzed for specific characteristics. Yet, even with the limitations, each of the aforementioned techniques provide important information on ENMs in different environmental samples.

Acknowledgments

J.G. Parsons is grateful for the generous support provided by a Departmental Grant from the Robert A. Welch Foundation (Grant No. BX-0048). M. Alcoutlabi would like to acknowledge the support from NSF PREM/award/under grant No. DMR-2122178: UTRGV-UMN Partnership for Fostering Innovation by Bridging Excellence in Research and Student Success. Partial support was provided by the Lloyd M. Bentsen, Jr. Endowed Chair in Engineering endowment at UTRGV.

References

Adediran, G. A., Lundberg, D., Almkvist, G., Pradas del Real, A. E., Klysubun, W., Hillier, S., Gustafsson, J. P., & Simonsson, M. (2021). *Micro and nano sized particles in leachates from agricultural soils: Phosphorus and sulfur speciation by X-ray micro-spectroscopy* (Vol. 189). Water Research. https://doi.org/10.1016/j.watres.2020.116585

Akaighe, N., MacCuspie, R. I., Navarro, D. A., Aga, D. S., Banerjee, S., Sohn, M., & Sharma, V. K. (2011). Humic acid-induced silver NPs formation under environmentally relevant conditions. *Environmental Science and Technology, 45*(9), 3895–3901. https://doi.org/10.1021/es103946g

Alanazi, A. K., Mezni, A., Saber, N. B., Sellami, B., Ibrahim, M. M., Alsawat, M., & Altalhi, T. (2019). Aquatic ecotoxicity effects of gold/titania based nanomaterials. *Journal of Biomaterials and Tissue Engineering*, 262–267. https://doi.org/10.1166/jbt.2019.1965

Amde, M., Tan, Z. Q., & Liu, J. (2019). Separation and size characterization of zinc oxide NPs in environmental waters using asymmetrical flow field-flow fractionation. *Talanta, 200*, 357–365. https://doi.org/10.1016/j.talanta.2019.03.074

Anaraki, N. I., Sadeghpour, A., Iranshahi, K., Toncelli, C., Cendrowska, U., Stellacci, F., Dommann, A., Wick, P., & Neels, A. (2020). New approach for time-resolved and dynamic investigations on NPs agglomeration. *Nano Research, 13*(10), 2847–2856. https://doi.org/10.1007/s12274-020-2940-4

António, D. C., Cascio, C., Jakšić, Jurašin, D., Lyons, D. M., Nogueira, A. J. A., Rossi, F., & Calzolai, L. (2015). Assessing silver NPs behaviour in artificial seawater by mean of AF4 and spICP-MS. *Marine Environmental Research, 111*, 162–169. https://doi.org/10.1016/j.marenvres.2015.05.006

Bali, R., & Harris, A. T. (2010). Biogenic synthesis of Au NPs using vascular plants. *Industrial and Engineering Chemistry Research, 49*(24), 12762–12772. https://doi.org/10.1021/ie101600m

Bao, D., Oh, Z. G., & Chen, Z. (2016). Characterization of silver NPs internalized by arabidopsis plants using single particle ICP-MS analysis. *Frontiers in Plant Science, 7*. https://doi.org/10.3389/fpls.2016.00032

Baysal, A., Saygin, H., & Ustabasi, G. S. (2019). Physicochemical transformation of ZnO and TiO2 NPs in sea water and its impact on bacterial toxicity. *Environmental Health Engineering and Management*, 73–80. https://doi.org/10.15171/ehem.2019.08

Beale, A. M., & Weckhuysen, B. M. (2010). EXAFS as a tool to interrogate the size and shape of mono and bimetallic catalyst NPs. *Physical Chemistry Chemical Physics, 12*(21), 5562–5574. https://doi.org/10.1039/b925206a

Bharadwaj, P., Beams, R., & Novotny, L. (2011). Nanoscale spectroscopy with optical antennas. *Chemical Science, 2*(1), 136–140. https://doi.org/10.1039/c0sc00440e

Bharadwaj, S., Pandey, A., Yagci, B., Ozguz, V., & Qureshi, A. (2018). Graphene nano-mesh-Ag-ZnO hybrid paper for sensitive SERS sensing and self-cleaning of organic pollutants. *Chemical Engineering Journal, 336*, 445–455. https://doi.org/10.1016/j.cej.2017.12.040

Bolea-Fernandez, E., Leite, D., Rua-Ibarz, A., Liu, T., Woods, G., Aramendia, M., Resano, M., & Vanhaecke, F. (2019). On the effect of using collision/reaction cell (CRC) technology in single-particle ICP-mass spectrometry (SP-ICP-MS). *Analytica Chimica Acta, 1077*, 95–106. https://doi.org/10.1016/j.aca.2019.05.077

Bone, A. J., Colman, B. P., Gondikas, A. P., Newton, K. M., Harrold, K. H., Cory, R. M., Unrine, J. M., Klaine, S. J., Matson, C. W., & Di Giulio, R. T. (2012). Biotic and abiotic interactions in aquatic microcosms determine fate and toxicity of Ag NPs: Part 2-toxicity and Ag speciation. *Environmental Science and Technology, 46*(13), 6925–6933. https://doi.org/10.1021/es204683m

Borowski, M. (1997). Size determination of small Cu-clusters by EXAFS. *Journal de Physique: IV, 7*(2). https://doi.org/10.1051/jp4/1997190. C2–C2.

Burachevskaya, M., Minkina, T., Mandzhieva, S., Bauer, T., Nevidomskaya, D., Shuvaeva, V., Sushkova, S., Kizilkaya, R., Gülser, C., & Rajput, V. (2021). Transformation of copper oxide and copper oxide NPs in the soil and their accumulation by Hordeum sativum. *Environmental Geochemistry and Health, 43*(4), 1655–1672. https://doi.org/10.1007/s10653-021-00857-7

Chen, Z. H., Kim, C., Zeng, X. B., Hwang, S. H., Jang, J., & Ungar, G. (2012). Characterizing size and porosity of hollow NPs: SAXS, SANS, TEM, DLS, and adsorption isotherms compared. *Langmuir, 28*(43), 15350–15361. https://doi.org/10.1021/la302236u

Chinnapongse, S. L., MacCuspie, R. I., & Hackley, V. A. (2011). Persistence of singly dispersed silver NPs in natural freshwaters, synthetic seawater, and simulated estuarine waters. *Science of the Total Environment, 409*(12), 2443–2450. https://doi.org/10.1016/j.scitotenv.2011.03.020

Collins, D., Luxton, T., Kumar, N., Shah, S., Walker, V. K., & Shah, V. (2012). Assessing the impact of copper and zinc oxide NPs on soil: A field study. *PLoS One, 7*(8). https://doi.org/10.1371/journal.pone.0042663

Comaschi, T., Balerna, A., & Mobilio, S. (2008). Temperature dependence of the structural parameters of gold NPs investigated with EXAFS. *Physical Review B, 77*(7). https://doi.org/10.1103/physrevb.77.075432

Cornelis, G., Ryan, B., McLaughlin, M. J., Kirby, J. K., Beak, D., & Chittleborough, D. (2011). Solubility and batch retention of CeO2 NPs in soils. *Environmental Science and Technology, 45*(7), 2777–2782. https://doi.org/10.1021/es103769k

Corredor, E., Testillano, P. S., Coronado, M. J., González-Melendi, P., Fernández-Pacheco, R., Marquina, C., Ibarra, M. R., De La Fuente, J. M., Rubiales, D., Pérez-De-Luque, A., & Risuẽo, M. C. (2009). NPs penetration and transport in living pumpkin plants: In situ subcellular identification. *BMC Plant Biology, 9*. https://doi.org/10.1186/1471-2229-9-45

Delay, M., Dolt, T., Woellhaf, A., Sembritzki, R., & Frimmel, F. H. (2011). Interactions and stability of silver NPs in the aqueous phase: Influence of natural organic matter (NOM) and ionic strength. *Journal of Chromatography A, 1218*(27), 4206–4212. https://doi.org/10.1016/j.chroma.2011.02.074

Deonarine, A., & Hsu-Kim, H. (2009). Precipitation of mercuric sulfide NPs in NOM-containing water: Implications for the natural environment. *Environmental Science and Technology, 43*(7), 2368–2373. https://doi.org/10.1021/es803130h

Dhandapani, P., Maruthamuthu, S., & Rajagopal, G. (2012). Bio-mediated synthesis of TiO 2 NPs and its photocatalytic effect on aquatic biofilm. *Journal of Photochemistry and Photobiology B: Biology, 110*, 43–49. https://doi.org/10.1016/j.jphotobiol.2012.03.003

Donovan, A. R., Adams, C. D., Ma, Y., Stephan, C., Eichholz, T., & Shi, H. (2016a). Detection of zinc oxide and cerium dioxide NPs during drinking water treatment by rapid single particle ICP-MS methods. *Analytical and Bioanalytical Chemistry, 408*(19), 5137–5145. https://doi.org/10.1007/s00216-016-9432-0

Donovan, A. R., Adams, C. D., Ma, Y., Stephan, C., Eichholz, T., & Shi, H. (2016b). Single particle ICP-MS characterization of titanium dioxide, silver, and gold NPs during drinking water treatment. *Chemosphere, 144*, 148–153. https://doi.org/10.1016/j.chemosphere.2015.07.081

Ellis, L. J. A., Valsami-Jones, E., Lead, J. R., & Baalousha, M. (2016). Impact of surface coating and environmental conditions on the fate and transport of silver NPs in the aquatic environment. *Science of the Total Environment, 568*, 95–106. https://doi.org/10.1016/j.scitotenv.2016.05.199

Farré, M., Sanchís, J., & Barceló, D. (2011). Analysis and assessment of the occurrence, the fate and the behavior of nanomaterials in the environment. *TrAC, Trends in Analytical Chemistry, 30*(3), 517–527. https://doi.org/10.1016/j.trac.2010.11.014

Flores, K., Turley, R. S., Valdes, C., Ye, Y., Cantu, J., Hernandez-Viezcas, J. A., Parsons, J. G., & Gardea-Torresdey, J. L. (2021). Environmental applications and recent innovations in single particle inductively coupled plasma mass spectrometry (SP-ICP-MS). *Applied Spectroscopy Reviews, 56*(1), 1–26. https://doi.org/10.1080/05704928.2019.1694937

Fréchette-Viens, L., Hadioui, M., & Wilkinson, K. J. (2019). Quantification of ZnO NPs and other Zn containing colloids in natural waters using a high sensitivity single particle ICP-MS. *Talanta, 200*, 156–162. https://doi.org/10.1016/j.talanta.2019.03.041

Gao, X., Spielman-Sun, E., Rodrigues, S. M., Casman, E. A., & Lowry, G. V. (2017). Time and NPs concentration affect the extractability of Cu from CuO NP-amended soil. *Environmental Science and Technology, 51*(4), 2226–2234. https://doi.org/10.1021/acs.est.6b04705

Gardea-Torresdey, J. L., Gomez, E., Peralta-Videa, J. R., Parsons, J. G., Troiani, H., & Jose-Yacaman, M. (2003). Alfalfa sprouts: A natural source for the synthesis of silver NPs. *Langmuir, 19*(4), 1357–1361. https://doi.org/10.1021/la020835i

Gardea-Torresdey, J. L., Parsons, J. G., Gomez, E., Peralta-Videa, J., Troiani, H. E., Santiago, P., & Yacaman, M. J. (2002). Formation and growth of Au NPs inside live alfalfa plants. *Nano Letters, 2*(4), 397–401. https://doi.org/10.1021/nl015673+

Gardea-Torresdey, J. L., Tiemann, K. J., Gamez, G., Dokken, K., Cano-Aguilera, I., Furenlid, L. R., & Renner, M. W. (2000). Reduction and accumulation of gold(III) by Medicago sativa alfalfa biomass: X-Ray absorption spectroscopy, pH, and temperature dependence. *Environmental Science and Technology, 34*(20), 4392–4396. https://doi.org/10.1021/es991325m

Gentile, L., Wang, T., Tunlid, A., Olsson, U., & Persson, P. (2018). Ferrihydrite NPs aggregation induced by dissolved organic matter. *The Journal of Physical Chemistry A, 122*(38), 7730–7738. https://doi.org/10.1021/acs.jpca.8b05622

Ghosh, M., Jana, A., Sinha, S., Jothiramajayam, M., Nag, A., Chakraborty, A., Mukherjee, A., & Mukherjee, A. (2016). Effects of ZnO NPs in plants: Cytotoxicity, genotoxicity, deregulation of antioxidant defenses, and cell-cycle arrest. *Mutation Research - Genetic Toxicology and Environmental Mutagenesis, 807*, 25–32. https://doi.org/10.1016/j.mrgentox.2016.07.006

Gogos, A., Thalmann, B., Voegelin, A., & Kaegi, R. (2017). Sulfidation kinetics of copper oxide NPs. *Environmental Science: Nano, 4*(8), 1733–1741. https://doi.org/10.1039/c7en00309a

Gondikas, A., Gallego-Urrea, J., Halbach, M., Derrien, N., & Hassellöv, M. (2020). Nanomaterial fate in seawater: A rapid sink or intermittent stabilization? *Frontiers in Environmental Science, 8*. https://doi.org/10.3389/fenvs.2020.00151

Gong, N., Shao, K., Feng, W., Lin, Z., Liang, C., & Sun, Y. (2011). Biotoxicity of nickel oxide NPs and bio-remediation by microalgae Chlorella vulgaris. *Chemosphere, 83*(4), 510–516. https://doi.org/10.1016/j.chemosphere.2010.12.059

Gribanyov, D., Zhdanov, G., Olenin, A., Lisichkin, G., Gambaryan, A., Kukushkin, V., & Zavyalova, E. (2021). Sers-based colloidal aptasensors for quantitative determination of influenza virus. *International Journal of Molecular Sciences, 22*(4), 1–13. https://doi.org/10.3390/ijms22041842

Gundlach-Graham, A., & Mehrabi, K. (2020). Monodisperse microdroplets: A tool that advances single-particle ICP-ms measurements. *Journal of Analytical Atomic Spectrometry, 35*(9), 1727–1739. https://doi.org/10.1039/d0ja00213e

Guo, H., Xing, B., Hamlet, L. C., Chica, A., & He, L. (2016). Surface-enhanced Raman scattering detection of silver NPs in environmental and biological samples. *Science of the Total Environment, 554*(555), 246–252. https://doi.org/10.1016/j.scitotenv.2016.02.084

Hansen, S. F., Hansen, O. F. H., & Nielsen, M. B. (2020). Advances and challenges towards consumerization of nanomaterials. *Nature Nanotechnology, 15*(12), 964–965. https://doi.org/10.1038/s41565-020-00819-7

Harada, M., Saijo, K., & Sakamoto, N. (2009). Characterization of metal NPs prepared by photoreduction in aqueous solutions of various surfactants using UV-vis, EXAFS and SAXS. *Colloids and Surfaces A: Physicochemical and Engineering Aspects, 349*(1–3), 176–188. https://doi.org/10.1016/j.colsurfa.2009.08.015

Haverkamp, R. G., & Marshall, A. T. (2009). The mechanism of metal NPs formation in plants: Limits on accumulation. *Journal of NPs Research, 11*(6), 1453–1463. https://doi.org/10.1007/s11051-008-9533-6

Hayder, M., Wojcieszek, J., Asztemborska, M., Zhou, Y., & Ruzik, L. (2020). Analysis of cerium oxide and copper oxide NPs bioaccessibility from radish using SP-ICP-MS. *Journal of the Science of Food and Agriculture, 100*(13), 4950–4958. https://doi.org/10.1002/jsfa.10558

Hernandez-Viezcas, J. A., Castillo-Michel, H., Peralta-Videa, J. R., & Gardea-Torresdey, J. L. (2016). Interactions between CeO_2 NPs and the desert plant mesquite: A spectroscopy approach. *ACS Sustainable Chemistry and Engineering, 4*(3), 1187–1192. https://doi.org/10.1021/acssuschemeng.5b01251

Hoque, M. E., Khosravi, K., Newman, K., & Metcalfe, C. D. (2012). Detection and characterization of silver NPs in aqueous matrices using asymmetric-flow field flow fractionation with inductively coupled plasma mass spectrometry. *Journal of Chromatography A, 1233*, 109–115. https://doi.org/10.1016/j.chroma.2012.02.011

House, S. D., Bonifacio, C. S., Timoshenko, J., Kunal, P., Wan, H., Duan, Z., Li, H., Yang, J. C., Frenkel, A. I., Humphrey, S. M., Crooks, R. M., & Henkelman, G. A. (2017).

Computationally assisted STEM and EXAFS characterization of tunable Rh/Au and Rh/Ag bimetallic NPs catalysts. *Microscopy and microanalysis*, 2030–2031. https://doi.org/10.1017/s1431927617010819

Ilie, A. G., Scarisoareanu, M., Morjan, M., Dutu, E., Badiceanu, M., & Mihailescu, I. (2016). Principal component analysis of Raman spectra for TiO_2 nanoparticle characterization. *Applied Surface Science, 573*, 93–103. https://doi.org/10.1016/j.apsusc.2017.01.193

Jiménez-Lamana, J., & Slaveykova, V. I. (2016). Silver NPs behaviour in lake water depends on their surface coating. *Science of the Total Environment, 573*, 946–953. https://doi.org/10.1016/j.scitotenv.2016.08.181

Kabelitz, A., Guilherme, A., Joester, M., Reinholz, U., Radtke, M., Bienert, R., Schulz, K., Schmack, R., Kraehnert, R., & Emmerling, F. (2015). Time-resolved in situ studies on the formation mechanism of iron oxide NPs using combined fast-XANES and SAXS. *CrystEngComm, 17*(44), 8463–8470. https://doi.org/10.1039/c5ce01585e

Kaegi, R., Voegelin, A., Ort, C., Sinnet, B., Thalmann, B., Krismer, J., Hagendorfer, H., Elumelu, M., & Mueller, E. (2013). Fate and transformation of silver NPs in urban wastewater systems. *Water Research, 47*(12), 3866–3877. https://doi.org/10.1016/j.watres.2012.11.060

Kaegi, R., Voegelin, A., Sinnet, B., Zuleeg, S., Hagendorfer, H., Burkhardt, M., & Siegrist, H. (2011). Behavior of metallic silver NPs in a pilot wastewater treatment plant. *Environmental Science and Technology, 45*(9), 3902–3908. https://doi.org/10.1021/es1041892

Keller, A. A., Wang, H., Zhou, D., Lenihan, H. S., Cherr, G., Cardinale, B. J., Miller, R., & Zhaoxia, J. I. (2010). Stability and aggregation of metal oxide NPs in natural aqueous matrices. *Environmental Science and Technology, 44*(6), 1962–1967. https://doi.org/10.1021/es902987d

Khaksar, M., Jolley, D. F., Sekine, R., Vasilev, K., Johannessen, B., Donner, E., & Lombi, E. (2015). In situ chemical transformations of silver NPs along the water-sediment continuum. *Environmental Science and Technology, 49*(1), 318–325. https://doi.org/10.1021/es504395m

Kim, J., Lee, S. K., Schellingerhout, D., Nahrendorf, M., Kim, K., Kim, J., & Kim, D. E. (2019). Spectroscopic assessment of gold NPs biodistribution using surface plasmon resonance phenomena. *ACS Biomaterials Science and Engineering, 5*(12), 6389–6394. https://doi.org/10.1021/acsbiomaterials.9b01079

Kumari, J., Kumar, D., Mathur, A., Naseer, A., Kumar, R. R., Thanjavur Chandrasekaran, P., Chaudhuri, G., Pulimi, M., Raichur, A. M., Babu, S., Chandrasekaran, N., Nagarajan, R., & Mukherjee, A. (2014). Cytotoxicity of TiO2 NPs towards freshwater sediment microorganisms at low exposure concentrations. *Environmental Research, 135*, 333–345. https://doi.org/10.1016/j.envres.2014.09.025

Lahr, R. H., & Vikesland, P. J. (2014). Surface-enhanced Raman spectroscopy (SERS) cellular imaging of intracellulary biosynthesized gold NPs. *In ACS Sustainable Chemistry and Engineering*, 2(Issue 7), 1599–1608. https://doi.org/10.1021/sc500105n. American Chemical Society.

Lee, K. C., Lin, S. J., Lin, C. H., Tsai, C. S., & Lu, Y. J. (2008). Size effect of Ag NPs on surface plasmon resonance. *Surface and Coatings Technology, 202*(22–23), 5339–5342. https://doi.org/10.1016/j.surfcoat.2008.06.080

Legg, B. A., Zhu, M., Comolli, L. R., Gilbert, B., & Banfield, J. F. (2014). Impacts of ionic strength on three-dimensional NPs aggregate structure and consequences for environmental transport and deposition. *Environmental Science and Technology, 48*(23), 13703–13710. https://doi.org/10.1021/es502654q

Leonardo, T., Farhi, E., Pouget, S., Motellier, S., Boisson, A. M., Banerjee, D., Rébeillé, F., Den Auwer, C., & Rivasseau, C. (2016). Silver accumulation in the green microalga Coccomyxa actinabiotis: Toxicity, in situ speciation, and localization investigated using synchrotron

XAS, XRD, and TEM. *Environmental Science and Technology, 50*(1), 359–367. https://doi.org/10.1021/acs.est.5b03306

León, A., Reuquen, P., Garín, C., Segura, R., Vargas, P., Zapata, P., & Orihuela, P. (2017). FTIR and Raman characterization of TiO_2 NPs coated with polyethylene glycol as carrier for 2-methoxyestradiol. *Applied Sciences, 7*(1), 49. https://doi.org/10.3390/app7010049

Li, X., Batchelor-Mcauley, C., & Compton, R. G. (2019). Silver NPs detection in real-world environments via particle impact electrochemistry. *ACS Sensors, 4*(2), 464–470. https://doi.org/10.1021/acssensors.8b01482

Li, W., He, Y., Wu, J., & Xu, J. (2012). Extraction and characterization of natural soil NPs from Chinese soils. *European Journal of Soil Science, 63*(5), 754–761. https://doi.org/10.1111/j.1365-2389.2012.01480.x

Liu, Z., Yan, Z., Jia, L., Song, P., Mei, L., Bai, L., & Liu, Y. (2017). Gold NPs decorated electrospun nanofibers: A 3D reproducible and sensitive SERS substrate. *Applied Surface Science, 403*, 29–34. https://doi.org/10.1016/j.apsusc.2017.01.157

Li, Z., Wang, J., & Li, D. (2016). Applications of Raman spectroscopy in detection of water quality. *Applied Spectroscopy Reviews, 51*(4), 313–337. https://doi.org/10.1080/05704928.2015.1131711

Long, C.l., Yang, Z.g., Yang, Y., Li, H.p., & Wang, Q. (2016). Determination of gold NPs in natural water using single particle-ICP-MS. *Journal of Central South University, 23*(7), 1611–1617. https://doi.org/10.1007/s11771-016-3215-9

López-Sánchez, J., Serrano, A., Del Campo, A., Abuín, M., Rodríguez De La Fuente, O., & Carmona, N. (2016). Sol-gel synthesis and micro-Raman characterization of ε-Fe2O3 micro- and NPs. *Chemistry of Materials, 28*(2), 511–518. https://doi.org/10.1021/acs.chemmater.5b03566

Lu, P. J., Huang, S. C., Chen, Y. P., Chiueh, L. C., & Shih, D. Y. C. (2015). Analysis of titanium dioxide and zinc oxide NPs in cosmetics. *Journal of Food and Drug Analysis, 23*(3), 587–594. https://doi.org/10.1016/j.jfda.2015.02.009

Luther, S., Borgfeld, N., Kim, J., & Parsons, J. G. (2012). Removal of arsenic from aqueous solution: A study of the effects of pH and interfering ions using iron oxide nanomaterials. *Microchemical Journal, 101*, 30–36. https://doi.org/10.1016/j.microc.2011.10.001

Majumdar, S., Peralta-Videa, J. R., Bandyopadhyay, S., Castillo-Michel, H., Hernandez-Viezcas, J. A., Sahi, S., & Gardea-Torresdey, J. L. (2014). Exposure of cerium oxide NPs to kidney bean shows disturbance in the plant defense mechanisms. *Journal of Hazardous Materials, 278*, 279–287. https://doi.org/10.1016/j.jhazmat.2014.06.009

Marshall, A. T., Haverkamp, R. G., Davies, C. E., Parsons, J. G., Gardea-Torresdey, J. L., & Van Agterveld, D. (2007). Accumulation of gold NPs in Brassic juncea. *International Journal of Phytoremediation, 9*(3), 197–206. https://doi.org/10.1080/15226510701376026

Ma, X., Wang, Q., Rossi, L., & Zhang, W. (2016). Cerium oxide NPs and bulk cerium oxide leading to different physiological and biochemical responses in Brassica rapa. *Environmental Science and Technology, 50*(13), 6793–6802. https://doi.org/10.1021/acs.est.5b04111

Meier, C., Voegelin, A., Pradas Del Real, A., Sarret, G., Mueller, C. R., & Kaegi, R. (2016). Transformation of silver NPs in sewage sludge during incineration. *Environmental Science and Technology, 50*(7), 3503–3510. https://doi.org/10.1021/acs.est.5b04804

Merrifield, R. C., Stephan, C., & Lead, J. (2017). Determining the concentration dependent transformations of Ag NPs in complex media: Using SP-ICP-ms and Au@Ag core-shell NPs as tracers. *Environmental Science and Technology, 51*(6), 3206–3213. https://doi.org/10.1021/acs.est.6b05178

Mogensen, K. B., & Kneipp, K. (2014). Size-dependent shifts of plasmon resonance in silver NPs films using controlled dissolution: Monitoring the onset of surface screening effects. *Journal of Physical Chemistry C, 118*(48), 28075–28083. https://doi.org/10.1021/jp505632n

Montaño, M. D., Olesik, J. W., Barber, A. G., Challis, K., & Ranville, J. F. (2016). Single particle ICP-MS: Advances toward routine analysis of nanomaterials. *Analytical and Bioanalytical Chemistry, 408*(19), 5053–5074. https://doi.org/10.1007/s00216-016-9676-8

Muccifora, S., Castillo-Michel, H., Barbieri, F., Bellani, L., Castiglione, M. R., Spanò, C., Pradas Del Real, A. E., Giorgetti, L., & Tassi, E. L. (2021). Synchrotron radiation spectroscopy and transmission electron microscopy techniques to evaluate Tio_2 nps incorporation, speciation, and impact on root cells ultrastructure of pisum sativum l. Plants. *Nanomaterials, 11*(4). https://doi.org/10.3390/nano11040921

Nastulyavichus, A. A., Kudryashov, S. I., Tolordava, E. R., Khaertdinova, L. F., Yushina, Y. K., Borodina, T. N., Gonchukov, S. A., & Ionin, A. A. (2021). Dynamic light scattering detection of silver NPs, food pathogen bacteria and their bactericidal interactions. *Laser Physics Letters, 18*(8). https://doi.org/10.1088/1612-202X/ac0d07

Nath, J., Dror, I., Landa, P., Vanek, T., Kaplan-Ashiri, I., & Berkowitz, B. (2018). Synthesis and characterization of isotopically-labeled silver, copper and zinc oxide NPs for tracing studies in plants. *Environmental Pollution, 242,* 1827–1837. https://doi.org/10.1016/j.envpol.2018.07.084

Ogunkunle, C. O., Bornmann, B., Wagner, R., Fatoba, P. O., Frahm, R., & Lützenkirchen-Hecht, D. (2019). Copper uptake, tissue partitioning and biotransformation evidence by XANES in cowpea (Vigna unguiculata L) grown in soil amended with nano-sized copper particles. *Environmental Nanotechnology, Monitoring and Management, 12.* https://doi.org/10.1016/j.enmm.2019.100231

Pandya, J. K., Dai, H., & He, L. (2021). An innovative filtration based Raman mapping technique for the size characterization of anatase titanium dioxide NPs. *Talanta, 224.* https://doi.org/10.1016/j.talanta.2020.121836

Parsons, J. G., Aldrich, M. V., & Gardea-Torresdey, J. L. (2002). Environmental and biological applications of extended X-ray absorption fine structure (EXAFS) and X-ray absorption near edge structure (XANES) spectroscopies. *Applied Spectroscopy Reviews, 37*(2), 187–222. https://doi.org/10.1081/ASR-120006044

Parsons, J. G., Lopez, M. L., Gonzalez, C. M., Peralta-Videa, J. R., & Gardea-Torresdey, J. L. (2010). Toxicity and biotransformation of uncoated and coated nickel hydroxide NPs on mesquite plants. *Environmental Toxicology and Chemistry, 29*(5), 1146–1154. https://doi.org/10.1002/etc.146

Peng, C., Duan, D., Xu, C., Chen, Y., Sun, L., Zhang, H., Yuan, X., Zheng, L., Yang, Y., Yang, J., Zhen, X., Chen, Y., & Shi, J. (2015). Translocation and biotransformation of CuO NPs in rice (Oryza sativa L.) plants. *Environmental Pollution, 197,* 99–107. https://doi.org/10.1016/j.envpol.2014.12.008

Petosa, A. R., Jaisi, D. P., Quevedo, I. R., Elimelech, M., & Tufenkji, N. (2010). Aggregation and deposition of engineered nanomaterials in aquatic environments: Role of physicochemical interactions. *Environmental Science and Technology, 44*(17), 6532–6549. https://doi.org/10.1021/es100598h

Pindar, S., & Dhawan, N. (2019). Carbothermal reduction of spent mobile phones batteries for the recovery of lithium, cobalt, and manganese values. *JOM, 71*(12), 4483–4491. https://doi.org/10.1007/s11837-019-03799-9

Quarato, M., Pinheiro, I., Vieira, A., Espiña, B., & Rodriguez-Lorenzo, L. (2021). Detection of silver NPs in seawater using surface-enhanced Raman scattering. *Nanomaterials, 11*(7). https://doi.org/10.3390/nano11071711

Raju, D., Mehta, U. J., & Ahmad, A. (2012). Phytosynthesis of intracellular and extracellular gold NPs by living peanut plant (Arachis hypogaea L.). *Biotechnology and Applied Biochemistry, 59*(6), 471–478. https://doi.org/10.1002/bab.1049

Ramallo-López, J. M., Giovanetti, L., Craievich, A. F., Vicentin, F. C., Marín-Almazo, M., José-Yacaman, M., & Requejo, F. G. (2007). XAFS, SAXS and HREM characterization of Pd NPs capped with n-alkyl thiol molecules. *Physica B: Condensed Matter, 389*(1), 150–154. https://doi.org/10.1016/j.physb.2006.07.044

Rathnayake, S., Unrine, J. M., Judy, J., Miller, A. F., Rao, W., & Bertsch, P. M. (2014). Multitechnique investigation of the ph dependence of phosphate induced transformations of ZnO NPs. *Environmental Science and Technology, 48*(9), 4757–4764. https://doi.org/10.1021/es404544w

Rick, A. R., & Arai, Y. (2011). Role of natural NPs in phosphorus transport processes in ultisols. *Soil Science Society of America Journal, 75*(2), 335–347. https://doi.org/10.2136/sssaj2010.0124nps

Roberts, D. R., Ford, R. G., & Sparks, D. L. (2003). Kinetics and mechanisms of Zn complexation on metal oxides using EXAFS spectroscopy. *Journal of Colloid and Interface Science, 263*(2), 364–376. https://doi.org/10.1016/S0021-9797(03)00281-9

Saleh, T. A. (2020). Trends in the sample preparation and analysis of nanomaterials as environmental contaminants. *Trends in Environmental Analytical Chemistry, 28*. https://doi.org/10.1016/j.teac.2020.e00101

Schavkan, A., Gollwitzer, C., Garcia-Diez, R., Krumrey, M., Minelli, C., Bartczak, D., Cuello-Nuñez, S., Goenaga-Infante, H., Rissler, J., Sjöström, E., Baur, G. B., Vasilatou, K., & Shard, A. G. (2019). Number concentration of gold NPs in suspension: SAXS and spICPMS as traceable methods compared to laboratory methods. *Nanomaterials, 9*(4). https://doi.org/10.3390/nano9040502

Schindler, T., Schmiele, M., Schmutzler, T., Kassar, T., Segets, D., Peukert, W., Radulescu, A., Kriele, A., Gilles, R., & Unruh, T. (2015). A combined SAXS/SANS study for the in situ characterization of ligand shells on small NPs: The case of ZnO. *Langmuir, 31*(37), 10130–10136. https://doi.org/10.1021/acs.langmuir.5b02198

Schwabe, F., Schulin, R., Limbach, L. K., Stark, W., Bürge, D., & Nowack, B. (2013). Influence of two types of organic matter on interaction of CeO_2 NPs with plants in hydroponic culture. *Chemosphere, 91*(4), 512–520. https://doi.org/10.1016/j.chemosphere.2012.12.025

Schwertfeger, D. M., Velicogna, J. R., Jesmer, A. H., Saatcioglu, S., McShane, H., Scroggins, R. P., & Princz, J. I. (2017). Extracting metallic NPs from soils for quantitative analysis: Method development using engineered silver NPs and SP-ICP-ms. *Analytical Chemistry, 89*(4), 2505–2513. https://doi.org/10.1021/acs.analchem.6b04668

Sciortino, L., Giannici, F., Martorana, A., Ruggirello, A. M., Liveri, V. T., Portale, G., Casaletto, M. P., & Longo, A. (2011). Structural characterization of surfactant-coated bimetallic cobalt/nickel nanoclusters by XPS, EXAFS, WAXS, and SAXS. *Journal of Physical Chemistry C, 115*(14), 6360–6366. https://doi.org/10.1021/jp200288y

Sekine, R., Marzouk, E. R., Khaksar, M., Scheckel, K. G., Stegemeier, J. P., Lowry, G. V., Donner, E., & Lombi, E. (2017). Aging of dissolved copper and copper-based NPs in five different soils: Short-term kinetics vs. long-term fate. *Journal of Environmental Quality, 46*(6), 1198–1205. https://doi.org/10.2134/jeq2016.12.0485

Servin, A. D., Castillo-Michel, H., Hernandez-Viezcas, J. A., Diaz, B. C., Peralta-Videa, J. R., & Gardea-Torresdey, J. L. (2012). Synchrotron micro-XRF and micro-XANES confirmation of the uptake and translocation of TiO_2 NPs in cucumber (Cucumis sativus) plants. *Environmental Science and Technology, 46*(14), 7637–7643. https://doi.org/10.1021/es300955b

Servin, A., De la Torre-Roche, R., Castillo-Michel, H., Pagano, L., Hawthorne, J., Musante, C., Pignatello, J., Uchimiya, M., & White, J. C. (2017). Exposure of agricultural crops to NPs CeO2 in biochar-amended soil. *Plant Physiology and Biochemistry, 110*, 147–157. https://doi.org/10.1016/j.plaphy.2016.06.003

Settimio, L., McLaughlin, M. J., Kirby, J. K., Langdon, K. A., Lombi, E., Donner, E., & Scheckel, K. G. (2014). Fate and lability of silver in soils: Effect of ageing. *Environmental Pollution, 191*, 151–157. https://doi.org/10.1016/j.envpol.2014.04.030

Shahwan, T., Abu Sirriah, S., Nairat, M., Boyacı, E., Eroğlu, A. E., Scott, T. B., & Hallam, K. R. (2011). Green synthesis of iron NPs and their application as a Fenton-like catalyst for the degradation of aqueous cationic and anionic dyes. *Chemical Engineering Journal, 172*(1), 258–266. https://doi.org/10.1016/j.cej.2011.05.103

Sikder, M., Lead, J. R., Chandler, G. T., & Baalousha, M. (2018). A rapid approach for measuring silver NPs concentration and dissolution in seawater by UV–Vis. *Science of the Total Environment, 618*, 597–607. https://doi.org/10.1016/j.scitotenv.2017.04.055

Silva, L. F. O., Dotto, G. L., Pinto, D., & Oliveira, M. L. S. (2021). NPs and interfaces with toxic elements in fluvial suspended sediment. *Marine Pollution Bulletin, 168*. https://doi.org/10.1016/j.marpolbul.2021.112405

Spielman-Sun, E., Lombi, E., Donner, E., Avellan, A., Etschmann, B., Howard, D., & Lowry, G. V. (2018). Temporal evolution of copper distribution and speciation in roots of Triticum aestivum exposed to CuO, Cu(OH)2, and CuS NPs. *Environmental Science and Technology, 52*(17), 9777–9784. https://doi.org/10.1021/acs.est.8b02111

Srabionyan, V. V., Bugaev, A. L., Pryadchenko, V. V., Avakyan, L. A., Van Bokhoven, J. A., & Bugaev, L. A. (2014). EXAFS study of size dependence of atomic structure in palladium NPs. *Journal of Physics and Chemistry of Solids, 75*(4), 470–476. https://doi.org/10.1016/j.jpcs.2013.12.012

Srabionyan, V. V., Bugaev, A. L., Pryadchenko, V. V., Makhiboroda, A. V., Rusakova, E. B., Avakyan, L. A., Schneider, R., Dubiel, M., & Bugaev, L. A. (2013). EXAFS study of changes in atomic structure of silver NPs in soda-lime glass caused by annealing. *Journal of Non-Crystalline Solids, 382*, 24–31. https://doi.org/10.1016/j.jnoncrysol.2013.09.025

Tamez, C., Hernandez, R., & Parsons, J. G. (2016). Removal of Cu (II) and Pb (II) from aqueous solution using engineered iron oxide NPs. *Microchemical Journal, 125*, 97–104. https://doi.org/10.1016/j.microc.2015.10.028

Thalmann, B., Voegelin, A., Morgenroth, E., & Kaegi, R. (2016). Effect of humic acid on the kinetics of silver NPs sulfidation. *Environmental Science: Nano, 3*(1), 203–212. https://doi.org/10.1039/c5en00209e

Tobler, D. J., Shaw, S., & Benning, L. G. (2009). Quantification of initial steps of nucleation and growth of silica NPs: An in-situ SAXS and DLS study. *Geochimica et Cosmochimica Acta, 73*(18), 5377–5393. https://doi.org/10.1016/j.gca.2009.06.002

Tomaszewska, E., Soliwoda, K., Kadziola, K., Tkacz-Szczesna, B., Celichowski, G., Cichomski, M., Szmaja, W., & Grobelny, J. (2013). Detection limits of DLS and UV-Vis spectroscopy in characterization of polydisperse NPs colloids. *Journal of Nanomaterials, 2013*. https://doi.org/10.1155/2013/313081

Tong, T., Hill, A. N., Alsina, M. A., Wu, J., Shang, K. Y., Kelly, J. J., Gray, K. A., & Gaillard, J. F. (2015). Spectroscopic characterization of TiO_2 polymorphs in wastewater treatment and sediment samples. *Environmental Science and Technology Letters, 2*(1), 12–18. https://doi.org/10.1021/ez5004023

Tsao, T., Chen, Y., Sheu, H., Tzou, Y., Chou, Y., & Wang, M. (2013). Separation and identification of soil NPs by conventional and synchrotron X-ray diffraction. *Applied Clay Science, 85*(1), 1–7. https://doi.org/10.1016/j.clay.2013.09.005

Tsukimura, K., Miyoshi, Y., Takagi, T., Suzuki, M., & Wada, S.i. (2021). Amorphous NPs in clays, soils and marine sediments analyzed with a small angle X-ray scattering (SAXS) method. *Scientific Reports, 11*(1). https://doi.org/10.1038/s41598-021-86573-9

Vance, M. E., Kuiken, T., Vejerano, E. P., McGinnis, S. P., Hochella, M. F., & Hull, D. R. (2015). Nanotechnology in the real world: Redeveloping the nanomaterial consumer products inventory. *Beilstein Journal of Nanotechnology, 6*(1), 1769–1780. https://doi.org/10.3762/bjnano.6.181

Vikesland, P. J., Rebodos, R. L., Bottero, J. Y., Rose, J., & Masion, A. (2016). Aggregation and sedimentation of magnetite NPs clusters. *Environmental Science: Nano, 3*(3), 567–577. https://doi.org/10.1039/c5en00155b

Vittori Antisari, L., Carbone, S., Gatti, A., Vianello, G., & Nannipieri, P. (2013). Toxicity of metal oxide (CeO_2, Fe_3O_4, SnO_2) engineered NPs on soil microbial biomass and their distribution in soil. *Soil Biology and Biochemistry, 60,* 87–94. https://doi.org/10.1016/j.soilbio.2013.01.016

Von Der Heyden, B., Roychoudhury, A., & Myneni, S. (2019). Iron-rich NPs in natural Aquatic environments. *Minerals, 9*(5). https://doi.org/10.3390/min9050287

Wang, R., Du, H., Wang, Y., Wang, D., Sun, Q., & Zhou, D. (2018). Retention of silver NPs and silver ion to natural soils: Effects of soil physicochemical properties. *Journal of Soils and Sediments, 18*(7), 2491–2499. https://doi.org/10.1007/s11368-018-1918-2

Wang, W., Martin, J. C., Fan, X., Han, A., Luo, Z., & Sun, L. (2012). Silica NPs and frameworks from rice husk biomass. *ACS Applied Materials and Interfaces, 4*(2), 977–981. https://doi.org/10.1021/am201619u

Wang, P., Menzies, N. W., Lombi, E., McKenna, B. A., Johannessen, B., Glover, C. J., Kappen, P., & Kopittke, P. M. (2013). Fate of ZnO NPs in soils and cowpea (Vigna unguiculata). *Environmental Science and Technology, 47*(23), 13822–13830. https://doi.org/10.1021/es403466p

Wang, D., Su, C., Zhang, W., Hao, X., Cang, L., Wang, Y., & Zhou, D. (2014). Laboratory assessment of the mobility of water-dispersed engineered NPs in a red soil (Ultisol). *Journal of Hydrology, 519,* 1677–1687. https://doi.org/10.1016/j.jhydrol.2014.09.053

Wang, D., Xie, Y., Jaisi, D. P., & Jin, Y. (2016). Effects of low-molecular-weight organic acids on the dissolution of hydroxyapatite NPs. *Environmental Science: Nano, 3*(4), 768–779. https://doi.org/10.1039/c6en00085a

Wang, Z. B., Yin, G. P., & Lin, Y. G. (2007). Synthesis and characterization of PtRuMo/C NPs electrocatalyst for direct ethanol fuel cell. *Journal of Power Sources, 170*(2), 242–250. https://doi.org/10.1016/j.jpowsour.2007.03.078

Wang, W., Zhang, K., Cai, Q., Mo, G., Xing, X. Q., Cheng, W. D., Chen, Z. J., & Wu, Z. H. (2010). Real-time SAXS and ultraviolet-visible spectral studies on size and shape evolution of gold NPs in aqueous solution. *European Physical Journal B, 76*(2), 301–307. https://doi.org/10.1140/epjb/e2010-00177-3

Wang, H., Zhao, X., Han, X., Tang, Z., Song, F., Zhang, S., Zhu, Y., Guo, W., He, Z., Guo, Q., Wu, F., Meng, X., & Giesy, J. P. (2018). Colloidal stability of Fe_3O_4 magnetic NPs differentially impacted by dissolved organic matter and cations in synthetic and naturally-occurred environmental waters. *Environmental Pollution, 241,* 912–921. https://doi.org/10.1016/j.envpol.2018.06.029

Wei, H., Leng, W., Song, J., Liu, C., Willner, M. R., Huang, Q., Zhou, W., & Vikesland, P. J. (2019). Real-time monitoring of ligand exchange kinetics on gold NPs surfaces enabled by hot spot-normalized surface-enhanced Raman scattering. *Environmental Science and Technology, 53*(2), 575–585. https://doi.org/10.1021/acs.est.8b03144

Yang, Q., Xu, W., Liu, G., Song, M., Tan, Z., Mao, Y., Yin, Y., Cai, Y., Liu, J., & Jiang, G. (2020). Transformation and uptake of silver NPs and silver ions in rice plant (: Oryza sativa L.): The effect of iron plaque and dissolved iron. *Environmental Science: Nano, 7*(2), 599–609. https://doi.org/10.1039/c9en01297d

Yan, W., Petkov, V., Mahurin, S. M., Overbury, S. H., & Dai, S. (2005). Powder XRD analysis and catalysis characterization of ultra-small gold NPs deposited on titania-modified SBA-15. *Catalysis Communications, 6*(6), 404–408. https://doi.org/10.1016/j.catcom.2005.04.004

Yeshchenko, O. A., Dmitruk, I. M., Alexeenko, A. A., Kotko, A. V., Verdal, J., & Pinchuk, A. O. (2012). Size and temperature effects on the surface plasmon resonance in silver NPs. *Plasmonics, 7*(4), 685–694. https://doi.org/10.1007/s11468-012-9359-z

Yilmaz, M., Yilmaz, A., Karaman, A., Aysin, F., & Aksakal, O. (2021). Monitoring chemically and green-synthesized silver NPs in maize seedlings via surface-enhanced Raman spectroscopy (SERS) and their phytotoxicity evaluation. *Talanta, 225*. https://doi.org/10.1016/j.talanta.2020.121952

Yin, Y., Yang, X., Zhou, X., Wang, W., Yu, S., Liu, J., & Jiang, G. (2015). Water chemistry controlled aggregation and photo-transformation of silver NPs in environmental waters. *Journal of Environmental Sciences, 34*, 116–125. https://doi.org/10.1016/j.jes.2015.04.005

Yoshida, K., Itoh, T., Tamaru, H., Biju, V., Ishikawa, M., & Ozaki, Y. (2010). Quantitative evaluation of electromagnetic enhancement in surface-enhanced resonance Raman scattering from plasmonic properties and morphologies of individual Ag nanostructures. *Physical Review B, 81*(11). https://doi.org/10.1103/physrevb.81.115406

Zeng, Y., Hu, R., Wang, L., Gu, D., He, J., Wu, S. Y., Ho, H. P., Li, X., Qu, J., Gao, B. Z., & Shao, Y. (2017). Recent advances in surface plasmon resonance imaging: Detection speed, sensitivity, and portability. *Nanophotonics, 6*(5), 1017–1030. https://doi.org/10.1515/nanoph-2017-0022

Zhang, Z., Guo, H., Deng, Y., Xing, B., & He, L. (2016). Mapping gold NPs on and in edible leaves: In situ using surface enhanced Raman spectroscopy. *RSC Advances, 6*(65), 60152–60159. https://doi.org/10.1039/c6ra11748a

Zhang, L., Li, X., He, R., Wu, L., Zhang, L., & Zeng, J. (2015). Chloride-induced shape transformation of silver NPs in a water environment. *Environmental Pollution, 204*, 145–151. https://doi.org/10.1016/j.envpol.2015.04.018

Zhang, P., Ma, Y., Zhang, Z., He, X., Zhang, J., Guo, Z., Tai, R., Zhao, Y., & Chai, Z. (2012). Biotransformation of ceria NPs in cucumber plants. *ACS Nano, 6*(11), 9943–9950. https://doi.org/10.1021/nn303543n

Zhang, W., Musante, C., White, J. C., Schwab, P., Wang, Q., Ebbs, S. D., & Ma, X. (2017). Bioavailability of cerium oxide NPs to Raphanus sativus L. in two soils. *Plant Physiology and Biochemistry, 110*, 185–193. https://doi.org/10.1016/j.plaphy.2015.12.013

Zhao, J., Wang, X., Hoang, S. A., Bolan, N. S., Kirkham, M. B., Liu, J., Xia, X., & Li, Y. (2021). Silver NPs in aquatic sediments: Occurrence, chemical transformations, toxicity, and analytical methods. *Journal of Hazardous Materials, 418*. https://doi.org/10.1016/j.jhazmat.2021.126368

Zhou, D., Abdel-Fattah, A. I., & Keller, A. A. (2012). Clay particles destabilize engineered NPs in aqueous environments. *Environmental Science and Technology, 46*(14), 7520–7526. https://doi.org/10.1021/es3004427

SECTION II

Biogeochemical cycle of ENMs before plant exposure explained at the molecular level

CHAPTER

4

Biophysicochemical transformations of ENMs in soil

Fabián Fernández-Luqueño[1], *Edgar Vázquez-Núñez*[2], *and Hermes Pérez-Hernandez*[3]

[1] Sustainability of Natural Resources and Energy Program, Cinvestav, Saltillo, Coahuila, Mexico; [2] Universidad de Guanajuato, Departamento de Ingenierías Química, Electrónica y Biomédica, División de Ciencias e Ingenierías, Campus León, León, Guanajuato, Mexico; [3] Universidad Autónoma de Chiapas, Facultad de Ciencias Agrícolas, Huehuetán, Chiapas, México

4.1 Introduction

Soils are a complex system where liquid, solid, and gaseous phases interact in an interdependent way (Silva & Lambers, 2018). Human activities have had a significant influence on soil characteristics; agriculture and forestry have negatively impacted the soil quality due to the escalated use of fertilizers to increase food production, pesticides to eliminate plagues, and new materials such as nanomaterials to improve soil quality and crop yield (Davari et al., 2020; Teixeira et al., 2021).

For these reasons, Engineered Nanomaterials (ENMs) can be found in terrestrial ecosystems, especially those manufactured for use in agriculture. The presence and persistence of ENMs in the soil can alter the quality of the soil microbiome, thus influencing key microbial processes like mineralization, nitrogen fixation, and plant growth-promoting activities (Khan, 2020, pp. 181–199).

Plant function can also be affected by the presence of nanomaterials in soil. In this context, adverse and positive effects of ENMs in the soil system have been reported. In order to accurately determine the conditions that allow a proper use of ENMs, further research is needed (Zhang et al., 2020).

Physicochemical Interactions of Engineered Nanoparticles and Plants
https://doi.org/10.1016/B978-0-323-90558-9.00012-7

4.2 Basic soil concepts

The two major categories of soils are organic and mineral soils. Organic soils form from plant debris, possess an intense dark color, are highly porous, and light in weight; these soils are named "peats" or "mucks" (Lavallee et al., 2020). Contrarily, mineral soils form from rocks or materials transported by wind, water, or landslides. In these soils, important amounts of silt, sand, and clay are commonly found (Totsche et al., 2018).

4.2.1 Soil horizons

Soil is a vertical sequence of layers called horizons, and the whole sequence constitutes the soil profile. The upper soil layer is termed A horizon, rich in decomposed, dark-colored, carbon-rich matter and microbial biomass mixed with minerals to form aggregate structures (Hartemink, 2016).

Below A lies B horizon, which is characterized by an accumulation of clay (particles less than 0.002 mm in diameter) structures linked together in various configurations as the horizon evolves (Hartemink et al., 2016).

In horizon C, the presence of humus is minor; the soil structure development is limited as well. This horizon is composed of unconsolidated parent material from which A and B horizons have formed. At some depth below the A, B, and C horizons, consolidated rock or the so-called R horizon is present (Mergelov et al., 2018). Fig. 4.1 schematizes the soil profiles described previously.

4.2.2 Physical characteristics of soil

4.2.2.1 Grain size and porosity

The grain size of soil particles and aggregate structures affects the transport of water, air, and nutrients. According to the size of the grain, they can be classified as clay, silt, or sand (<0.002 mm, 0.002–0.005 mm and, 0.05–2 mm, respectively) (Yang et al., 2015). Soil texture refers to the relative proportions of sand, silt, and clay particles, irrespective of chemical or mineralogical composition. Sandy soils are called coarse-textured, and clay-rich soils are called fine-textured. Loam is a textural class representing about one-fifth of clay, with sand and silt equally sharing the remainder (Saha et al., 2017). The space between soil particles (pores) is called porosity, which reflects the capacity of soil to hold air and water. Also, permeability describes the ease of transport of fluids and their dissolved components. In general, average pore size decreases because of agricultural practices and other human activities (Dignac et al., 2017).

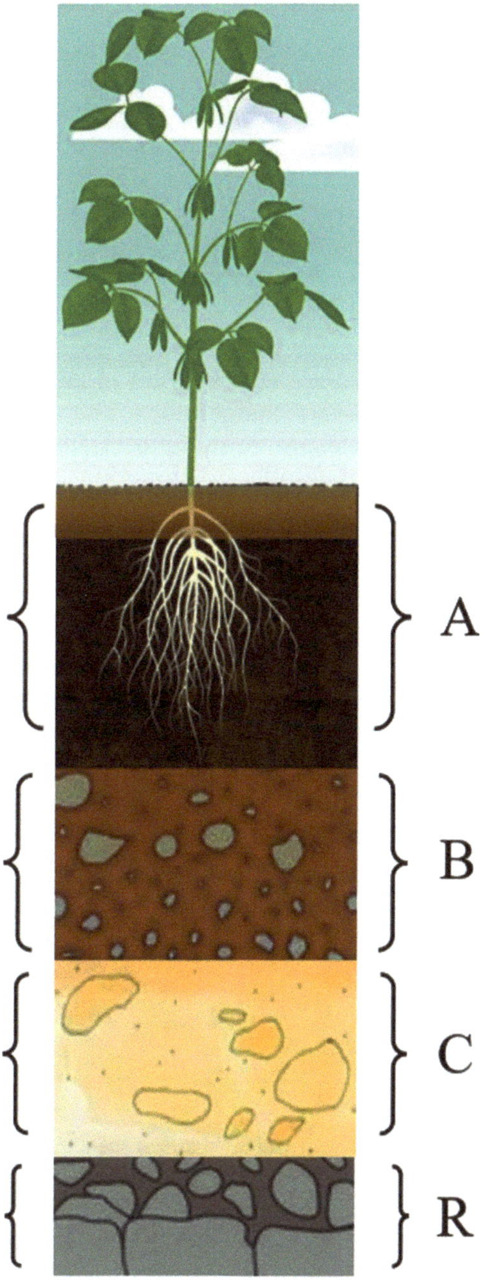

FIGURE 4.1 Schematic drawing of the soil profiles.

4.2.2.2 *Water in soil*

Soils whose horizon A is dense and unstructured increase the fraction of precipitation that will undergo surface runoff and have a high potential for erosion and flooding (Chen et al., 2018). In general, subsurface runoff processes are characteristic of soils in humid regions, whereas surface runoff is characteristic of arid regions. Any landscape is significantly altered by cultivation or urbanization (Owuor et al., 2016).

4.2.3 Chemical characteristics

4.2.3.1 *Mineral content*

Bulk soil consists of mineral particles composed of arrays of silicate ions (SiO_4^{4-}) combined with various positively charged ions. The most common mineral found in Earth's crust is feldspar, an aluminosilicate that contains sodium, potassium, or calcium in addition to aluminum ions. Weathering breaks crystals of feldspars and other silicate minerals and releases chemical compounds such as bases, silica, and oxides of iron and aluminum (Fe_2O_3 and alumina [Al_2O_3]) (Crundwell, 2014; White, 2018).

The solid particles are chemically reactive because of electrically charged sites on their surface, which can bind a dissolved ion or molecule to form a stable unit in a process called surface complexation (Kubicki et al., 2007). Ions such as Ca^{2+}, Mg^{2+}, Na^{+}, and NO_3^- do not tend to adsorb strongly, which makes these essential plant nutrients easily replaceable. Soils leached by freshwater release hydrogen ions (H^+) that increase mineral solubility. This may result in the liberation of, for example, Al^{3+} (aluminum), a toxic ion that is able to displace nutrients such as Ca^{2+} (Faulkner & Richardson, 2020, pp. 41–72). Soils display their acidity by decreasing the content of acid-soluble minerals and an increase in insoluble minerals (Barsova et al., 2019).

4.2.3.2 *Organic content*

Organic matter produced by organisms is the second major component of soils. Total organic matter in soil is called humus. This soil component plays a crucial role in controlling soil acidity, cycling of nutrients, and the detoxification of hazardous compounds (Yang et al., 2021). It has been demonstrated that in humus, molecules containing hydrogen ions dissociate in freshwater to form species that bear a net negative charge, which can interact with toxic metal ions and effectively remove them from further interaction with the environment. On the other hand, the uncharged portion of humic substances can react with synthetic organic compounds such as pesticides, fertilizers, solid and liquid waste materials, and degradation products (Fedotov & Shoba, 2015; Nardi et al., 2017).

4.2.4 Living organisms in soil

Soils are biological environments with life at all size scales, microfauna (body size <0.1 mm), mesofauna (up to 2 mm), and macrofauna (up to 20 mm). The most abundant soil microorganisms are unicellular microfauna. For example, it has been reported that 500 billion bacteria, 10 billion actinomycetes, and nearly one billion fungi can be found in 1 kg soil (Orlov, 2020; Weber et al., 2018). The multicellular animal population can be as high as 500 million in a kilogram of soil, and the most abundant are nematodes, followed by mesofauna including mites. Regarding larger soil animal species, earthworms, millipedes, centipedes, and insects make up most of the rest of the soil species (Dervash et al., 2018). These living organisms may influence stability and transformation of ENMs in soils.

4.3 Soil as the main recipient of ENMs

According to Keller et al. (2013), coatings, paints, and pigments represent the major segment releasing ENMs to the environment, followed by electronics and optics, cosmetics, energy and environment, catalysts, automotive, and medical segments (80,500, 48,700, 48,000, 43,700, 37,500, 23,500, and 13,400 metric tons/year, respectively). In a recent study, the concentration of nanomaterials in the environment was estimated using stochastic models covering the release of ENMs and subsequent transport between giving and receiving compartments, ENMs in economic circulation, and ENMs present in nature and technical sinks from 1950 to 2015 (Giese et al., 2018). The study also resulted in an attempt to distinguish predicted concentrations derived from ENMs use and after-use release.

Soil is the environmental compartment where most ENMs will end up via atmospheric deposition, inundation from rivers, sewage sludge application in agriculture, waste disposal, and targeted applications (Giese et al., 2018; McKee & Filser, 2016). Conversely, they represent a likely source of ENMs in natural soil systems. Thus, it is essential to identify their potential impact on the food chain and groundwater, as well as mechanisms of transport and fate (Vázquez Núñez & de la Rosa-Álvarez, 2018).

4.4 Interactions of ENMs with soil components: analysis of the molecular reactions

4.4.1 Interactions of ENMs with abiotic soil components

Soils are composed of minerals, water, air, and organic matter. Fig. 4.2 displays the microstructure of (A) coarse soils; (B) fine soils; (C) well-

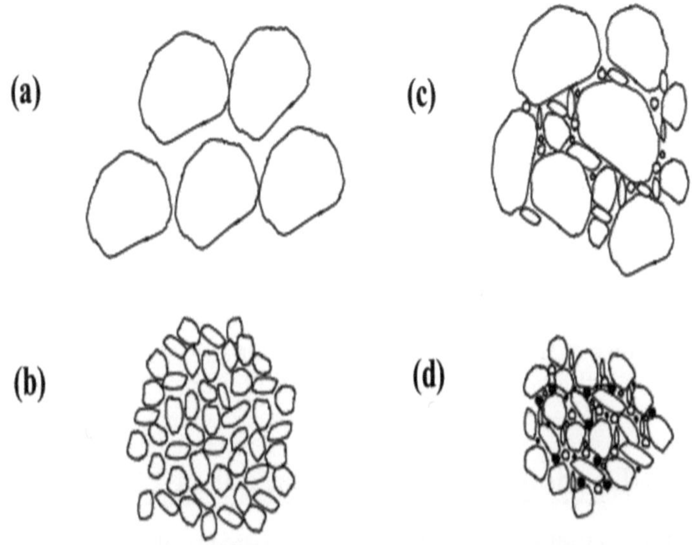

FIGURE 4.2 Microstructure of (a) coarse soils; (b) fine soils; (c) well-sorted; and (d) poorly sorted soils.

sorted; and (D) poorly sorted soils. Coarse and well-sorted soils provide pore spaces that allow for a higher hydraulic conductivity than fine and poorly sorted soils. In the former, pore spaces are blocked by small particles (Fig. 4.1A and D). For this reason, it would be expected that ENMs present in the soil solution would be more easily transported in coarse and well-sorted soils. The same occurs with compactness as a higher value is related to changes in pore geometry and volume of structural soil pores (Richard et al., 2001).

Depending on the degree of ENMs aggregation, they may be excluded from the soil solution and separated by gravity. In this case, aggregates may also clog soil pores. Temperature is a factor potentially affecting ENMs' interaction with soil media. To the best of our knowledge, no studies have reported the effect of temperature on ENMs dissolution in the soil solution phase. However, it has been demonstrated that ZnO NPs dissolution in seawater was negatively correlated with temperature (Wong & Leung, 2014). There is a need to study other ENMs and their stability in different media at different temperatures.

Abiotic soil matrix is mainly composed of silicates, which are assemblies containing silicon oxides. Silicates (Fig. 4.3) are organized in different crystalline structures depending on the number of silicon and oxygen atoms taking part in the structural units that form large networks. These networks display specific properties such as texture, porosity, and others at the macro- and micro-levels.

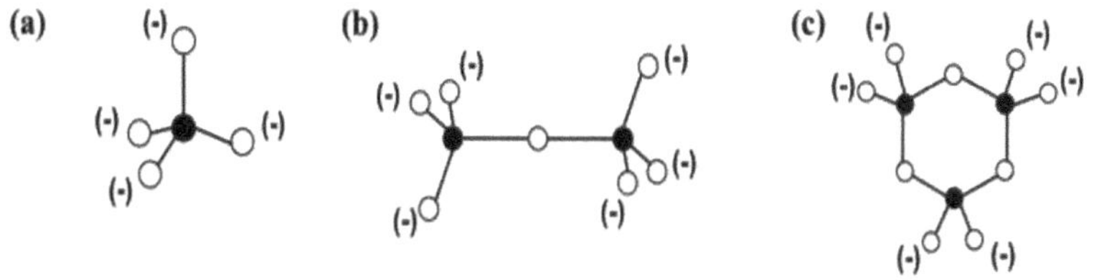

FIGURE 4.3 Structural silicate units: (a) SiO_4^{4-}; (b) $Si_2O_7^{6-}$; (c) $Si_3O_9^{6-}$. Black dots represent Si; white dots correspond to O. Negative oxygen charges are displayed.

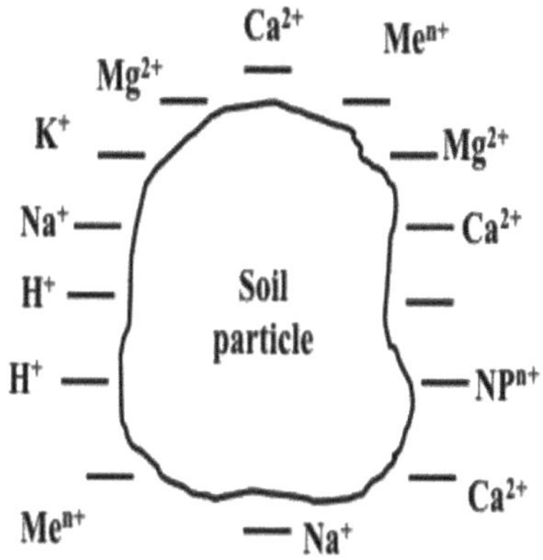

FIGURE 4.4 Representation of a soil particle showing negative charges due to oxygen and different cations that are eventually attracted towards the particle.

As Fig. 4.3 displays, silicates contain oxygen units that are negatively charged depending on soil pH. Fig. 4.4 is a representation of a soil particle with negative surface charges that can attract and bind different ions. Conversely, specific ions are repelled and maintained in the soil solution. It would be expected that negatively charged soil particles would attract positively charged ENMs. Surprisingly, charged and uncharged Ag NPs were almost equally sorbed onto silt particles (Klitzke et al., 2015). However, it is unknown whether the same phenomenon occurs with other particle sizes such as clay and sand.

4.4.2 Interactions of ENMs with biotic soil components

The biotic soil fraction is where most ENMs biotransformation occurs because of the highly variable diversity of modifications that soil may suffer due to the action of living organisms. Biotic soil fraction is transformed relatively faster as compared to the abiotic fraction. Living organisms are not the only ones responsible for this phenomenon of rapid decomposition of organic matter, which eventually forms the humic components in soils (humic acids, fulvic acids, and humin). Table 4.1 displays the main functional groups contained in macromolecules present in humic soil components.

As shown in Table 4.1, functional groups contain moieties that can attract and eventually bind to the metals present in the ENMs because of the free electrons in O, N, and S. It has been proven that some ENMs can bind to soil organic matter. For example, it was demonstrated that Ag NPs preferentially bound small organic matter molecules to long-chain-aromatic DOM (Klitzke et al., 2015); in the above-mentioned study it was also determined that DOM (dissolved organic matter) was adsorbed onto Ag NPs, reducing their dissolution and oxidation, and increasing their colloidal stability. Similarly, DOM increased colloidal stability in PVP-Ag NPs and Citrate-Ag NPs (Gunsolus et al., 2015). In addition, hydrophilic and hydrophobic DOM differentially affected Ag and TiO_2 NPs aggregation (Zehlike et al., 2019). In summary, these ENMs with hydrophilic DOM resulted in larger NP aggregates, consequently reducing their colloidal stability. Moreover, when Ca^{2+} interacted with the hydrophilic system DOM-NPs, further aggregation and NPs instability occurred.

TABLE 4.1 Main functional groups present in humic soil components (de Melo et al., 2016; Dinu et al., 2021; Gong et al., 2020).

Humic soil component	Functional groups	Formula
Humic acids	Carboxylic acids	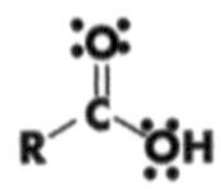
	Phenolic groups	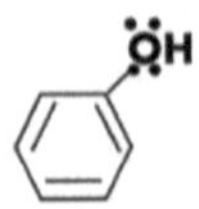
	Ether groups	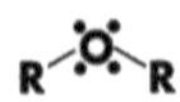
	Ester groups	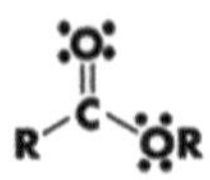

TABLE 4.1 Main functional groups present in humic soil components (de Melo et al., 2016; Dinu et al., 2021; Gong et al., 2020).—cont'd

Humic soil component	Functional groups	Formula
	Amine	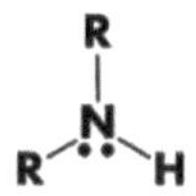
	Ketone	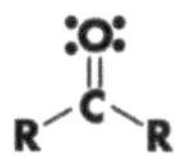
	Hydroxyl groups	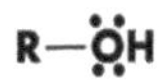
	Imine groups	
Fulvic acids	Carboxylic acids	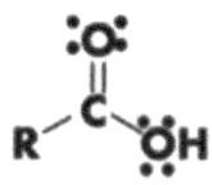
	Phenolic groups	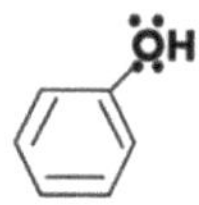
	Aldehyde	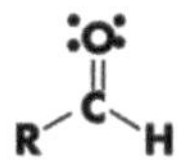
	Ester groups	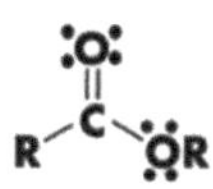

Continued

TABLE 4.1 Main functional groups present in humic soil components (de Melo et al., 2016; Dinu et al., 2021; Gong et al., 2020).—cont'd

Humic soil component	Functional groups	Formula
	Amine groups	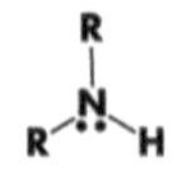
Humin	Carboxylic acids	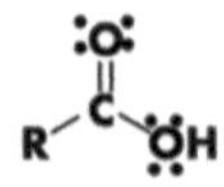
	Hydroxyl	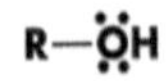
	Sugars (cellulose/ glucose units)	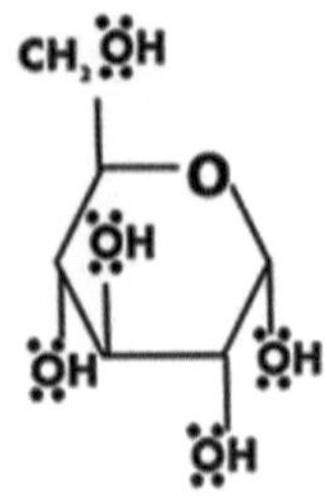

ENMs' interaction with DOM is also affected by the nature of capping, which displays differential affinity toward this organic matter; for example, the stability of Ag NPs capped with either citrate or polyvinylpyrrolidone was evaluated and it was finally proposed that DOM competes more successfully with citrate than PVP because citrate capping is less strongly bound to the Ag NPs surface than PVP (Gunsolus et al., 2015). These experiments were performed at pH 7.5, where citrate groups are negatively charged, and PVP moieties are neutral. Even when competition occurs between citrate replacement by DOM and repulsion of both species, their results indicated that DOM could replace citrate and provide more colloidal stability compared to PVP-AgNPs.

Other important biotic components are root exudates, a mixture of organic acids, sugars, amino acids, proteins, phytosiderophores, enzymes, hormones, and mucilage (Canarini et al., 2019). Notably, organic acids react with minerals, salts, and ENMs and can mobilize metallic species (Zhang et al., 2012). Moreover, others have proven these types of interactions. Zhao et al. (2019) also demonstrated that the metabolite profile of soils amended with SiO_2, TiO_2, and Fe_3O_4 NPs was significantly affected. These researchers found out that concentrations of levoglucosan, allo-inositol, and methyl phosphate, among others, increased in ENMs amended soils, probably indicating an impact on microbial growth and plant growth regulation. In the following section, specific reactions will be discussed. In addition, the role of microorganisms will be detailed in a separate chapter.

4.4.3 Transformation of ENMs as a result of biophysicochemical interaction with soil components

Different studies have tried to explain the transformation of NPs when they enter the vascular plant system. A study evaluated the uptake of CeO_2 by cucumber plants in hydroponics. In this study, it was determined that within the first steps of NPs transformation, citric acid present in root exudates dissolved CeO_2 NPs. In addition, ascorbic acid may act as a reducing agent to produce Ce(III), which is later complexed to PO_4^{3-}, forming $CePO_4$. Moreover, Ce(III) is also bound to carboxylate moieties to form Ce-OOC-like species (Zhang et al., 2012). This same process likely occurs in soil solution. However, more research is needed in order to corroborate this hypothesis.

Using isothermal titration calorimetry, different thermodynamic parameters relevant to the interaction of synthetic root exudates and Cu NPs were determined. According to the results, dissolution, oxidation, and reduction resulted in the transformation of Cu NPs to Cu^{2+} and Cu^{+} (Huang et al., 2017). Also, a similar study evaluated the growth of wheat on sand media and the effects of fulvic acids and root exudates on plant response to CuO NPs. In addition, soluble Cu complexes including 2′-deoxymugineic acid-Cu and gluconate-Cu were identified (Hortin et al., 2019).

Thus, carboxylic acids play an essential role in solubilizing and mobilizing ENMs. In summary, some potential reactions and processes occurring between ENMs and soil components (excluding microorganisms) include:

$$\text{ENM} + \text{R–COOH} \rightarrow \text{RCOO}^-\text{M}^+$$

$$\text{M}^+ + \text{charged soil particle} \rightarrow \text{M-soil particle}$$

$$\text{ENM-Citrate} + \text{small DOM} \rightarrow \text{ENM-small DOM} + \text{citrate}$$

$$\text{CeO}_4 + \text{citric acid} \rightarrow \text{dissolved Ce(IV) or Ce(IV) complexed with citrate}$$

$$\text{Ce(IV)} + \text{ascorbic acid} \rightarrow \text{Ce(III)}$$

$$\text{Ce(III)} + \text{PO}_4{}^{3-} \rightarrow \text{CePO}_4$$

$$\text{Ce(III)} + \text{R-COO}^- \rightarrow \text{R–COO–Ce}$$

$$\text{CuO ENMs} + \text{fulvic acids} \rightarrow \text{Cu-fulvate} + \text{Cu(II)} + \text{Cu(I)}$$

Cu(II) + Cu(I) + "2′-deoxymugineic acid → Cu(II)- 2′-deoxymuginate + Cu(I)- 2′-deoxymuginate

Cu(II) + Cu(I) + gluconic acid →Cu(II)-gluconate + Cu(I)-gluconate

4.4.4 Effects of ENMs on soil properties

ENMs properties affect soils as well as soil properties affect ENMs. However, in both cases, few data are available. A study reported positive effects of nano alumina and nano copper on different soil properties, including compaction, crack intensity factor, volumetric shrinkage strain, and expansive volumetric strain (Taha & Taha, 2012). Also, the presence of NPs reduced the development of desiccation cracks with no negative consequences on hydraulic conductivity. The effects were mainly the result of ENMs filling small pores between soil particles. Soil pH may also change. In maize plants grown for 4 weeks in presence of NPs, it was determined that Fe_3O_4 NPs decreased soil pH value from 5.35 to 5.13 (L. Zhao et al., 2019). Unfortunately, very little data on this subject have been reported. Thus, it is necessary to investigate this area and provide data that may serve for decision-making purposes.

4.5 Bioavailability and stability of ENMs affected by physicochemical soil properties

The potential environmental and human health risks of the ENMs have been shown through several studies, but currently there is a general lack of consensus regarding their transportation, stability, and transformation in the soil ecosystem (Hou et al., 2017; Liu et al., 2021; Ng et al., 2018). Besides, information regarding the toxicity of ENMs on soil organisms and bioavailability is still scarce (Gómez-Sagasti et al., 2019). It has been demonstrated that polystyrene nanomaterials were transported through desert soil, black soil, and red soil. Thus, it is important to identify specific soil characteristics that influence transport and retention of ENMs in soils (Wu, Lyu et al., 2020).

According to recent studies, at least 25 factors grouped in four sections (ENMs properties, aqueous phase chemistry, properties of porous media, and flow conditions) affect nanomaterials transport in porous media, including soils (Liu et al., 2020). Regarding ENMs properties, these factors include particle surface charge, particle size, particle shape, particle density, and particle concentration. In addition, ionic strength, ion type, pH, and temperature are important when considering aqueous phase chemistry. Relevant porous media factors and properties include medium type, medium size, medium surface roughness, moisture content, organic matter content, mineral content, clay particles, plastic particles, bacteria cells, and biofilms. In addition, factors related to flow conditions involve flow velocity, flow direction, and flow interruption.

4.5.1 pH

The pH buffering capacity of the soil (its ability to resist pH change upon lime addition) mainly depends on clay and organic matter content. Soil pH is an important factor in the distinct mobility of polystyrene NPs in soils (Wu, Lyu et al., 2020). It has to be noted that varying pH influences the surface charges of soil minerals such as Fe/Al oxides. Besides, clays with amphoteric Al-OH and Si-OH functional groups can also develop pH-dependent surface charge and therefore increase the mobility of polystyrene NPs. It should be noticed that soil pH influenced the surface charges of polystyrene NPs, changing the retention and transport in soil column experiments. These results indicated that polystyrene NPs are most likely to move far in alkaline soil with low Fe/Al oxide content and thereby impose collateral risks on groundwater (Wu, Lyu et al., 2020).

In order to understand the life cycle of CuO NPs in the environment and their potential effect on human health, an experiment was carried out using a soil column. In this study, it was determined that the transport of CuO NPs was significantly affected when columns were taken to pH values of 6, 8, and 9; ionic strength levels of 1, 10, and 50 mM; and humic acid concentrations of 0.1–10 mg-C/mL (Wu, Fang et al., 2020). pH changes may also affect the transport of nanomaterials in porous media by changing the surface charges of NPs and thus NPs aggregation and deposition, while severe pH changes could affect the structure and composition of ENMs (Liu et al., 2020).

4.5.2 Texture

Soil texture (STex) is a critical soil characteristic that refers to the relative percentages of sand, silt, and clay-sized particles that make up the mineral fraction of the soil. There are 12 soil textural classes based upon these relative proportions. STex determines numerous soil properties such as the rate at which water drains through a saturated soil, soil aeration, susceptibility to erosion (erodibility), organic matter levels, tillage and fertility management, cation exchange capacity, the pH buffering capacity, and how much water is available to plants and organisms.

A 3-month experiment was performed to understand the effects of different concentrations of nZVI on soil microbial properties in a clay-loam versus a sandy-loam soil. Data on microbial biomass, activity, and functional and structural diversity evidenced that soil microbial properties from sandy-loam soil were more vulnerable to the presence of nZVI than clay-loam soil (Gómez-Sagasti et al., 2019).

It is well known that clay particles are a critical factor in the transport of ENMs. For example, in was found that clay particles destabilize both positively and negatively charged ENMs, in the form of the clay-facilitated aggregation (heteroaggregation), which inhibited the transport of NMs by physical straining (Liu et al., 2020). The study of the transport and long-term release of poly vinylpyrrolidone-coated silver NPs (PVP-Ag NPs) was carried out in a quartz sand column. In this case, various sand grain sizes, input concentrations of PVP-Ag, and flow rates were selected. Release tests showed that the long-term release of retained PVP-Ag NPs in the quartz sand was mainly in NPs rather than ions form. Besides, their studies indicated that sand grain size, input concentration, and flow rate significantly affected the transport and long-term release behavior of ENMs, i.e., NPs and ions (Hou et al., 2017).

Finally, it is important to mention that some soil aggregates such as nano-maghemite and amorphous Mn-oxide reduced the mobility of As in soil by accelerating the formation and precipitation of secondary Fe/Mn/Al oxyhydroxides (Vítková et al., 2015).

4.5.3 Water holding capacity (WHC)

The ENMs differ in the rate of transformations depending on their aggregative state and other characteristics due to soil properties or constituents, such as pH, organic matter, or water content. They also mediate the dissolution processes of metal-based ENMs, being a potential source of free metal ions (Rajput et al., 2020). Besides, the bioavailability of ENMs depends on their transformation in soil.

4.5.4 Organic matter

Organic matter can enhance the transport of ENMs in soils because it improves ENMs stability and reduces aggregate size (Qian et al., 2020). On the other hand, humic acids (HA) and Ca^{2+} display a prominent influence on the transport and long-term release behavior of Ag NPs in saturated quartz sand (Hou et al., 2017).

In a long-term study, ZnO-NPs and Fe_xO_y-NPs were mixed with natural sediments in the presence of HA at different pH values. Results indicated that a fraction (3.58%–7.97%) of the Zn was dissolved and eluted at the outlet when the pH was maintained at 6. In the case of Fe_xO_y–NPs, in the presence of HA, extensive retention of particles was observed during injection at both pHs (6 and 8) (Kumar & Raychoudhury, 2020). This study concluded that after the release of ZnO-NPs and Fe_xO_y-NPs mixed with the soil, significant concentrations of Zn and Fe were expected to remain in the porous media. However, a significant amount of retained NPs were likely to be detached from the soil, causing a risk of groundwater contamination in the longer run.

Organic matter is a ubiquitous component in the ecosystems which is adsorbed onto the surface of soil particles or the surface of NMs. Therefore, organic matter increases NPs sorption, precipitation, and transportation behaviors between soil aggregates (Qian et al., 2020). This affinity could be originated by electrostatic, hydrogen bonding, and hydrophobic interactions (Navarro et al., 2008).

4.5.5 Soil biodiversity

Soils are important reservoirs of biological diversity that underpins key processes and moderate ecosystem service provision. For example, highly abundant and diverse communities of organisms are supported in soils. These organisms vary in body size from a few micrometers to several meters (Nielsen et al., 2015).

The ENMs fate, in terms of dispersion, might be triggered by parameters such as pH, osmolarity, and organic matter, which can interact with the specific properties of the ENMs themselves, thus affecting uptake and toxicity in target or nontarget organisms (Corsi et al., 2014). The outcome of such interactions is also affected by the biological status of the organism, for instance, its ability to face and react to such exposure (Corsi et al., 2018).

Diverse studies have been carried out to determine the effect of ENMs on soil organisms. For instance, it was observed that earthworms did not detect low concentrations of Fe-NPs. For this reason, they did not limit soil consumption, resulting in high amounts of Fe in their bodies (Pérez-Hernández et al., 2020). It was also demonstrated that bioaccumulation of Zn and Fe in earthworms (*Eisenia foetida*) increased on par with dose when the soil was amended with $ZnO-$, TiO_2-, or Fe_2O_3-NPs (Valerio-Rodríguez et al., 2020). The effect of TiO_2 NPs has also been studied on the bacterial community structure and earthworms in arable soil (Sánchez-López et al., 2019). These authors found that the application of oats reduced the effect of TiO_2 NP applied to the soil. However, as the oats were mineralized by the soil microorganisms, the effect of the nanoparticles increased again over time. This implied that the oats had a collateral effect regarding adsorptions and delivery of ENMs, e.g., NPs.

Contradictory results were obtained when researchers studied the collateral effect of ENMs on nontarget organisms such as plants, mesofauna, macrofauna, and soil microbiota (Pérez-Hernández et al., 2020). Considering that NPs may change the physicochemical and biological properties of the soils, these authors recommended long-term in situ field trials. At the same time, land application of NPs should be limited to scientific experiments to fill knowledge gaps, not jeopardize global food production or the environment and worldwide human health.

4.5.6 Effect of soil type on ENMs bioavailability

There are more than hundred thousand kinds of soils worldwide, which contain different minerals and can interact in different ways with ENMs. It was found that iron and aluminum oxides were crucial in controlling the transport of polystyrene NPs in desert, black, and red soils (Wu, Fang et al., 2020). The mineralogy of these three soils was composed of quartz, goethite, hematite, montmorillonite, kaolinite, and albite; it was also found that higher amounts of Fe/Al oxides caused higher nano-polystyrene retention in red soils > black soil > desert soil (Wu, Lyu et al., 2020).

Aerated soils are predominantly oxidizing environments, while carbon-rich sediments may deplete oxygen, resulting in metal-based ENMs reduction. Besides, tidal zones with dynamic redox environments may well encounter the cycling of ENMs between different redox states, with different effects and transformation routes (Lowry et al., 2012). These could include chemical, physical, or biological transformations coupled with macromolecules such as proteins and natural organic matter.

4.6 Bioavailability and stability of ENMs as affected by biotic soil components

More than a decade (13 years) ago, the first article reporting the effects and nanotoxicology of NPs on the environment was published (Klaine et al., 2008). Nowadays, we have reliable information on the fate, behavior, and toxicological effects of ENMs and the interaction with biotic and abiotic factors in the soil. Different studies have suggested that microorganisms, along with other factors, are mediators in the process of ENMs dissolution. This results in the release of free metallic ions and consequently, both the ionic and dissolved forms may display greater availability and potentially cause environmental damage (Peng et al., 2020). In this section, we discuss the effects of soil biotic factors (bacteria, fungi, and other microorganisms) involved in the bioavailability and stability of ENMs. However, it is important to include the effects of soil mesofauna and macrofauna as promoters of these processes.

4.6.1 Biotic soil components potentially interacting with ENMs

According to Ameen et al. (2021), in recent years, the study of the interaction of ENMs with soil components has increased. Most studies have focused on plants, macroinvertebrates, bacteria, fungi, and nematodes. Particularly, microorganisms secrete enzymes, which play an essential role in the mineralization of nutrients, the transformation of OM, and degradation of pollutants, including ENMs. Nevertheless, it is known that the contact and mechanisms of action of ENMs and organisms depend on soil properties, buffering capacity, natural organic matter, nanoparticle aggregation and immobilization, nanoparticle deposition, and environmental corona formation around the NPs (Ameen et al., 2021).

4.6.2 ENMs biotransformation mediated by soil biotic components: bacteria, fungi, microfauna, and plant metabolites

After ENMs are released into the environment, they undergo various transformation processes that change their physicochemical properties. Within the transformation processes recognized in ENMs are (a) physical (i) agglomeration/aggregation, (ii) adsorption, and (iii) deposition, (b) (i) chemical sulfurization, (ii) dissolution, and (iii) redox reactions, and (c) interaction of NPs with macromolecules (Ahmed et al., 2021). Nevertheless, as previously mentioned, the migration, stability, transformation, and bioavailability of ENMs are uncertain. Even more, they are affected by the reduction process mediated by microorganisms and plants (root exudates). Several studies have explored the biocompatibility and toxicity of ENMs in different bio models. For example, Peng et al. (2017) evaluated the fate of CuO and Cu NPs in the soil-plant system (*Oryza sativa*). According to their results, the combination of humic acid with CuO and Cu resulted in Cu_2S and Cu associated with goethite.

Luo et al. (2017) performed experiments and computational simulations at the molecular scale and demonstrated that the bacterial strain *Shewanella oneidensis* MR-1 transformed hematite to magnetite. In this process, the breakage of iron–oxygen bonds in hematite and the concomitant formation of Fe complexing microstructures for biogenic magnetite production occurred. In terms of soil decontamination, the importance of microorganisms for the degradation, elimination, and transformation of heavy metals and contaminating organic compounds has been demonstrated. In the case of selenium, a trace element that is toxic in high concentrations, its transformation in the presence of *Azospirillum brasilense*, considered a plant growth-promoting bacterium, has been evaluated. The bacterium demonstrated the ability to transform selenium NPs into insoluble Se^0 particles formed extracellularly under aerobic conditions (Vogel et al., 2018).

In the context of the biotransformation of ENMs mediated by fungi, several researchers have proposed different methodologies to transform toxic ENMs. For example, it has been demonstrated that *Mariannaea* sp. HJ reduces Se (IV) to Se (0) intracellularly. In this process, the presence of proteins influences the detoxification process, which could be considered natural stabilizing agents (Zhang et al., 2019). On the other hand, it has been found that the rice rhizospheric fungus (*Aspergillus flavus*) is capable of biotransforming soluble arsenic (As) into immobilized arsenic particles. Furthermore, the study showed that As particles were not toxic to both fungi and bacteria. This was concluded because null growth effects in *Azospirillum brasilense* sp7 (bacteria), *Piriformospora indica* (fungus), and *Dictyostelium discoideum* AX2 (slime mold) were observed (Mohd et al., 2019).

It has been documented that root exudate composition is a key factor in the transformation of ENMs, i.e., NPs and, consequently, their translocation. In this regard, the literature reports that both root exudates and microbial communities produce nanosized minerals that contribute to the binding of soil carbon. In this same line, the authors reported that roots contribute to the dissolution and availability of nanosized soil minerals (Yu, 2018,pp. 143–166). Other authors suggested that low molecular weight organic acids such as succinic acid and malic acid and some reducing substances such as phenols were essential to transform CeO_2. However, the xylem of plant species (monocots and dicots) influences translocation (Zhang et al., 2012).

4.6.3 Effects of ENMs on soil fauna

Many studies have focused on evaluating the damage caused by NMs on bacteria, fungi, algae, collembola, isopods, and earthworms, among others. However, the experiments have used high concentrations, mostly in the levels of 1000 and 3000 mg kg^{-1} of dry soil, a situation that is unlikely to occur in natural soils (Pérez-Hernández et al., 2020). Also, it is important to emphasize that most research, in terms of nanotoxicology, has focused on single biological materials. Few researchers have reported and discussed different aspects of nano-biological-ecological interactions (He et al., 2018). This section will focus on nanobiointeractions and some aspects of ecological interaction in the soil system.

The toxic effects of various ENMs in model organisms have been discussed (Pérez-Hernández et al., 2020). Regarding the adverse effects that ENMs, i.e., NPs cause on soil fauna, three main events may be considered: (1) The ions released from NPs are internalized in cells, resulting in a degradation of ATP and reduction of cell duplication; (2) ions produce ROS and damage cell organs; and (3) NPs accumulate and dissolve in cell membranes leading to permeability alteration and consequently protein loss (Singh et al., 2019).

In a meta-analysis study with information collected over 12 years, valuable knowledge was presented on the releases of Ag NPs and their environmental risk. This study examined the behavior of Ag NPs along the "source – wastewater treatment plant (WWTP)–soil–receptor" pathway. According to the analysis, it was concluded that Ag in bulk and

Ag NPs from WWTPs accumulate in agricultural soils. However, the mass flow of Ag NPs entering the soil only represents a small fraction of the total Ag, with low concentrations compared to the values of the total concentrations measured in soils treated with mud. Another result is that Ag NPs are transformed into Ag_2S in WWTPs before being applied to agricultural soils. Therefore, it is estimated that the final concentrations of $\leq 6\ \mu g\ L^{-1}$ do not represent any harm to organisms and plants (Wang et al., 2018).

Other studies have shown conclusive results on the toxicity to the soil macrofauna. For instance, a study evaluating Ag effects on the earthworm-microbe-soil-plant system found that earthworms (*Eisenia fetida*) exposed to Ag NPs (10 mg kg^{-1}) suffered oxidative stress and decreased weight. In plants, NPs reduced the availability of nutrients and, in microorganisms, they reduced microbial growth (Das et al., 2018). Regarding the enzymatic change in earthworms, Galaktionova et al. (2019) reported that Zn and Cu NPs decreased urease and catalase activity after combined exposure of Zn-Cu NPs (50 mg kg^{-1}), while invertase activity was sensitive to 200 mg kg^{-1}.

Studies on soil decontamination suggest that *E. fetida* can be used in soil remediation. It has been demonstrated that earthworms decreased the bioavailability of Zn in soils treated with ZnO NPs, but not in soil amended with ZnS NPs (bioavailability was higher). The authors suggest that the decrease in soil pH mediated by worms increases the dissolution rate of the ZnS NPs (Bao et al., 2020). Therefore, the extractable Zn in the soil treated with ZnS NP increased in the presence of earthworms. Other results indicate that 400 mg Zn NPs kg^{-1} (Zn and ZnS NPs) induced toxicity on *E. fetida*. The effects were verified with survival and growth data.

There are many gaps in the knowledge about Cu NPs in the intestinal tract of soil organisms. Valuable information regarding toxicity in collembolans exposed to NPs of Cu and ionic copper has demonstrated that both NMs affected the intestinal microbial community of springtails. However, Cu NPs only affected microbial diversity. Furthermore, both NMs altered C and N isotopes in collembola and consequently their metabolism (Ding et al., 2020).

It has been documented that two species of nematodes (*Panagrellus redivivus* and *Folsomia candida*) exposed to ZnO NPs had different responses (Kiss et al., 2018). *P. redivivus* grown in artificial soil showed higher mortality compared to *F. candida* in natural soil. It has been stated that organic matter, the dissolution of Zn, and the aggregation of NPs influence organisms' toxicity (Kiss et al., 2018).

Some factors and mechanisms that occur with ENMs after having contact with some abiotic components of the soil have been studied, like pH, soil texture, ionic strength, temperature, soil minerals, and organic matter. Nevertheless, the contact with plants and soil organisms has generated a series of discussions that, to some extent, are controversial. For instance, conclusions have suggested the use of nanomaterials as promoters of plant or animal growth. However, the few reports at the field level suggest that the combined action of soil and water physicochemical components and biotic components should be studied in greater depth and detail.

The concentrations of ENMs and the living species are considered the most critical factors in identifying adverse effects. However, the absence or presence of organic matter and soil type (texture and pH) has been a key factor in transforming ENMs and generating toxic effects in organisms.

Therefore, it is expected that with the advance in the methodologies for the precise detection of ENMs in the environment and complete studies that integrate soil, plants, and organisms, the knowledge gaps can be reduced to such extent that the suggestions for the use of NMs may have greater scientific rigor.

4.7 Final remarks

During the past decade, nanomaterials' development and application have inevitably led to the emergence of ENMs in the environment. The evidence has demonstrated the toxicity of ENMs, raising concerns about their environmental fate, transport, and exposure. ENMs reach the environment via solid wastes, transport, water effluents, production, and manufacturing processes. However, the available information about their effects on the ecosystems is limited. Recently, more studies have investigated the transport of ENMs through porous media, like soil. Soil characteristics, such as organic matter, pH, and porosity, have been observed as enhancers of transportability of nanomaterials under laboratory conditions. Currently, new studies have focused on the fate of ENMs in natural environments. These studies have allowed elucidating the behavior of NMs under oxidation, transformation, and dissolution conditions. Regarding the biological components and their interaction with ENMs, there is an important interest in analyzing and elucidating the effect of ENMs with soil biota and plants, so that several studies have proposed methods to determine hazard effects of nanomaterials to biological components. These studies have been carried out either under laboratory or field conditions. In addition, they have provided meaningful information to predict the possible adverse effects on ecosystems including their incorporation to the trophic chain, bioaccumulation, bioaugmentation, and direct toxicological effect to microorganisms and plants. The knowledge of soil characteristics where the ENMs are discharged may facilitate the elucidation mechanisms of uptake, translocation, transformation, and interaction with biological components of the system, i.e., microorganisms and plants.

References

Ahmed, B., Rizvi, A., Ali, K., Lee, J., Zaidi, A., Khan, M. S., & Musarrat, J. (2021). Nanoparticles in the soil—plant system: A review. *Environmental Chemistry Letters*, 1—65.

Ameen, F., Alsamhary, K., Alabdullatif, J. A., & ALNadhari, S. (2021). A review on metal-based nanoparticles and their toxicity to beneficial soil bacteria and fungi. *Ecotoxicology and Environmental Safety, 213*, 112027. https://doi.org/10.1016/j.ecoenv.2021.112027

Bao, S., Huang, M., Tang, W., Wang, T., Xu, J., & Fang, T. (2020). Opposite effects of the earthworm *Eisenia fetida* on the bioavailability of Zn in soils amended with ZnO and ZnS nanoparticles. *Environmental Pollution, 260*, 114045. https://doi.org/10.1016/j.envpol.2020.114045

Barsova, N., Motuzova, G., Kolchanova, K., Stepanov, A., Karpukhin, M., Minkina, T., & Mandzhieva, S. (2019). The effect of humic substances on Cu migration in the soil profile. *Chemistry and Ecology, 35*(1), 86—101. https://doi.org/10.1080/02757540.2018.1540613

Canarini, A., Kaiser, C., Merchant, A., Richter, A., & Wanek, W. (2019). Root exudation of primary metabolites: Mechanisms and their roles in plant responses to environmental stimuli. *Frontiers in Plant Science, 10*. https://doi.org/10.3389/fpls.2019.00157

Chen, H., Zhang, X., Abla, M., Lü, D., Yan, R., Ren, Q., Ren, Z., Yang, Y., Zhao, W., Lin, P., Liu, B., & Yang, X. (2018). Effects of vegetation and rainfall types on surface runoff and soil erosion on steep slopes on the Loess Plateau, China. *Catena, 170*, 141—149. https://doi.org/10.1016/j.catena.2018.06.006

Corsi, I., Cherr, G. N., Lenihan, H. S., Labille, J., Hassellov, M., Canesi, L., Dondero, F., Frenzilli, G., Hristozov, D., Puntes, V., Della Torre, C., Pinsino, A., Libralato, G., Marcomini, A., Sabbioni, E., & Matranga, V. (2014). Common strategies and technologies for the ecosafety assessment and design of nanomaterials entering the marine environment. *ACS Nano, 8*(10), 9694—9709. https://doi.org/10.1021/nn504684k

Corsi, I., Winther-Nielsen, M., Sethi, R., Punta, C., Della Torre, C., Libralato, G., Lofrano, G., Sabatini, L., Aiello, M., Fiordi, L., Cinuzzi, F., Caneschi, A., Pellegrini, D., & Buttino, I. (2018). Ecofriendly nanotechnologies and nanomaterials for environmental applications: Key issue and consensus recommendations for sustainable and ecosafe nanoremediation. *Ecotoxicology and Environmental Safety, 154*, 237—244. https://doi.org/10.1016/j.ecoenv.2018.02.037

Crundwell, F. K. (2014). The mechanism of dissolution of minerals in acidic and alkaline solutions: Part II application of a new theory to silicates, aluminosilicates and quartz. *Hydrometallurgy, 149*, 265—275. https://doi.org/10.1016/j.hydromet.2014.07.003

Das, P., Barua, S., Sarkar, S., Chatterjee, S. K., Mukherjee, S., Goswami, L., Das, S., Bhattacharya, S., Karak, N., & Bhattacharya, S. S. (2018). Mechanism of toxicity and transformation of silver nanoparticles: Inclusive assessment in earthworm-microbe-soil-plant system. *Geoderma, 314*, 73—84.

Davari, M., Gholami, L., Nabiollahi, K., Homaee, M., & Jafari, H. J. (2020). Deforestation and cultivation of sparse forest impacts on soil quality (case study: West Iran, Baneh). *Soil and Tillage Research, 198*, 104504. https://doi.org/10.1016/j.still.2019.104504

de Melo, B. A. G., Motta, F. L, & Santana, M. H. A. (2016). Humic acids: Structural properties and multiple functionalities for novel technological developments. *Materials Science and Engineering, C 62*, 967—974. https://doi.org/10.1016/j.msec.2015.12.001

Dervash, M., Bhat, R., Mushtaq, N., & Singh, V. (2018). Dynamics and importance of soil mesofauna. *International Journal of Advance Research in Science and Engineering, 7*.

Dignac, M. F., Derrien, D., Barré, P., Barot, S., Cécillon, L., Chenu, C., Chevallier, T., Freschet, G. T., Garnier, P., Guenet, B., Hedde, M., Klumpp, K., Lashermes, G., Maron, P. A., Nunan, N., Roumet, C., & Basile-Doelsch, I. (2017). Increasing soil carbon storage: Mechanisms, effects of agricultural practices and proxies. A Review. *Agronomy for Sustainable Development, 37*(2). https://doi.org/10.1007/s13593-017-0421-2

Ding, J., Liu, J., Chang, X. B., Zhu, D., & Lassen, S. B. (2020). Exposure of CuO nanoparticles and their metal counterpart leads to change in the gut microbiota and resistome of collembolans. *Chemosphere, 258*. https://doi.org/10.1016/j.chemosphere.2020.127347

Dinu, R., Montes, S., Orange, F., & Mija, A. (2021). Reprocessable humins thermosets and composites. *Composites Science and Technology, 207*, 108655. https://doi.org/10.1016/j.compscitech.2021.108655

Faulkner, S. P., & Richardson, C. J. (2020). *Physical and chemical characteristics of freshwater wetland soils* (pp. 41–72). Informa UK Limited. https://doi.org/10.1201/9781003069850-5

Fedotov, G. N., & Shoba, S. A. (2015). On the nature of humic substances. *Eurasian Soil Science, 48*(12), 1292–1299. https://doi.org/10.1134/S1064229315120066

Galaktionova, L., Gavrish, I., & Lebedev, S. (2019). Bioeffects of Zn and Cu nanoparticles in soil systems. *Toxicology and Environmental Health Sciences, 11*(4), 259–270. https://doi.org/10.1007/s13530-019-0413-5

Giese, B., Klaessig, F., Park, B., Kaegi, R., Steinfeldt, M., Wigger, H., von Gleich, A., & Gottschalk, F. (2018). Risks, release and concentrations of engineered nanomaterial in the environment. *Scientific Reports, 8*, 1–18. https://doi.org/10.1038/s41598-018-19275-4

Gómez-Sagasti, M. T., Epelde, L., Anza, M., Urra, J., Alkorta, I., & Garbisu, C. (2019). The impact of nanoscale zero-valent iron particles on soil microbial communities is soil dependent. *Journal of Hazardous Materials, 364*, 591–599. https://doi.org/10.1016/j.jhazmat.2018.10.034

Gong, G. Q., Zhao, Y. F., Zhang, Y. J., Deng, B., Liu, W. X., Wang, M., Yuan, X., & Xu, L. W. (2020). Establishment of a molecular structure model for classified products of coal-based fulvic acid. *Fuel, 267*, 117210. https://doi.org/10.1016/j.fuel.2020.117210

Gunsolus, I. L., Mousavi, M. P. S., Hussein, K., Bühlmann, P., & Haynes, C. L. (2015). Effects of humic and fulvic acids on silver nanoparticle stability, dissolution, and toxicity. *Environmental Science and Technology, 49*(13), 8078–8086. https://doi.org/10.1021/acs.est.5b01496

Hartemink, A. E. (2016). The definition of soil since the early 1800s. *In Advances in Agronomy, 137*, 73–126. https://doi.org/10.1016/bs.agron.2015.12.001. Academic Press Inc.

He, X., Fu, P., Aker, W. G., & Hwang, H.-M. (2018). Toxicity of engineered nanomaterials mediated by nano–bio–eco interactions. *Journal of Environmental Science and Health, Part C, 36*(1), 21–42. https://doi.org/10.1080/10590501.2017.1418793

Hortin, J. M., Anderson, A. J., Britt, D. W., Jacobson, A. R., & McLean, J. E. (2019). Soil-derived fulvic acid and root exudates, modified by soil bacteria, alter CuO nanoparticle-induced root stunting of wheat: Via Cu complexation. *Environmental Science: Nano, 6*(12), 3638–3652. https://doi.org/10.1039/c9en00728h

Hou, J., Zhang, M., Wang, P., Wang, C., Miao, L., Xu, Y., You, G., Lv, B., Yang, Y., & Liu, Z. (2017). Transport, retention, and long-term release behavior of polymer-coated silver nanoparticles in saturated quartz sand: The impact of natural organic matters and electrolyte. *Environmental Pollution, 229*, 49–59. https://doi.org/10.1016/j.envpol.2017.05.059

Huang, Y., Zhao, L., & Keller, A. A. (2017). Interactions, transformations, and bioavailability of nano-copper exposed to root exudates. *Environmental Science and Technology, 51*(17), 9774–9783. https://doi.org/10.1021/acs.est.7b02523

Keller, A. A., McFerran, S., Lazareva, A., & Suh, S. (2013). Global life cycle releases of engineered nanomaterials. *Journal of Nanoparticle Research, 15*(6). https://doi.org/10.1007/s11051-013-1692-4

Khan, S. T. (2020). *Interaction of engineered nanomaterials with soil microbiome and plants: Their impact on plant and soil health* (pp. 181–199). Springer Science and Business Media LLC. https://doi.org/10.1007/978-3-030-33996-8_10

Kiss, L. V., Hrács, K., Nagy, P. I., & Seres, A. (2018). Effects of zinc oxide nanoparticles on Panagrellus redivivus (nematoda) and Folsomia candida (collembola) in various test media.

International Journal of Environmental Research, 12(2), 233–243. https://doi.org/10.1007/s41742-018-0086-y

Klaine, S. J., Alvarez, P. J. J., Batley, G. E., Fernandes, T. F., Handy, R. D., Lyon, D. Y., Mahendra, S., McLaughlin, M. J., & Lead, J. R. (2008). Nanomaterials in the environment: Behavior, fate, bioavailability, and effects. *Environmental Toxicology and Chemistry, 27*(9), 1825–1851. https://doi.org/10.1897/08-090.1

Klitzke, S., Metreveli, G., Peters, A., Schaumann, G. E., & Lang, F. (2015). The fate of silver nanoparticles in soil solution - sorption of solutes and aggregation. *Science of The Total Environment, 535*, 54–60. https://doi.org/10.1016/j.scitotenv.2014.10.108

Kubicki, J. D., Kwon, K. D., Paul, K. W., & Sparks, D. L. (2007). Surface complex structures modelled with quantum chemical calculations: Carbonate, phosphate, sulphate, arsenate and arsenite. *European Journal of Soil Science, 58*(4), 932–944. https://doi.org/10.1111/j.1365-2389.2007.00931.x

Kumar, A., & Raychoudhury, T. (2020). Long-term fate of ZnO-FexOy mix-nanoparticles through the saturated porous media under constant head condition. *Science of The Total Environment, 721*, 137669. https://doi.org/10.1016/j.scitotenv.2020.137669

Lavallee, J. M., Soong, J. L., & Cotrufo, M. F. (2020). Conceptualizing soil organic matter into particulate and mineral-associated forms to address global change in the 21st century. *Global Change Biology, 26*(1), 261–273. https://doi.org/10.1111/gcb.14859

Liu, Y., Shao, H., Liu, J., Cao, R., Shang, E., Liu, S., & Li, Y. (2021). Transport and transformation of microplastics and nanoplastics in the soil environment: A critical review. *Soil Use and Management, 37*(2), 224–242. https://doi.org/10.1111/sum.12709

Liu, G., Zhong, H., Ahmad, Z., Yang, X., & Huo, L. (2020). Transport of engineered nanoparticles in porous media and its enhancement for remediation of contaminated groundwater. *Critical Reviews in Environmental Science and Technology, 50*(22), 2301–2378. https://doi.org/10.1080/10643389.2019.1694823

Lowry, G. V., Gregory, K. B., Apte, S. C., & Lead, J. R. (2012). *Transformations of nanomaterials in the environment* (pp. 6893–6899). Environmental Science & Technology. https://doi.org/10.1021/es300839e

Luo, H. W., Zhang, X., Chen, J. J., Yu, H. Q., & Sheng, G. P. (2017). Probing the biotransformation of hematite nanoparticles and magnetite formation mediated by: Shewanella oneidensis MR-1 at the molecular scale. *Environmental Science: Nano, 4*(12), 2395–2404. https://doi.org/10.1039/c7en00767a

McKee, M. S., & Filser, J. (2016). Impacts of metal-based engineered nanomaterials on soil communities. *Environmental Science: Nano, 3*(3), 506–533. https://doi.org/10.1039/c6en00007j

Mergelov, N., Mueller, C. W., Prater, I., Shorkunov, I., Dolgikh, A., Zazovskaya, E., Shishkov, V., Krupskaya, V., Abrosimov, K., Cherkinsky, A., & Goryachkin, S. (2018). Alteration of rocks by endolithic organisms is one of the pathways for the beginning of soils on Earth. *Scientific Reports, 8*(1). https://doi.org/10.1038/s41598-018-21682-6

Mohd, S., Kushwaha, A. S., Shukla, J., Mandrah, K., Shankar, J., Arjaria, N., Saxena, P. N., Khare, P., Narayan, R., Dixit, S., Siddiqui, M. H., Tuteja, N., Das, M., Roy, S. K., & Kumar, M. (2019). Fungal mediated biotransformation reduces toxicity of arsenic to soil dwelling microorganism and plant. *Ecotoxicology and Environmental Safety, 176*, 108–118. https://doi.org/10.1016/j.ecoenv.2019.03.053

Nardi, S., Ertani, A., & Francioso, O. (2017). Soil–root cross-talking: The role of humic substances. *Journal of Plant Nutrition and Soil Science, 180*(1), 5–13. https://doi.org/10.1002/jpln.201600348

Navarro, E., Baun, A., Behra, R., Hartmann, N. B., Filser, J., Miao, A. J., Quigg, A., Santschi, P. H., & Sigg, L. (2008). Environmental behavior and ecotoxicity of engineered nanoparticles to algae, plants, and fungi. *Ecotoxicology, 17*(5), 372–386. https://doi.org/10.1007/s10646-008-0214-0

Ng, E. L., Huerta Lwanga, E., Eldridge, S. M., Johnston, P., Hu, H. W., Geissen, V., & Chen, D. (2018). An overview of microplastic and nanoplastic pollution in agroecosystems. *Science of The Total Environment, 627*, 1377–1388. https://doi.org/10.1016/j.scitotenv.2018.01.341

Nielsen, U. N., Wall, D. H., & Six, J. (2015). Soil biodiversity and the environment. *Annual Review of Environment and Resources, 40*, 63–90. https://doi.org/10.1146/annurev-environ-102014-021257

Orlov, D. S. (2020). *Humic substances of soils and general theory of humification.*

Owuor, S. O., Butterbach-Bahl, K., Guzha, A. C., Rufino, M. C., Pelster, D. E., Díaz-Pinés, E., & Breuer, L. (2016). Groundwater recharge rates and surface runoff response to land use and land cover changes in semi-arid environments. *Ecological Processes, 5*, 1–21. https://doi.org/10.1186/s13717-016-0060-6

Peng, C., Tong, H., Shen, C., Sun, L., Yuan, P., He, M., & Shi, J. (2020). Bioavailability and translocation of metal oxide nanoparticles in the soil-rice plant system. *Science of The Total Environment, 713*, 136662. https://doi.org/10.1016/j.scitotenv.2020.136662

Peng, C., Xu, C., Liu, Q., Sun, L., Luo, Y., & Shi, J. (2017). Fate and transformation of CuO nanoparticles in the soil-rice system during the life cycle of rice plants. *Environmental Science and Technology, 51*(9), 4907–4917. https://doi.org/10.1021/acs.est.6b05882

Pérez-Hernández, H., Fernández-Luqueño, F., Huerta-Lwanga, E., Mendoza-Vega, J., & Álvarez-Solís José, D. (2020). Effect of engineered nanoparticles on soil biota: Do they improve the soil quality and crop production or jeopardize them? *Land Degradation and Development, 31*(16), 2213–2230. https://doi.org/10.1002/ldr.3595

Qian, Y., Qin, C., Chen, M., & Lin, S. (2020). Nanotechnology in soil remediation – applications vs. implications. *Ecotoxicology and Environmental Safety, 201*. https://doi.org/10.1016/j.ecoenv.2020.110815

Rajput, V., Minkina, T., Mazarji, M., Shende, S., Sushkova, S., Mandzhieva, S., Burachevskaya, M., Chaplygin, V., Singh, A., & Jatav, H. (2020). Accumulation of nanoparticles in the soil-plant systems and their effects on human health. *Annals of Agricultural Sciences, 65*(2), 137–143. https://doi.org/10.1016/j.aoas.2020.08.001

Richard, G., Cousin, I., Sillon, J., Bruand, A., & Guérif, J. (2001). Effect of compaction on the porosity of a silty soil: Influence on unsaturated hydraulic properties. *European Journal of Soil Science, 52*(1), 49–58.

Saha, J. K., Selladurai, R., Coumar, M. V., Dotaniya, M. L., Kundu, S., & Patra, A. K. (2017). (pp. 11–36). Springer Nature. https://doi.org/10.1007/978-981-10-4274-4_2.

Sánchez-López, K. B., De Los Santos-Ramos, F. J., Gómez-Acata, E. S., Luna-Guido, M., Navarro-Noya, Y. E., Fernández-Luqueño, F., & Dendooven, L. (2019). TiO2 nanoparticles affect the bacterial community structure and *Eisenia fetida* (Savigny, 1826) in an arable soil. *PeerJ, 2019*(7). https://doi.org/10.7717/peerj.6939

Silva, L. C., & Lambers, H. (2018). Soil–plant–atmosphere interactions: Ecological and biogeographical considerations for climate-change research. *In Developments in Soil Science, 35*, 29–60 (Elsevier).

Singh, J., Vishwakarma, K., Ramawat, N., Rai, P., Singh, V. K., Mishra, R. K., Kumar, V., Tripathi, D. K., & Sharma, S. (2019). Nanomaterials and microbes' interactions: A contemporary overview. *3 Biotech, 9*(3). https://doi.org/10.1007/s13205-019-1576-0

Taha, M. R., & Taha, O. M. E. (2012). Influence of nano-material on the expansive and shrinkage soil behavior. *Journal of Nanoparticle Research, 14*(10), 1–13.

Teixeira, H. M., Bianchi, F. J. J. A., Cardoso, I. M., Tittonell, P., & Peña-Claros, M. (2021). Impact of agroecological management on plant diversity and soil-based ecosystem services in pasture and coffee systems in the Atlantic forest of Brazil. *Agriculture, Ecosystems and Environment, 305*. https://doi.org/10.1016/j.agee.2020.107171

Totsche, K. U., Amelung, W., Gerzabek, M. H., Guggenberger, G., Klumpp, E., Knief, C., Lehndorff, E., Mikutta, R., Peth, S., Prechtel, A., Ray, N., & Kögel-Knabner, I. (2018). Microaggregates in soils. *Journal of Plant Nutrition and Soil Science, 181*(1), 104–136. https://doi.org/10.1002/jpln.201600451

Valerio-Rodríguez, M. F., Trejo-Téllez, L. I., Aguilar-González, M.Á., Medina-Pérez, G., Zúñiga-Enríquez, J. C., Ortegón-Pérez, A., & Fernández-Luqueño, F. (2020). Effects of ZnO, TiO 2 or Fe 2 O 3 nanoparticles on the body mass, reproduction, and survival of *Eisenia fetida*. *Polish Journal of Environmental Studies, 29.*

Vázquez Núñez, E., & de la Rosa-Álvarez, G. (2018). Environmental behavior of engineered nanomaterials in terrestrial ecosystems: Uptake, transformation and trophic transfer. *Current Opinion in Environmental Science and Health, 6*, 42–46. https://doi.org/10.1016/j.coesh.2018.07.011

Vítková, M., Komárek, M., Tejnecký, V., & Šillerová, H. (2015). Interactions of nano-oxides with low-molecular-weight organic acids in a contaminated soil. *Journal of Hazardous Materials, 293*, 7–14. https://doi.org/10.1016/j.jhazmat.2015.03.033

Vogel, M., Fischer, S., Maffert, A., Hübner, R., Scheinost, A. C., Franzen, C., & Steudtner, R. (2018). Biotransformation and detoxification of selenite by microbial biogenesis of selenium-sulfur nanoparticles. *Journal of Hazardous Materials, 344*, 749–757. https://doi.org/10.1016/j.jhazmat.2017.10.034

Wang, P., Lombi, E., Menzies, N. W., Zhao, F. J., & Kopittke, P. M. (2018). Engineered silver nanoparticles in terrestrial environments: A meta-analysis shows that the overall environmental risk is small. *Environmental Science: Nano, 5*(11), 2531–2544. https://doi.org/10.1039/C8EN00486B

Weber, J., Chen, Y., Jamroz, E., & Miano, T. (2018). Preface: Humic substances in the environment. *Journal of Soils and Sediments, 18*(8), 2665–2667. https://doi.org/10.1007/s11368-018-2052-x

White, A. F. (2018). Chemical weathering rates of silicate minerals in soils. *Chemical Weathering Rates of Silicate Minerals*, 407–462.

Wong, S. W., & Leung, K. M. (2014). Temperature-dependent toxicities of nano zinc oxide to marine diatom, amphipod and fish in relation to its aggregation size and ion dissolution. *Nanotoxicology, 8*(Suppl. 1), 24–35.

Wu, H., Fang, H., Xu, C., Ye, J., Cai, Q., & Shi, J. (2020). Transport and retention of copper oxide nanoparticles under unfavorable deposition conditions caused by repulsive van der Waals force in saturated porous media. *Environmental Pollution, 256*, 113400. https://doi.org/10.1016/j.envpol.2019.113400

Wu, X., Lyu, X., Li, Z., Gao, B., Zeng, X., Wu, J., & Sun, Y. (2020). Transport of polystyrene nanoplastics in natural soils: Effect of soil properties, ionic strength and cation type. *Science of The Total Environment, 707*, 136065. https://doi.org/10.1016/j.scitotenv.2019.136065

Yang, F., Tang, C., & Antonietti, M. (2021). Natural and artificial humic substances to manage minerals, ions, water, and soil microorganisms. *Chemical Society Reviews, 50*(10), 6221–6239. https://doi.org/10.1039/d0cs01363c

Yang, X., Zhang, Q., Li, X., Jia, X., Wei, X., & Shao, M. (2015). Determination of soil texture by laser diffraction method. *Soil Science Society of America Journal, 79*(6), 1556–1566. https://doi.org/10.2136/sssaj2015.04.0164

Yu, G. (2018). *Root exudates and microbial communities drive mineral dissolution and the formation of nano-size minerals in soils: Implications for soil carbon storage* (pp. 143–166). Springer Science and Business Media LLC. https://doi.org/10.1007/978-3-319-75910-4_5

Zehlike, L., Peters, A., Ellerbrock, R. H., Degenkolb, L., & Klitzke, S. (2019). Aggregation of TiO2 and Ag nanoparticles in soil solution – effects of primary nanoparticle size and dissolved organic matter characteristics. *Science of The Total Environment, 688*, 288–298. https://doi.org/10.1016/j.scitotenv.2019.06.020

Zhang, P., Guo, Z., Zhang, Z., Fu, H., White, J. C., & Lynch, I. (2020). Nanomaterial transformation in the soil–plant system: Implications for food safety and application in agriculture. *Small, 16.*

Zhang, P., Ma, Y., Zhang, Z., He, X., Zhang, J., Guo, Z., Tai, R., Zhao, Y., & Chai, Z. (2012). Biotransformation of ceria nanoparticles in cucumber plants. *ACS Nano, 6*(11), 9943–9950. https://doi.org/10.1021/nn303543n

Zhang, H., Zhou, H., Bai, J., Li, Y., Yang, J., Ma, Q., & Qu, Y. (2019). Biosynthesis of selenium nanoparticles mediated by fungus Mariannaea sp. HJ and their characterization. *Colloids and Surfaces A: Physicochemical and Engineering Aspects, 571*, 9–16. https://doi.org/10.1016/j.colsurfa.2019.02.070

Zhao, L., Zhang, H., White, J. C., Chen, X., Li, H., Qu, X., & Ji, R. (2019). Metabolomics reveals that engineered nanomaterial exposure in soil alters both soil rhizosphere metabolite profiles and maize metabolic pathways. *Environmental Science: Nano, 6*(6), 1716–1727. https://doi.org/10.1039/c9en00137a

Further reading

Shah, T., Lateef, S., & Noor, M. A. (2020). Carbon and nitrogen cycling in agroecosystems: An overview. *Carbon and Nitrogen Cycling in Soil*, 1–15.

Shah, S. N. A., Shah, Z., Hussain, M., & Khan, M. (2017). Hazardous effects of titanium dioxide nanoparticles in ecosystem. *Bioinorganic Chemistry and Applications, 2017*. https://doi.org/10.1155/2017/4101735

CHAPTER

5

Biophysicochemical transformations of ENMs in water

Sheng Yin[1,2], Yulu Ge[1,2], Mariana Marcos-Hernandez[1,2], and Dino Villagran[1,2]

[1] The University of Texas at El Paso, Department of Chemistry and Biochemistry, El Paso, TX, United States; [2] Nanosystems Engineering Research Center for Nanotechnology-Enabled Water Treatment (NEWT), United States

5.1 Introduction

The rational design and composition of one or more nanomaterials in order to yield selected properties define the term: engineered nanomaterials (ENMs). In lockstep with the increasing range of applications for ENMs is the increasing pursuit to assess their toxicological effects in the environment (Selck et al., 2016). Nano environmental and nanotoxicological research is driven by questions regarding the environmental risk posed by ENMs (Selck et al., 2016). In order to assess this risk, multiple factors must be taken into account. Some of these include: the increased production and output discharge of ENMs (Sun et al., 2014); their suspected behavior in the environment and in biological systems (Luoma et al., 2014); their detrimental biological effects (Fabrega et al., 2011); the complexity of risk assessment (Laux et al., 2018); and their fate and bioavailability (Lead et al., 2018) (Gottschalk et al., 2013), among others. Once ENMs enter aqueous systems, their fate and bioavailability depend upon numerous transport and reaction mechanisms. Bioavailability of ENMs can be defined as the available fraction of the total ambient material that can be taken up by an organism through all possible

https://doi.org/10.1016/B978-0-323-90558-9.00011-5

sources including food, water, and environmental media. Bioaccumulation is directly related to the amount of ENMs available for uptake in the source (e.g., water or diet). Therefore, the accumulation of metals or ENMs onto or inside the tissues of an organism resulting from this uptake is what defines bioaccumulation. Metal-based ENMs can then undergo degradation or dissolution inside organisms or in the environment leading to metal leaching from the ENMs. Bioavailability and bioaccumulation must take into account both the metals coming from the ENMs and the engineered nanomaterial itself. Several interrelated factors affect both the bioavailability and bioaccumulation of the ENMs in the environment and in organisms (Luoma et al., 2014). These are:

Concentration: The amount of available product or byproduct drives the uptake and accumulation of the ENMs;
Environmental matrix: The water, soil, or food matrices drive secondary reactions and transformations of ENMs;
Properties and composition of ENMs: Metal composition, aggregation, dissolution, reactivity, surface coating, shape, size, charge, among other properties;
Organisms: Biological traits such as physiology, ion regulation, uptake processes and the type of environment where the organism lives and how it obtains food and oxygen, as well as its longevity and development.

Different combinations of these factors must be considered in order to estimate the toxicity and fate of a particular type of ENMs.

The study of the toxicity, transport, fate, and accumulation of ENMs in different environments (i.e., soils, aquatic, atmospheric, urban areas, among others) is an ongoing research field (Klaine et al., 2008). Nanoscale properties in ENMs such as high reactivity, dispersibility, and high surface area might influence their behavior in the environment (Lead et al., 2018). A relevant question might be: do ENMs pose a greater health risk to ecosystems due to these properties? To evaluate this risk, it is necessary to associate the concentration, the possible ecotoxicological effects, as well as the possible interactions of the ENMs with the environment where released. Some of the ENMs found in urban atmospheres come from gasoline vehicles, industrial and domestic activities, as well as air transport. Other ENMs in the air are formed from the breaking down of bigger particles or the assembly of smaller ones (Klaine et al., 2008). In soil and bedrock, natural metal oxide/hydroxides nanoparticles, clays, organic matter, as well as other minerals, influence soil structural behavior, and development. Existing soil particles and colloids might interact and transform with ENMs that then can be mobilized into plants and/or groundwater (Shrivastava et al., 2019). Finally, aquatic environments are the ultimate sink where ENMs accumulate and settle (Selck et al., 2016).

Interactions, transformations, and the physical and chemical processes of ENMs in water matrices are the focus of this chapter. Specifically, the following aspects are covered: (a) how ENMs enter aqueous systems; (b) the basic chemical, physical, and biological transformations of ENMs in aqueous systems, and factors that influence these transformations; and (c) the bioavailability of ENMs in aqueous systems.

5.2 How are ENMs released in aqueous systems?

ENMs and their related byproducts enter aquatic environments during manufacture. The leaching occurs during different types of applications, specifically when used for water treatment/management processes (Tolaymat et al., 2017), as shown in Fig. 5.1.

Point sources of ENMs result from direct release from industrial processes into aqueous systems, while the diffuse sources are ENMs released from industrial products. For instance, paints containing ENMs are collected in the sewer system, preliminarily captured or treated through wastewater treatment plants, and eventually released into aquatic environments (Gottschalk & Nowack, 2011). The ENMs released into water bodies during manufacturing are estimated to be between 0.1 and 2.0 of the total percent (Gottschalk & Nowack, 2011; Keller et al., 2013). Similarly, consumer products (i.e., sensors, packaging, and plastics) can liberate ENMs into aquatic environments and this accounts for 5% of their total release, while the water runoff from the textile industry accounts for approximately 95% of the total (Keller et al., 2013). In addition, since 63%

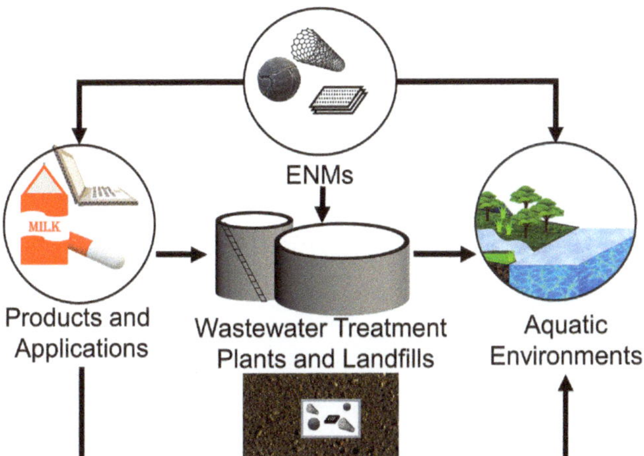

FIGURE 5.1 Release of ENMs into aquatic environments.

−91% of ENMs are placed in landfills, it is necessary to estimate how much ENMs can be transported in the leachate and enter aquatic environments in the future (Keller et al., 2013).

ENMs can be released to aquatic environments directly through sewage, effluents, and river influx (Gottschalk et al., 2009; Tolaymat et al., 2017). Although nanoscale separation and removal techniques have been developed, they are still not widely used in conventional wastewater treatment processes. That is to say, ENMs-mediated environmental remediation can increase the possibility of direct release (Liu et al., 2014; Shevlin et al., 2018). For example, nano TiO_2 released into aqueous environments through sewage spills was found to be over 100 μg/L in some water samples, which estimated a total discharge of 1.1−3.8 tons of TiO_2 into surface waters in the USA through sewage spills per year (Loosli et al., 2019). On another hand, the extensive use of cosmetics, coatings, and detergents results in significant ZnO and TiO_2 release into water (Gottschalk et al., 2009). Thus, to a certain extent, the production volumes of specific ENMs can provide relevant data to predict their concentration in the aqueous systems (Bundschuh et al., 2018).

Moreover, the water cycle can drive the spreading and transport of ENMs from the soil and atmosphere into aqueous systems (Hochella et al., 2019), see Fig. 5.2. Both, ENMs and natural nanomaterials can be mobilized through rivers, groundwater, and glaciers. In addition, precursors of ENMs such as sulfuric acid, gaseous ammonia, and oxidized organic compounds can react with small particles contained in the anthropogenic exhaust emissions in the atmosphere to form new nanomaterials (Kulmala et al., 2014; Merikanto et al., 2009). These

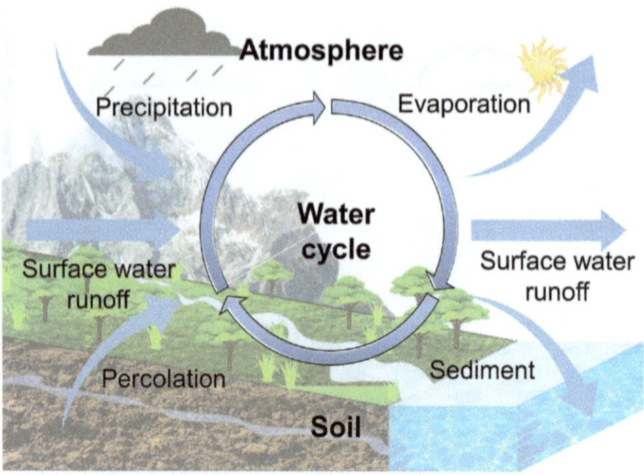

FIGURE 5.2 The water cycle.

nanomaterials grow into larger particles, and finally enter surface water, such as lakes and rivers by dry or wet deposition/precipitation (John et al., 2017). The runoff of surface and ground water and aggregated precipitation release ENMs into the soil, while evaporation processes bring ENMs to the atmosphere.

5.3 Fate and transformation of ENMs in aqueous system

After entering the aqueous environments, ENMs face different fates depending on various factors present in the corresponding environment. These factors include the physical and chemical properties of ENMs and of the water matrix. In this section, we will describe the transformation and fate of ENMs through physical and chemical transformations.

5.3.1 Physical fate and transformation

In most laboratory experiments, nanoparticles are well dispersed through continuous energy supply (i.e., shaking, stirring, and/or sonication), or with the help of solvents (Laux et al., 2018). Different to laboratory conditions, ENMs enter natural waters in the form of aggregates, suspensions, or sedimentation due to their neutral or weak electrostatic charge and the lack of continuous energy supply (Lin, Tian, et al., 2010), see Fig. 5.3.

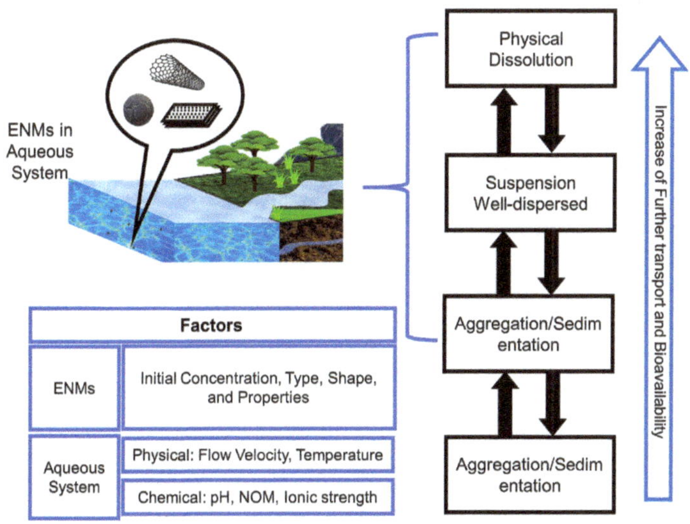

FIGURE 5.3 Physical fate and transformation of ENMs in aqueous systems.

Surface charged ENMs do not aggregate at a considerable rate and can be dispersed through electrostatic repulsion forces. This enhances their transportation and bioavailability since well-dispersed ENMs are easier to remove when compared to aggregates and/or dispersed ENMs (Lin, Tian, et al., 2010). For example, nano copper and copper (II) hydroxide ($Cu(OH)_2$) (100−1000 nm) particles will aggregate quickly in natural water matrices and form stable phases in three stages (Conway et al., 2015). In the beginning, the well-dispersed nano Cu and nano $Cu(OH)_2$ particles will aggregate to form 5 to 10 μm particles in seconds. After 10 min, larger aggregates gradually sediment while new aggregates start to grow. At the end, a stable phase with a diameter in the range of 700−2000 nm is formed (Conway et al., 2015). During the aggregation process, the interface free energy will decrease due to the reduction of overall specific surface area of ENMs, and this leads to decrease in reactivity (Lin, Tian, et al., 2010). Compared with nano Cu and $Cu(OH)_2$ particles, nano copper oxide (CuO) (20−100 nm) exhibits a different performance. Nano-CuO aggregates are smaller in water due to better dispersion. Sedimentation of these particles hardly occurs in freshwater and storm runoff (Conway et al., 2015).

The physical aggregation, suspension, or sedimentation process of the ENMs is affected by their initial concentration, particle shape, charge, and other properties. However, when they form stable suspensions, ENMs with sizes between 1 and 1000 nm are usually classified as colloidal particles. At this point, factors and parameters of water system including temperature, flow velocity (Lv et al., 2016), chemical composition, natural organic matter (NOM), inorganic species (Laux et al., 2018; Surette & Nason, 2016; Warren, 2000), and pH (Lin, Tian, et al., 2010) have to be taken into consideration.

5.3.1.1 *Physical properties of ENMs*

The initial ENMs concentration has a direct effect on their fate. Using nano TiO_2 as an example, the diameters of aggregates are around 1034 nm in lake water after 7 h for a diluted TiO_2 concentration (10 mg/L), while for an initial concentration of 100 mg/L larger diameters of aggregates around 1374 nm are shown (Lv et al., 2016). Therefore, the higher the ENMs initial concentration the higher the probability to form colloids with enhanced aggregation rates (Lv et al., 2016; Petosa et al., 2010).

The aggregation process can also be affected when different types of ENMs exist in the same system/environment. The coexistence of nano ZnO and nano TiO_2 in water, for instance, will enhance each of their individual stability. Both nanomaterials will repel each other in water due to electrostatic repulsions. This lowers the chance of forming colloids and aggregates compared to their individual tendencies, and thus, their sedimentation rates decrease by 30.7%–49.1% (ZnO) and 42.4% (TiO_2) compared to their respective individual aggregation (Fang et al., 2017).

The shape and size of ENMs influence their aggregation processes in water (Hotze et al., 2010). Nano-hematite (Fe_2O_3), which is used in optoelectrical materials, magnetic applications, and in water remediation, is widely distributed in aqueous environments (Ashraf et al., 2020; Keller et al., 2013). It has been reported that as the size of nano-Fe_2O_3 increases, the aggregation rates become slower. This is because different sizes of nano-Fe_2O_3 show different point of zero charge (PZC) values. For example, 12 nm Fe_2O_3 nanoparticles have a PZC at pH = 7.8, while nanoparticles of 32 and 65 nm diameter have PZC at pH = 8.2, and 8.8, respectively Also, sizes of 32 and 65 nm present a PZC at pH = 8.2, and 8.8, respectively. In neutral conditions, smaller particles are less charged, resulting in lower electrostatic double-layer repulsion forces, and these have a higher tendency to aggregate (He et al., 2008). Nano TiO_2 shows similar behavior. After 30 days, 95% of the initial concentration of 10×40 nm rods of rutile and 50 nm anatase at pH 7 remained dispersed in water, while only 5% of 10 nm anatase was present after only 4 days (Liu et al., 2011).

Surface functionalization and coatings (i.e., polymers, polyelectrolytes, and surfactants) of the ENMs can also influence the aggregation process. In general, there are two ways to affect the aggregation: (a) the surface functionalization and coating will change the coulombic attraction and hydrophobic interactions between particles and water, leading to the inhibition/enhancement of the aggregation process; and (b) surfactants can change the surface charge, electrostatic repulsion, and interface energy between particles and solvents through adsorption or formation of covalent bonds thereby affecting the ENMs aggregation (Hotze et al., 2010). Carbon nanotubes (CNTs) hardly form stable suspensions in natural waters due to their hydrophobicity and relatively large particle size. In addition, the lack of continuous energy supply (such as sonication or stirring, common in laboratory settings) results in poor dispersion and aggregation in field waters (Lin, Liu, et al., 2010). In contrast, functionalization of the CNT surface with oxygen-containing functional groups, such as carboxylic acid (-COOH) and hydroxy groups (-OH), increases their hydrophilicity and reduces their aggregation (Laux et al., 2018; Smith et al., 2009). For example, multi-walled carbon nanotubes (MWCNTs) (7.5 mg/L) with a surface oxygen content of 9.5% hardly aggregate within 120 min in pH 6 in a 64 nM NaCl solution. Under the same conditions, when the surface oxygen content decreases to 7.6%, small aggregates are observed and when the oxygen content is decreased even further to 4.3%, the larger aggregates appear (Smith et al., 2009). Nano metal oxides show similar behavior. Hydrophobic nanocrystals of TiO_2 and Fe_3O_4 modified with short-chain carboxylic acids show a considerable increase in dispersion in aqueous systems compared to nanoparticles not functionalized (Vo et al., 2009). In addition, the compression of the electric double layer or the adsorption of counter ions can change the stability of nanoparticles. This can be observed with coated gold nanoparticles, and this results from the changes in electrostatic repulsion (Surette & Nason, 2016).

5.3.1.2 Water parameters and properties

The properties of the media, and in this discussion, water, directly affect the properties, stability, and fate of nanoparticles. Water temperature is an important factor which is positively correlated with the diameters of ENMs aggregates. For example, in a 50 mg/L TiO_2 dispersion, the sedimentation rate of ultrapure water samples at 10°C is 16%. As the temperature rises to 35°C, the proportion of aggregates increases to 19%, and it keeps increasing until the temperature is raised to 60°C (Lv et al., 2016). TiO_2 ENMs are more likely to be efficiently transported and bioavailable in low temperature aqueous environments. This phenomenon can be explained by Brownian motion, which is directly proportional to temperature. Lower temperatures result in lower thermal energy, which decreases the collision rate of the ENMs and slows down Brownian motion (Lv et al., 2016). Another physical parameter that needs to be taken into account is the flow velocity of the water body. When flow velocity is increased through shaking, stirring, or sonication, the aggregation process is usually disturbed. In other words, the water flow contributes to the homogeneous dispersion of ENMs. In a study with nano TiO_2, when the flow speed of 0.32 m/s (150 rpm) was introduced into ultrapure water, the diameters of aggregates decreased by 50%. In ultrapure water with a continuous flow velocity of 0.32 m/s for 7 h, the sedimentation ratio was only around 3.2%, which was much lower than 26% of the same period in static water (i.e., without water flow, 0 m/s). In addition, with the increase in flow velocity, the aggregates of ENMs decreased in size and were able to redisperse into aqueous systems (Lv et al., 2016). Thus, ENMs in aqueous systems with higher flow velocity and at low temperature will have higher stability and better dispersion, leading to transformations that can occur over long-distances and will affect their regional bioavailability.

Coexisting compounds including natural organic matter (NOM) and inorganic species in aqueous systems also impact the aggregation of ENMs (Shevlin et al., 2018). NOM is a complex mixture of organic compounds widely distributed in natural terrestrial and aquatic environments, produced by the decomposition of animals, plants, and microorganisms. It is mainly composed of a humic acid part, such as humic acid and fulvic acid, and a nonhumic part, such as fresh proteins and amino acids (Darko et al., 2014; Hotze et al., 2010). NOM can be adsorbed by ENMs to form a coating on the surface to facilitate dispersion in aqueous systems via electrostatic, steric, and hydrophobic interactions (Zhang & Nosaka, 2014). NOM adsorbed or coated on neutral or weakly negatively charged ENMs can therefore stabilize suspensions. At the same time, NOM can neutralize positively charged ENMs and enhance the ENMs aggregation and deposition when their surface charge changes from positive to negative. For example, the suspension of citrate-stabilized gold nanoparticles (Ci-Au-NP, negatively charged) shows classical Derjaguin–Landau–Verwey–Overbeek (DLVO-type) behavior with regards to the aggregation of the dispersion. This behavior is driven by van der Waals forces and electrical double-layer forces (Nason et al., 2012). The addition of four different types of NOM affects the Ci-Au-NP aggregation processes and leads to different levels of stabilization. This is due to the various adsorption processes between the surface of Ci-Au-NP and NOM. Different concentrations and types of NOM contribute to the ENMs aggregation process: large molecular weight NOM (Suwannee River Humic Acid, SRHA) or high concentrations (>5 mg/L) with low molecular weight NOM (Suwannee River Fulvic Acid, SRFA and Pony Lake Fulvic Acid, PLFA) are able to stabilize dispersions of Ci-Au-NP (Nason et al., 2012). TiO_2 shows a similar behavior when NOM is added to the solution. Upon addition of SRHA and SRFA NOM to nano TiO_2 dispersions, their stability increases under certain suitable pH values. This is due to favorable electrostatic, steric, and hydrophobic interactions. However, in the absence of NOM, nano TiO_2 nanoparticles aggregate spontaneously under controlled laboratory conditions (Danielsson et al., 2018; Loosli et al., 2013). In typical aqueous environments, negatively charged NOM will be adsorbed on positively charged or neutral nano TiO_2. When the pH value is lower than the PZC of nano TiO_2, the TiO_2 aggregates are positively charged and tend to break into a greater number of free TiO_2 particles that are able to adsorb more NOM. This leads to the formation of stable aggregates that can be better dispersed. When the pH is equal to or higher than the PZC, the charge will be neutralized or reversed, resulting in fast aggregation and deposition (Loosli et al., 2013). CNTs show similar behavior. They can be charged when they adsorb NOM (such as Suwannee River natural organic matter (SRNOMSRHA), SRHA, SRFA) under suitable pH values, therefore forming stable suspensions in water via steric and electrostatic forces (Hyung & Kim, 2008). In summary, NOM can change the physical and chemical properties of ENMs by surface adsorption, affecting their aggregation and suspension properties.

Inorganic matter in aqueous systems is another factor that affects the aggregation processes of ENMs. The aggregation processes are enhanced with the increase in the ionic strength on the aqueous solution. Metal oxides-based ENMs, such as nano ZnO (Bian et al., 2011; Li et al., 2017), nano CuO (Conway et al., 2015), nano Au (Nason et al., 2012), and nano

TiO_2 (French et al., 2009), tend to aggregate with higher ionic strength values, while stable suspensions are formed and stabilized at lower ionic strength. For example, after 24 h, suspended nano ZnO will gradually form aggregates in the size range of 2500–3000 nm; when adding 50 mM NaCl, the aggregates size increases to 4500–5500 nm within the same time scale (Li et al., 2017). This phenomenon is mainly due to the enhanced ionic strength of the ENMs in their electrical double layer, leading to a faster aggregation (Li et al., 2017; Nason et al., 2012). Moreover, the type of ions also affects their aggregation. Divalent cations, such as Ca^{2+} and Mg^{2+}, increase aggregation when compared to monovalent ions, such as Na^+ and K^+. This is due to the strong compressive effect on electrical double layer and the bridging effect by divalent ions (Li et al., 2017; Miao et al., 2016). Anions, such as SO_4^{2-} and NO_3^-, promote the aggregation of ENMs (such as nano-TiO_2), while phosphate, PO_4^{3-}, slows down these processes (He et al., 2017).

5.3.2 Chemical fate and transformation

The chemical and physical transformations of ENMs in water bodies include dissolution, redox reactions, complexation, precipitation, and photochemistry (Wagner et al., 2014). These processes involve the breaking and formation of chemical bonds, and energy transfer. Therefore, the chemical properties of the aqueous system, such as pH and coexisting substances, have a direct impact on the chemical fate and transformation of the ENMs.

5.3.2.1 Dissolution and redox reactions

The dissolution process gradually decreases the size of ENMs, and can eventually form new species and precipitates (Hortin et al., 2020; Shevlin et al., 2018; Wagner et al., 2014). The extent of the dissolution process depends on the properties of ENMs (i.e., composition, structure, size, etc.), and the solution parameters (i.e., pH, coexisting substances/ions, ENMs concentration, temperature, etc.).

Nanometal oxides, such as CuO and ZnO, can be dissolved in water to form their corresponding cations (Cu^{2+} and Zn^{2+}). As shown in Eqs. (5.1) and (5.2), ENMs are more likely to dissolve at lower pH values. For example, nano CuO is stable (at a concentration up to 10 mM) at pH 11; while at pH = 4, nano CuO will rapidly dissolve and aggregate (Adeleye et al., 2014). Similarly, when nano ZnO dissolves in water, the concentration of Zn^{2+} decreases from 5.00 × 10 M at pH 7 to 7.50 × 10 M when pH increases to 9 (Miao et al., 2010).

The dissolution of silver nanoparticles starts with redox reactions. The surface of nano Ag is first oxidized to silver oxide (Ag_2O) (Eq. 5.1), leaving Ag core in the inside. This composite can then be dissolved as in Eq. (5.2).

Further oxidation of the core and dissolution of the oxide result in a large amount of surface Ag^+ released into water. In neutral water, the surface of Ag_2O is protonated to yield hydroxyl terminated Ag-OH (Eq. 5.3), which can be further protonated in acidic conditions (Eq. 5.2) (Peretyazhko et al., 2014).

$$4\,Ag + O_2 \rightarrow 2\,Ag_2O \tag{5.1}$$

$$M_2O_x + 2x\,H_2O \rightarrow 2M^{x+} + 2x\,H_2O\ (Acidic\ conditions) \tag{5.2}$$

$$M_2O_x + 2x\,H_2O \rightarrow 2M^{x+} + 2x\,OH^-\ (Neutral\ conditions) \tag{5.3}$$

The size, shape, and structure of ENMs influence the dissolution process. For example, under neutral conditions within a 2 h time frame, the solubility of 50 nm ZnO ENMs was 8.59 ± 0.45 mg/L, which was higher than that for 100 nm ZnO ENMs (with 7.85 ± 0.27 mg/L) (Avramescu et al., 2017). In another side, the dissolution of 30 nm Fe_2O_3 rhombohedral crystals was enhanced due to exposed (and likely reactive) crystal facets that presented sharp edges and surface defects. Yet, in general, amorphous structures are easier to dissolve compared to crystalline materials (Echigo et al., 2012).

The temperature of the water body is another factor in all dissolution processes. Depending on whether the dissolution process is exothermic or endothermic, temperature changes will reduce or increase the solubility of ENMs (Avramescu et al., 2017).

Moreover, when the concentration of ENMs decreases, the dissolution processes are enhanced (Baalousha et al., 2016). For example, at high concentration of Ag ENMs (1500 μg/L), only $10 \pm 2.5\%$ Ag was dissolved in seawater during a period of 4 days, while at low concentration (25 μg/L), $83 \pm 4\%$ of Ag dissolved in only 2 days (Sikder et al., 2018).

$$Ag^+ + 2Cl^- \rightarrow AgCl_2^- \tag{5.4}$$

$$AgCl_2^- + Cl^- \rightarrow AgCl_3^{2-} \tag{5.5}$$

The solubility constant (K_{sp}) is the equilibrium constant for the dissolution of a solid dissolved into an aqueous solution at a specific temperature. For the reaction

$$A(s) \rightleftharpoons bB(aq) + cC(aq)$$

the K_{sp} for compound A is shown in Eq. (5.6):

$$K_{sp} = [B]^b\ [C]^c \tag{5.6}$$

For AgCl, $K_{sp} = 1.77 \times 10^{-10}$ at 25°C, while the equilibrium constant (K) of Eqs. (5.4) and (5.5) are 10^5 and 10^{10} times larger than AgCl K_{sp} (Skompska et al., 2010). When Ag^+ is released into water, it is encapsulated with different anions including Cl^-, forming ionic complexes such as $AgCl_2$ and $AgCl_3^{2-}$, as shown in Eqs. (5.4) and (5.5).

5.3.2.2 Precipitation

Opposite to the dissolution processes, the dissolved metal salts can precipitate out of solution and form new compounds with a lower K_{sp} value. Sulfidation is a common precipitation process. For example, metal oxides (i.e., ZnO, Fe_3O_4, MnO_2) are sensitive to sulfides. Sulfidation reactions can reduce the toxicity of metal oxides by preventing them from lowering their mobility in aqueous environments (Fan et al., 2017; Levard et al., 2011; Levard et al., 2013; Ma et al., 2013). When nano ZnO is dissolved in water, Zn^{2+} will react with sulfide and rapidly reprecipitate to yield 2.5–5 nm ZnS due to its lower K_{sp} value (1.6×10^{-24}) compared to ZnO (1.2×10^{-17}) (Wu et al., 2015). When the sulfide concentration increases, the sulfidation conversion rate can reach 100% (Levard et al., 2011; Ma et al., 2013). However, when trace amounts of sulfide are present in the solution, a metal sulfide shell can form around the metal oxide core. In addition, some common ions such as halides and carbonates also can cause the dissolution and reprecipitation of certain ENMs. For example, phosphate can reprecipitate dissolved nano ZnO into hopeite ($Zn_3(PO_4)_2 \cdot 4H_2O$, $K_{sp} = 9.0 \times 10^{-33}$) under weakly acidic conditions (pH = 6); while under pH = 8, a $Zn_3(PO_4)_2$ shell is formed on the surface of nano ZnO (Rathnayake et al., 2014). Similarly, nano-magnesium oxide (MgO) and calcium oxide (CaO) will dissolve to form hydroxides in water. However, when they are in contact with phosphates, the phosphate salts will precipitate as metal phosphates, for example, $Mg_3(PO_4)_2$ ($K_{sp} = 1 \times 10^{-25}$); and calcium phosphate, $Ca_3(PO_4)_2$ ($K_{sp} = 2 \times 10^{-29}$) (Li et al., 2018). The formed particles will have different physical properties as well as different types of interactions between themselves and the water body. These interactions might include aggregation, suspension, or sedimentation and will be either bioconsumed or stay in the environment for a long term.

5.3.2.3 Photochemical transformations

Certain ENMs can be excited by exposure to specific wavelengths of light resulting in optical degradation, oxidation, and reduction. The active intermediates and byproducts of ENMs within the photochemical processes affect their chemical and physical properties, which impact their transport and fate (Montaño et al., 2014). For example, when nano TiO_2 particles are irradiated with wavelengths of around 390 nm, electron-hole (e^-–h^+) pairs and free radicals (such as $\cdot OH$, $HO_2^{-\cdot}$) can be generated on its surface (Lee & Park, 2013; J. Zhang & Nosaka, 2014). During this process, nano TiO_2 is considered to be a photocatalyst, which can react with contaminants coated or adsorbed on the particle surfaces. For example, humic acid in water can be first adsorbed on the surface of nano TiO_2. Once exposed to sunlight, humic acid can be photodegraded to smaller organic compounds with the help of nano TiO_2 (Babel et al., 2017; Zhang et al., 2008). Arsenic (As) and arsenite (As(III)) can be photocatalytically oxidized to arsenate (As(V)) by nano TiO_2 (Dutta et al., 2005). Other metal oxides that can also act as photocatalysts in water are zinc oxide (ZnO), tin oxide (SnO_2), and cerium oxide (CeO_2) (Khan et al., 2015). Sunlight displays different effects on metal nanoparticles in water. For example, sunlight irradiation accelerates the transformation of Ag nanoparticles to Ag^+ in aqueous systems. Recent studies have shown that solar irradiation can decrease the size of Ag nanoparticles and change their morphology with no aggregation. For example, sunlight can lead to the formation of an oxide shell and enhance the oxidation of nano Ag nanospheres to Ag^+.

ENM morphology also plays a role in photooxidation processes. For example, Ag nanocubes showed slower photooxidation rates when compared to spherical Ag particles because particles underwent a morphological change before the photochemical process. It has been reported that the edges and corners of the cubes initially became blunt (Grillet et al., 2013). In addition, Ag^+ can be reduced back to Ag metal with the help of dissolved organic matter under sunlight (Yin et al., 2012; Yongguang Yin et al., 2015). Compared to Ag, Au nanoparticles are more resistant to oxidation or dissolution. Thus, Au nanoparticles in water are more likely to behave as photocatalysts instead of undergoing redox reactions (Yin et al., 2012).

Carbonaceous ENMs in water are also affected by photochemical processes. For example, when high energy photons (wavelengths ranges: 254–400 nm) excite O_2 molecules to the highly reactive singlet state, singlet oxygen oxidizes the CNTs surface and different oxygen functional groups can be introduced (Deline et al., 2020). When single-walled carbon nanotubes (SWCNTs) are exposed to UV light (254 nm), photo hydroxylation occurs on the SWCNTs surface to form CNT–OH (Alvarez et al., 2008). In addition, photooxidation can also introduce other oxygen-containing functional groups on the surface of CNTs including ethers (C–O–C), epoxides, and carboxyl groups (–COOH) (Parekh et al., 2006). Research shows that, when exposed to sunlight, nitrogen-containing carbon dots (NCDs) can be photobleached. Over long-term exposure to •OH radicals, NCDs can be depolymerized at very slow rates. Subsequently, they can be oxidized to carboxylic acids to finally result in volatile compounds (CO or CO_2), while the nitrogen atoms are oxidized into nitrate ions (Frank et al., 2020). Other carbon allotropes, such as fullerenes, also have photochemical activity in aquatic environments. When aqueous buckminsterfullerene (C_{60}) clusters are exposed to 300–400 nm sunlight, the size of clusters will decrease with a color change from brown to yellow. This may be due to the formation of soluble products that result from sunlight irradiation. These products are eventually mineralized or volatilized to CO_2 and volatile organic compounds and will be removed from aqueous systems (Hou & Jafvert, 2009).

5.4 ENMs bioavailability and biological fate

The interactions of ENMs with biotic and abiotic components in aquatic environments and their potential to cause harm are still poorly understood. These current uncertainties are driving concerns on the risk they may pose to human and environmental health (Luoma & Rainbow, 2005). In general, the processes that govern bioavailability and bioaccumulation of ENMs in any given circumstance are likely to be the result of the combined influence of the following: concentration; physical and chemical properties of the ENMs, the environment exposure route and the biology and functional ecology of the involved organism. Below we discuss the impact of nanosilver, ceria, and carbon-based ENMs toward (1) aquatic environments; (2) biosystem accumulation; and (3) their corresponding biofate.

5.4.1 Silver nanoparticles (AgNPs)

Silver nanoparticles are the most widely used nanomaterials in technology, medicine, and consumer products. This is mainly due to their antimicrobial, optical, electrical, and magnetic properties. However, the high surface-to-area ratio enhances the toxicity of nanosized Ag when compared to the bulk counterpart (Council, 2012). AgNPs can enter the aquatic environment accidentally or incidentally through washing of treated clothes and the release from products, such as cosmetics, toothpaste, soaps, and food containers. The measured concentrations in natural waters for AgNPs are in the ng/L range (Benn & Westerhoff, 2008). Because of easy access into aquatic environments, AgNPs can exert toxic effects on aquatic organisms, and this is mostly caused by the dissolution of the Ag ion, as previously mentioned (Asharani et al., 2008; Griffitt et al., 2009). When AgNPs enter aquatic media, the nanoparticles can either penetrate into an organism or be retained at the external surfaces where they can cause damage. The toxicity of AgNPs can be attributed to either direct surface effects or to indirect effects of dissolved silver species (i.e., Ag^+, AgCl (aq) and AgOH (aq)). Nanosilver can be taken up by different cells and become internalized. Due to the absence of Ag-specific cell walls in higher organisms, AgNPs and their aggregates that are smaller than the largest pore can pass through and reach the plasma membrane causing toxicity in some fish species and other aquatic animals (Lee et al., 2007). On the other hand, released Ag^+ can bind onto cell membranes through particle-specific interaction and affect permeability and respiratory functioning of the cells by decreasing the Na^+/K^+ ATPase activity (Morones et al., 2005). Most plants, bacteria, and fungi are surrounded by semipermeable cell walls that serve as barriers that prevent the entry of bigger AgNPs. However, the toxicity of AgNPs is mainly due to their fast dissolution that leads to Ag^+ release. Additionally, AgNPs are also able to disrupt cell membranes and increase permeability, which eventually causes cell death (Dorobantu et al., 2015).

High levels of Ag accumulation from AgNPs have been reported in the liver, gills, kidneys, intestines, and muscles of fish. For example, it was found that the smaller AgNPs are more likely to aggregate inside the gills of rainbow trout (Scown et al., 2010). A 10-day exposure to commercial silver particles of three nominal sizes (10 nm, 35 nm, and 600–1600 nm) and bulk silver nitrate showed that 10 nm nanoparticles were found in the highest concentration within gill tissues (0.61 ± 0.07 μg/g) while bulk silver was most highly concentrated in liver (1.63 ± 0.18 μg/g). AgNPs enter in the organism as monomeric NPs without apparent build-up accumulation at the surface and in the gills with a weak bio-concentration. In addition, like most nanoparticles, AgNPs are likely to aggregate depending on concentration and type of organic matter and solids in suspension. Research shows that AgNPs appeared in municipal effluents as small noncharged aggregates (11.7 ± 0.14 nm) (Bruneau et al., 2016). A significant increase in Ag concentration was measured in gills and livers of fish exposed to AgNPs in diluted (10%) municipal waste-water. Among these, the liver accumulated more Ag (1.0–1.6 mg/kg) than gills (0.3–0.6 mg/kg) which suggest a systemic distribution of AgNPs in fish. Total Ag concentration in liver for AgNP exposure was 0.3 mg/kg and in exposure media the total Ag concentration was 26 μg/L after 96 h.

Another possible way for AgNPs to bioaccumulate is through the aggregation on the surface of the membrane and the release of free metal ions within the surface layers. In this case, a high Ag^+ concentration nano environment can be created by the dissolution of adsorbed Ag within the superficial microlayer of the membrane (Liu & Hurt, 2010; Wu et al., 2010). An early-life stage test on Japanese medaka (*Oryzias latipes*) showed that exposure to 100–1000 μg/L homogeneously dispersed AgNPs for 70 days induced a variety of morphological malformations such as edema, spinal abnormalities, finfold abnormalities, heart malformations, and eye defects. The nanoparticles could be absorbed and diffused into the yolk sac and the body via passive diffusion. The acute toxicity tests showed that the 48 h LC_{50} of AgNPs in adult medaka was 1.03 mg/L, 100% mortality of embryonic medaka under 2.0 mg/L AgNPs exposure and 0.8 mg/L in larval-juvenile medaka. In addition, embryotoxicity of nanoscale silver in embryonic freshwater fish caused significant growth retardation and various types of malformations in medaka embryos after the AgNPs exposure (Kashiwada, 2006).

The toxicity and bioavailability of nanomaterials to algae also involves adsorption to cell surfaces and disruption to membrane transport. Algae are 10–100 times more sensitive than fish cells to both AgNPs and $AgNO_3$ exposure (Yue et al., 2017). The toxicity of AgNPs to algae is mediated solely by dissolved silver while in fish cells, AgNPs were found to induce toxicity by both nanoparticles uptake and dissolved silver. This is mainly because the cell wall of algae is a barrier for particle uptake. Instead of accessing to intercellular proteins, AgNPs may adhere to the algal surface from which Ag^+ may dissolve and, as such, be taken up in the cells. Research shows AgNPs are toxic to the freshwater alga. *Chlamydomonas reinhardtii* exhibits a decreased photosynthetic yield in a concentration-dependent manner with respect to nanoparticles (Navarro, Piccapietra, et al., 2008). The results showed about 1% of the AgNPs are present as Ag^+ ions, and the toxicity of $AgNO_3$ was 18 times higher than that of AgNPs (in terms of EC50). However, when compared as a function of Ag^+ concentration, the toxicity of AgNPs appeared to be much higher than that of $AgNO_3$. There is also evidence showing the possibility of cell permeability toward AgNPs. *Ochromonas danica* experience both direct uptake of AgNPs and indirect uptake of Ag^+ ions. The direct uptake effects from AgNPs result in altered protein function due to the formation of protein-corona species. Additionally, Ag^+ ions are capable of binding with extracellular proteins, disrupting nutrient acquisition (Huang et al., 2019).

5.4.2 Ceria (CeO_2) nanoparticles

Cerium is considered the most abundant rare earth metal found in the Earth's crust. Its corresponding nanoparticulated metal oxide is intensively used. Some applications include: biomedical sensors, antioxidants, catalysts, as well as diesel fuel and glass additives (Abbasi et al., 2011; Hedrick & Sinha, 1994; Reinhardt & Winkler, 2002). Cerium pollution in aqueous media is mainly associated with leaching from landfills coming from electronic device waste and wastewater discharges and partially accumulated in the sludge used for soil amendments (Limbach et al., 2008). In nature, the most common oxidation states of cerium are Ce(III) and Ce(IV) and these form hexa-aquo complexes. The redox processes (Ce (III) and Ce (IV)) strongly influence particle aggregation and dissolution due to their different solubilities in aqueous environments, which also influence the exposure pathways and therefore the potential bio fate and bioaccumulation (Dahle et al., 2015). CeO_2 NPs are more toxic than bulk cerium oxide and may induce cell death, oxidative stress, and DNA damage (Pulido-Reyes et al., 2015). They are usually absorbed and accumulated by the biota and diluted in the constructed food web, potentially causing an irreversible disturbance in the sustainability of the aquatic system.

Compared to other metal oxide NPs (i.e., AgNPs and ZnNPs) that readily dissolve in freshwater and release ionic forms upon degradation, CeO_2 NPs are much less reactive and present as irregular ceramic-like aggregates in insoluble forms. This relatively lower solubility results in less accumulation in mussels when compared with ZnO NPs (Xia et al., 2008). Moreover, the tendency of CeO_2 NPs to aggregate may lead to different exposure routes which relate directly to the NPs internalization and bio-uptake. The tightly cohered CeO_2 NP aggregates show significant uptake of cerium in the liver and gills of zebrafish exposed to 500 μg/L CeO_2 but little uptake in fish exposed to 5000 μg/L CeO_2 (Johnston et al., 2010). In the case of Daphnia, bio-uptake can happen through two possible pathways: by adsorption/accumulation of large NPs aggregates on the cuticle, and/or by internalization of CeO_2 NPs into the cells. Exposure of Daphnia to 10 mg/L of CeO_2 NPs for 48 h results in a significant amount of accumulation of particles onto the distal spine and onto specific areas of carapace, which is responsible for different mechanical trapping of CeO_2 NPs and their aggregates (Artells et al., 2013). On the other hand, the nanoparticle size also influences the long-term fate of cerium ENMs. When two different sizes (3.8 and 185 nm) of CeO_2 NPs were exposed to freshwater over 9 months, the accumulation of cerium in the floc was significantly greater for small particles (256 ± 57 mg) compared to the large particles (39.9 ± 4.2 mg). Small CeO_2 nanoparticles were taken up significantly more by these aquatic plants, and the internal cerium mass was more than twice the external concentration (23.3 ± 1.6 mg interior vs. 9.5 ± 0.9 mg exterior). Large particles, in contrast, were approximately twice as likely to be found on the exterior than the interior of the aquatic plant *Egeria densa* (3.0 ± 0.5 mg interior vs. 6.0 ± 3.0 mg exterior) (Geitner et al., 2018).

5.4.3 Carbon-based nanomaterials (CNMs)

The main classes of CNMs that can be highlighted are buckyballs or spherical fullerenes, CNTs, graphene, and carbon black. The origin of these materials is diverse, some are released naturally into the environment as a result of volcanic eruptions and forest fires, while others are produced during anthropogenic combustion processes, or manufactured (Dickens et al., 2004; Navarro, Baun, et al., 2008). CNMs are used in a wide variety of applications, such as water treatment, medicine, optics, electronic engineering, photovoltaic devices, automotive industry, sports equipment, cosmetics, among others (Bakry et al., 2007; Gupta & Saleh, 2013). CNMs enter aquatic systems through direct inputs, including sewage and water effluents, as well as indirect inputs, which are associated to runoff events or atmospheric depositions. In aquatic ecosystems, CNMs can be accumulated in river sediments, or remain suspended in the water columns and transported to marine systems. Highly stable single and double carbon bonds within CNMs result in highly persistent ENMs in the environment. Moreover, the interactions with natural organic matter (NOM) can stabilize CNMs in aqueous suspension (Hyung et al., 2007). Although evidence of water contamination by CNMs is rare, C60 has been detected in water samples from wastewater treatment plants in the range of ng/L (Bäuerlein et al., 2017). In addition, different shapes and sizes of CNMs may influence the uptake and accumulation in aquatic organisms. Research shows that after a 24 h exposure in moderately hard freshwater, smaller outer diameter (8–15 nm) MWCNTs are 250% more likely to accumulate in *D. magna* than those with a larger outer diameter (20–30 nm) (Cano et al., 2017). Besides, decreased growth and reproduction were observed in *Daphnia magna* exposed to MWCNTs due to physical blockage of the intestinal gut, suggesting that such toxicity was the result of a physical rather than chemical mode of action (Alloy & Roberts, 2011). Carbon nanotubes are also reported to form a coating on the surfaces of the organisms which contribute to the adverse effects on the respiratory systems. SWCNT aggregate in the surface mucus of the gill rather than inside the gill tissue, which may eventually cause gill injury in a dose-dependent manner (Smith et al., 2007). Studies on fish suggest that C60 fullerenes are toxic in the milligram per liter range (between 460 μg/L and 7.9 mg/L). However, the obtained LC50 values (during 48 h) are highly dependent on the method of preparation of the material and the addition of dispersants (Lovern & Klaper, 2006). Exposure of juvenile largemouth bass (*Micropterus salmonids*) to dissolved C60 fullerenes (in THF) for up to 48 h showed an increased amount of lipid peroxidation products in the brain (Oberdörster, 2004). In the case of graphene oxide, short-term exposure (14 days) at 2.5 and 5 mg/L on the eastern oyster (*Crassostrea virginica*) induced elevated lipid peroxidation and changes in glutathione-s-transferase (GST) levels in the gills tissues and digestive gland (Khan et al., 2019). Unlike metal nanoparticles, the insoluble nature of carbon-based nanomaterials makes it difficult to measure the exact delivery dose.

5.5 Summary

Over the last few decades, the use and application of ENMs-containing products have increased exponentially. Large amounts of ENMs are constantly released in waste products and they inevitably end in water sources. These materials can reach aqueous systems directly by industrial processes, wastewater, and daily common usage, or from soil runoff and by being part of the water cycle. Once ENMs enter aqueous environments, they will form aggregates, suspensions, or sedimentations depending on their types and their properties. We have discussed here the processes of dissolution, aggregation, and precipitation, as a function of concentration, and other physical and chemical aspects of their fate in aquatic environments. Environmental factors of the aqueous systems such as their mobility in aqueous environments (flow velocity), temperature, pH and the presence of coexisting substances directly affect ENMS stability and persistence. In addition to this, ENMs in aqueous media can be exposed to sunlight and take part in photochemical transformations that can yield new materials and contaminants with different chemical and physical properties. All these transformations can result in different species that are available to uptake or interact with biological systems. Finally, in this chapter, we discussed the biological fate of Ag, CeO_2, CNMs, and their potential bioaccumulation in animal organs and cells.

References

Abbasi, Z., Haghighi, M., Fatehifar, E., & Saedy, S. (2011). Synthesis and physicochemical characterization of nanostructured Pt/CeO_2 catalyst used for total oxidation of toluene. *International Journal of Chemical Reactor Engineering, 9*. https://doi.org/10.1515/1542-6580.2690

Adeleye, A. S., Conway, J. R., Perez, T., Rutten, P., & Keller, A. A. (2014). Influence of extracellular polymeric substances on the long-term fate, dissolution, and speciation of copper-based nanoparticles. *Environmental Science and Technology, 48*(21), 12561–12568. https://doi.org/10.1021/es5033426

Alloy, M. M., & Roberts, A. P. (2011). Effects of suspended multi-walled carbon nanotubes on daphnid growth and reproduction. *Ecotoxicology and Environmental Safety, 74*(7), 1839–1843. https://doi.org/10.1016/j.ecoenv.2011.06.020

Alvarez, N. T., Kittrell, C., Schmidt, H. K., Hauge, R. H., Engel, P. S., & Tour, J. M. (2008). Selective photochemical functionalization of surfactant-dispersed single wall carbon nanotubes in water. *Journal of the American Chemical Society, 130*(43), 14227–14233. https://doi.org/10.1021/ja804164y

Artells, E., Issartel, J., Auffan, M., Borschneck, D., Thill, A., Tella, M., Brousset, L., Rose, J., Bottero, J. Y., & Thiéry, A. (2013). Exposure to cerium dioxide nanoparticles differently affect swimming performance and survival in two daphnid species. *PLoS One, 8*(8). https://doi.org/10.1371/journal.pone.0071260

Asharani, P. V., Lian Wu, Y., Gong, Z., & Valiyaveettil, S. (2008). Toxicity of silver nanoparticles in zebrafish models. *Nanotechnology, 19*(25), 255102. https://doi.org/10.1088/0957-4484/19/25/255102

Ashraf, M., Khan, I., Usman, M., Khan, A., Shah, S. S., Khan, A. Z., Saeed, K., Yaseen, M., Ehsan, M. F., Tahir, M. N., & Ullah, N. (2020). Hematite and magnetite nanostructures for green and sustainable energy harnessing and environmental pollution control: A review. *Chemical Research in Toxicology, 33*(6), 1292–1311. https://doi.org/10.1021/acs.chemrestox.9b00308

Avramescu, M. L., Rasmussen, P. E., Chénier, M., & Gardner, H. D. (2017). Influence of pH, particle size and crystal form on dissolution behaviour of engineered nanomaterials. *Environmental Science and Pollution Research, 24*(2), 1553–1564. https://doi.org/10.1007/s11356-016-7932-2

Baalousha, M., Sikder, M., Prasad, A., Lead, J., Merrifield, R., & Chandler, G. T. (2016). The concentration-dependent behaviour of nanoparticles. *Environmental Chemistry, 13*(1), 1–3. https://doi.org/10.1071/EN15142

Babel, S., Sekartaji, P. A., & Sudrajat, H. (2017). TiO_2 as an effective nanocatalyst for photocatalytic degradation of humic acid in water environment. *Journal of Water Supply: Research and Technology - AQUA, 66*(1), 25–35. https://doi.org/10.2166/aqua.2016.102

Bakry, R., Vallant, R. M., Najam-ul-Haq, M., Rainer, M., Szabo, Z., Huck, C. W., & Bonn, G. K. (2007). Medicinal applications of fullerenes. *International Journal of Nanomedicine, 2*(4), 639–649.

Bäuerlein, P. S., Emke, E., Tromp, P., Hofman, J. A. M. H., Carboni, A., Schooneman, F., de Voogt, P., & van Wezel, A. P. (2017). Is there evidence for man-made nanoparticles in the Dutch environment? *Science of the Total Environment, 576*, 273–283. https://doi.org/10.1016/j.scitotenv.2016.09.206

Benn, T. M., & Westerhoff, P. (2008). Nanoparticle silver released into water from commercially available sock fabrics. *Environmental Science and Technology, 42*(11), 4133–4139. https://doi.org/10.1021/es7032718

Bian, S. W., Mudunkotuwa, I. A., Rupasinghe, T., & Grassian, V. H. (2011). Aggregation and dissolution of 4 nm ZnO nanoparticles in aqueous environments: Influence of pH, ionic strength, size, and adsorption of humic acid. *Langmuir, 27*(10), 6059–6068. https://doi.org/10.1021/la200570n

Bruneau, A., Turcotte, P., Pilote, M., Gagné, F., & Gagnon, C. (2016). Fate of silver nanoparticles in wastewater and immunotoxic effects on rainbow trout. *Aquatic Toxicology, 174*, 70–81. https://doi.org/10.1016/j.aquatox.2016.02.013

Bundschuh, M., Filser, J., Lüderwald, S., McKee, M. S., Metreveli, G., Schaumann, G. E., Schulz, R., & Wagner, S. (2018). Nanoparticles in the environment: Where do we come from, where do we go to? *Environmental Sciences Europe, 30*(1). https://doi.org/10.1186/s12302-018-0132-6

Cano, A. M., Maul, J. D., Saed, M., Shah, S. A., Green, M. J., & Cañas-Carrell, J. E. (2017). Bioaccumulation, stress, and swimming impairment in Daphnia magna exposed to multiwalled carbon nanotubes, graphene, and graphene oxide. *Environmental Toxicology and Chemistry, 36*(8), 2199–2204. https://doi.org/10.1002/etc.3754

Conway, J. R., Adeleye, A. S., Gardea-Torresdey, J., & Keller, A. A. (2015). Aggregation, dissolution, and transformation of copper nanoparticles in natural waters. *Environmental Science and Technology, 49*(5), 2749–2756. https://doi.org/10.1021/es504918q

Dahle, J. T., Livi, K., & Arai, Y. (2015). Effects of pH and phosphate on CeO_2 nanoparticle dissolution. *Chemosphere, 119*, 1365–1371. https://doi.org/10.1016/j.chemosphere.2014.02.027

Danielsson, K., Persson, P., Gallego-Urrea, J. A., Abbas, Z., Rosenqvist, J., & Jonsson, C. M. (2018). Effects of the adsorption of NOM model molecules on the aggregation of TiO_2 nanoparticles in aqueous suspensions. *NanoImpact, 10*, 177–187. https://doi.org/10.1016/j.impact.2018.05.002

Darko, B., Jiang, J. Q., Kim, H., Machala, L., Zboril, R., & Sharma, V. K. (2014). Advances made in understanding the interaction of ferrate(VI) with natural organic matter in water.

In *Water reclamation and sustainability* (pp. 183–197). Elsevier Inc. https://doi.org/10.1016/B978-0-12-411645-0.00008-0

Deline, A. R., Frank, B. P., Smith, C. L., Sigmon, L. R., Wallace, A. N., Gallagher, M. J., Goodwin, D. G., Durkin, D. P., & Howard Fairbrother, D. (2020). Influence of oxygen-containing functional groups on the environmental properties, transformations, and toxicity of carbon nanotubes. *Chemical Reviews, 120*(20), 11651–11697. https://doi.org/10.1021/acs.chemrev.0c00351

Dickens, A. F., Gélinas, Y., Masiello, C. A., Wakeham, S., & Hedges, J. I. (2004). Reburial of fossil organic carbon in marine sediments. *Nature, 427*(6972), 336–339. https://doi.org/10.1038/nature02299

Dorobantu, L. S., Fallone, C., Noble, A. J., Veinot, J., Ma, G., Goss, G. G., & Burrell, R. E. (2015). Toxicity of silver nanoparticles against bacteria, yeast, and algae. *Journal of Nanoparticle Research, 17*(4). https://doi.org/10.1007/s11051-015-2984-7

Dutta, P. K., Pehkonen, S. O., Sharma, V. K., & Ray, A. K. (2005). Photocatalytic oxidation of arsenic (III): Evidence of hydroxyl radicals. *Environmental Science and Technology, 39*(6), 1827–1834. https://doi.org/10.1021/es0489238

Echigo, T., Aruguete, D. M., Murayama, M., & Hochella, M. F. (2012). Influence of size, morphology, surface structure, and aggregation state on reductive dissolution of hematite nanoparticles with ascorbic acid. *Geochimica et Cosmochimica Acta, 90*, 149–162. https://doi.org/10.1016/j.gca.2012.05.008

Fabrega, J., Zhang, R., Renshaw, J. C., Liu, W. T., & Lead, J. R. (2011). Impact of silver nanoparticles on natural marine biofilm bacteria. *Chemosphere, 85*(6), 961–966. https://doi.org/10.1016/j.chemosphere.2011.06.066

Fang, J., Shijirbaatar, A., Lin, D.h., Wang, D.j., Shen, B., Sun, P.d., & Zhou, Z.q. (2017). Stability of co-existing ZnO and TiO_2 nanomaterials in natural water: Aggregation and sedimentation mechanisms. *Chemosphere, 184*, 1125–1133. https://doi.org/10.1016/j.chemosphere.2017.06.097

Fan, D., Lan, Y., Tratnyek, P. G., Johnson, R. L., Filip, J., O'Carroll, D. M., Nunez Garcia, A., & Agrawal, A. (2017). Sulfidation of iron-based materials: A review of processes and implications for water treatment and remediation. *Environmental Science and Technology, 51*(22), 13070–13085. https://doi.org/10.1021/acs.est.7b04177

Frank, B. P., Sigmon, L. R., Deline, A. R., Lankone, R. S., Gallagher, M. J., Zhi, B., Haynes, C. L., & Fairbrother, D. H. (2020). Photochemical transformations of carbon dots in aqueous environments. *Environmental Science and Technology, 54*(7), 4160–4170. https://doi.org/10.1021/acs.est.9b07437

French, R. A., Jacobson, A. R., Kim, B., Isley, S. L., Penn, L., & Baveye, P. C. (2009). Influence of ionic strength, pH, and cation valence on aggregation kinetics of titanium dioxide nanoparticles. *Environmental Science and Technology, 43*(5), 1354–1359. https://doi.org/10.1021/es802628n

Geitner, N. K., Cooper, J. L., Avellan, A., Castellon, B. T., Perrotta, B. G., Bossa, N., Simonin, M., Anderson, S. M., Inoue, S., Hochella, M. F., Richardson, C. J., Bernhardt, E. S., Lowry, G. V., Ferguson, P. L., Matson, C. W., King, R. S., Unrine, J. M., Wiesner, M. R., & Hsu-Kim, H. (2018). Size-based differential transport, uptake, and mass distribution of ceria (CeO_2) nanoparticles in wetland mesocosms. *Environmental Science and Technology, 52*(17), 9768–9776. https://doi.org/10.1021/acs.est.8b02040

Gottschalk, F., & Nowack, B. (2011). The release of engineered nanomaterials to the environment. *Journal of Environmental Monitoring, 13*(5), 1145–1155. https://doi.org/10.1039/c0em00547a

Gottschalk, F., Sonderer, T., Scholz, R. W., & Nowack, B. (2009). Modeled environmental concentrations of engineered nanomaterials (TiO_2, ZnO, Ag, CNT, fullerenes) for different regions. *Environmental Science and Technology, 43*(24), 9216–9222. https://doi.org/10.1021/es9015553

Gottschalk, F., Sun, T., & Nowack, B. (2013). Environmental concentrations of engineered nanomaterials: Review of modeling and analytical studies. *Environmental Pollution, 181*, 287–300. https://doi.org/10.1016/j.envpol.2013.06.003

Griffitt, R. J., Hyndman, K., Denslow, N. D., & Barber, D. S. (2009). Comparison of molecular and histological changes in zebrafish gills exposed to metallic nanoparticles. *Toxicological Sciences, 107*(2), 404–415. https://doi.org/10.1093/toxsci/kfn256

Grillet, N., Manchon, D., Cottancin, E., Bertorelle, F., Bonnet, C., Broyer, M., Lermé, J., & Pellarin, M. (2013). Photo-oxidation of individual silver nanoparticles: A real-time tracking of optical and morphological changes. *The Journal of Physical Chemistry C, 117*(5), 2274–2282. https://doi.org/10.1021/jp311502h

Gupta, V. K., & Saleh, T. A. (2013). Sorption of pollutants by porous carbon, carbon nanotubes and fullerene- an overview. *Environmental Science and Pollution Research, 20*(5), 2828–2843. https://doi.org/10.1007/s11356-013-1524-1

He, H., Cheng, Y., Yang, C., Zeng, G., Zhu, C., & Yan, Z. (2017). Influences of anion concentration and valence on dispersion and aggregation of titanium dioxide nanoparticles in aqueous solutions. *Journal of Environmental Sciences, 54*, 135–141. https://doi.org/10.1016/j.jes.2016.06.009

Hedrick, J. B., & Sinha, S. P. (1994). Cerium-based polishing compounds: Discovery to manufacture. *Journal of Alloys and Compounds, 207–208*(C), 377–382. https://doi.org/10.1016/0925-8388(94)90243-7

He, Y. T., Wan, J., & Tokunaga, T. (2008). Kinetic stability of hematite nanoparticles: The effect of particle sizes. *Journal of Nanoparticle Research, 10*(2), 321–332. https://doi.org/10.1007/s11051-007-9255-1

Hochella, M. F., Mogk, D. W., Ranville, J., Allen, I. C., Luther, G. W., Marr, L. C., McGrail, B. P., Murayama, M., Qafoku, N. P., Rosso, K. M., Sahai, N., Schroeder, P. A., Vikesland, P., Westerhoff, P., & Yang, Y. (2019). Natural, incidental, and engineered nanomaterials and their impacts on the earth system. *Science, 363*(6434). https://doi.org/10.1126/science.aau8299

Hortin, J. M., Anderson, A. J., Britt, D. W., Jacobson, A. R., & McLean, J. E. (2020). Copper oxide nanoparticle dissolution at alkaline pH is controlled by dissolved organic matter: Influence of soil-derived organic matter, wheat, bacteria, and nanoparticle coating. *Environmental Science: Nano, 7*(9), 2618–2631. https://doi.org/10.1039/d0en00574f

Hotze, E. M., Phenrat, T., & Lowry, G. V. (2010). Nanoparticle aggregation: Challenges to understanding transport and reactivity in the environment. *Journal of Environmental Quality, 39*(6), 1909–1924. https://doi.org/10.2134/jeq2009.0462

Hou, W. C., & Jafvert, C. T. (2009). Photochemical transformation of aqueous C clusters in sunlight. *Environmental Science and Technology, 43*(2), 362–367. https://doi.org/10.1021/es802465z

Huang, B., Wei, Z. B., Yang, L. Y., Pan, K., & Miao, A. J. (2019). Combined toxicity of silver nanoparticles with hematite or plastic nanoparticles toward two freshwater algae. *Environmental Science and Technology, 53*(7), 3871–3879. https://doi.org/10.1021/acs.est.8b07001

Hyung, H., Fortner, J. D., Hughes, J. B., & Kim, J. H. (2007). Natural organic matter stabilizes carbon nanotubes in the aqueous phase. *Environmental Science and Technology, 41*(1), 179–184. https://doi.org/10.1021/es061817g

Hyung, H., & Kim, J. H. (2008). Natural organic matter (NOM) adsorption to multi-walled carbon nanotubes: Effect of NOM characteristics and water quality parameters. *Environmental Science and Technology, 42*(12), 4416–4421. https://doi.org/10.1021/es702916h

John, A. C., Küpper, M., Manders-Groot, A. M. M., Debray, B., Lacome, J. M., & Kuhlbusch, T. A. J. (2017). Emissions and possible environmental Implication of engineered nanomaterials (ENMs) in the atmosphere. *Atmosphere, 8*(5). https://doi.org/10.3390/atmos8050084

Johnston, B. D., Scown, T. M., Moger, J., Cumberland, S. A., Baalousha, M., Linge, K., Van Aerle, R., Jarvis, K., Lead, J. R., & Tyler, C. R. (2010). Bioavailability of nanoscale metal oxides TiO_2, CeO_2, and ZnO to fish. *Environmental Science and Technology, 44*(3), 1144–1151. https://doi.org/10.1021/es901971a

Kashiwada, S. (2006). Distribution of nanoparticles in the see-through medaka (Oryzias latipes). *Environmental Health Perspectives, 114*(11), 1697–1702. https://doi.org/10.1289/ehp.9209

Keller, A. A., McFerran, S., Lazareva, A., & Suh, S. (2013). Global life cycle releases of engineered nanomaterials. *Journal of Nanoparticle Research, 15*(6). https://doi.org/10.1007/s11051-013-1692-4

Khan, B., Adeleye, A. S., Burgess, R. M., Russo, S. M., & Ho, K. T. (2019). Effects of graphene oxide nanomaterial exposures on the marine bivalve, Crassostrea virginica. *Aquatic Toxicology, 216*. https://doi.org/10.1016/j.aquatox.2019.105297

Khan, M. M., Adil, S. F., & Al-Mayouf, A. (2015). Metal oxides as photocatalysts. *Journal of Saudi Chemical Society, 19*(5), 462–464. https://doi.org/10.1016/j.jscs.2015.04.003

Klaine, S. J., Alvarez, P. J. J., Batley, G. E., Fernandes, T. F., Handy, R. D., Lyon, D. Y., Mahendra, S., McLaughlin, M. J., & Lead, J. R. (2008). Nanomaterials in the environment: Behavior, fate, bioavailability, and effects. *Environmental Toxicology and Chemistry, 27*(9), 1825–1851. https://doi.org/10.1897/08-090.1

Kulmala, M., Petäjä, T., Ehn, M., Thornton, J., Sipilä, M., Worsnop, D. R., & Kerminen, V. M. (2014). Chemistry of atmospheric nucleation: On the recent advances on precursor characterization and atmospheric cluster composition in connection with atmospheric new particle formation. *Annual Review of Physical Chemistry, 65*, 21–37. https://doi.org/10.1146/annurev-physchem-040412-110014

Laux, P., Riebeling, C., Booth, A. M., Brain, J. D., Brunner, J., Cerrillo, C., Creutzenberg, O., Estrela-Lopis, I., Gebel, T., Johanson, G., Jungnickel, H., Kock, H., Tentschert, J., Tlili, A., Schäffer, A., Sips, A. J. A. M., Yokel, R. A., & Luch, A. (2018). Challenges in characterizing the environmental fate and effects of carbon nanotubes and inorganic nanomaterials in aquatic systems. *Environmental Science: Nano, 5*(1), 48–63. https://doi.org/10.1039/c7en00594f

Lead, J. R., Batley, G. E., Alvarez, P. J. J., Croteau, M. N., Handy, R. D., McLaughlin, M. J., Judy, J. D., & Schirmer, K. (2018). Nanomaterials in the environment: Behavior, fate, bioavailability, and effects—an updated review. *Environmental Toxicology and Chemistry, 37*(8), 2029–2063. https://doi.org/10.1002/etc.4147

Lee, K. J., Nallathamby, P. D., Browning, L. M., Osgood, C. J., & Nancy Xu, X. H. (2007). In vivo imaging of transport and biocompatibility of single silver nanoparticles in early development of zebrafish embryos. *ACS Nano, 1*(2), 133–143. https://doi.org/10.1021/nn700048y

Lee, S. Y., & Park, S. J. (2013). TiO_2 photocatalyst for water treatment applications. *Journal of Industrial and Engineering Chemistry, 19*(6), 1761–1769. https://doi.org/10.1016/j.jiec.2013.07.012

Levard, C., Hotze, E. M., Colman, B. P., Dale, A. L., Truong, L., Yang, X. Y., Bone, A. J., Brown, G. E., Tanguay, R. L., Di Giulio, R. T., Bernhardt, E. S., Meyer, J. N., Wiesner, M. R., & Lowry, G. V. (2013). Sulfidation of silver nanoparticles: Natural antidote to their toxicity. *Environmental Science and Technology, 47*(23), 13440–13448. https://doi.org/10.1021/es403527n

Levard, C., Reinsch, B. C., Michel, F. M., Oumahi, C., Lowry, G. V., & Brown, G. E. (2011). Sulfidation processes of PVP-coated silver nanoparticles in aqueous solution: Impact on dissolution rate. *Environmental Science and Technology, 45*(12), 5260–5266. https://doi.org/10.1021/es2007758

Limbach, L. K., Bereiter, R., Müller, E., Krebs, R., Gälli, R., & Stark, W. J. (2008). Removal of oxide nanoparticles in a model wastewater treatment plant: Influence of agglomeration

and surfactants on clearing efficiency. *Environmental Science and Technology, 42*(15), 5828–5833. https://doi.org/10.1021/es800091f
Lin, D., Liu, N., Yang, K., Xing, B., & Wu, F. (2010). Different stabilities of multiwalled carbon nanotubes in fresh surface water samples. *Environmental Pollution, 158*(5), 1270–1274. https://doi.org/10.1016/j.envpol.2010.01.020
Lin, D., Tian, X., Wu, F., & Xing, B. (2010). Fate and transport of engineered nanomaterials in the environment. *Journal of Environmental Quality, 39*(6), 1896–1908. https://doi.org/10.2134/jeq2009.0423
Liu, X., Chen, G., & Su, C. (2011). Effects of material properties on sedimentation and aggregation of titanium dioxide nanoparticles of anatase and rutile in the aqueous phase. *Journal of Colloid and Interface Science, 363*(1), 84–91. https://doi.org/10.1016/j.jcis.2011.06.085
Liu, J., & Hurt, R. H. (2010). Ion release kinetics and particle persistence in aqueous nano-silver colloids. *Environmental Science and Technology, 44*(6), 2169–2175. https://doi.org/10.1021/es9035557
Liu, Y., Tourbin, M., Lachaize, S., & Guiraud, P. (2014). Nanoparticles in wastewaters: Hazards, fate and remediation. *Powder Technology, 255*, 149–156. https://doi.org/10.1016/j.powtec.2013.08.025
Li, R., Wang, J. J., Zhang, Z., Awasthi, M. K., Du, D., Dang, P., Huang, Q., Zhang, Y., & Wang, L. (2018). Recovery of phosphate and dissolved organic matter from aqueous solution using a novel CaO-MgO hybrid carbon composite and its feasibility in phosphorus recycling. *Science of the Total Environment, 642*, 526–536. https://doi.org/10.1016/j.scitotenv.2018.06.092
Li, X., Yoneda, M., Shimada, Y., & Matsui, Y. (2017). Effect of surfactants on the aggregation and sedimentation of zinc oxide nanomaterial in natural water matrices. *Science of the Total Environment, 581–582*, 649–656. https://doi.org/10.1016/j.scitotenv.2016.12.175
Loosli, F., Le Coustumer, P., & Stoll, S. (2013). TiO_2 nanoparticles aggregation and disaggregation in presence of alginate and Suwannee River humic acids. pH and concentration effects on nanoparticle stability. *Water Research, 47*(16), 6052–6063. https://doi.org/10.1016/j.watres.2013.07.021
Loosli, F., Wang, J., Rothenberg, S., Bizimis, M., Winkler, C., Borovinskaya, O., Flamigni, L., & Baalousha, M. (2019). Sewage spills are a major source of titanium dioxide engineered (nano)-particle release into the environment. *Environmental Science: Nano, 6*(3), 763–777. https://doi.org/10.1039/c8en01376d
Lovern, S. B., & Klaper, R. (2006). Daphnia magna mortality when exposed to titanium dioxide and fullerene (C60) nanoparticles. *Environmental Toxicology and Chemistry, 25*(4), 1132–1137. https://doi.org/10.1897/05-278R.1
Luoma, S. N., Khan, F. R., & Croteau, M. N. (2014). Bioavailability and bioaccumulation of metal-based engineered nanomaterials in aquatic environments. Concepts and processes. *Frontiers of Nanoscience, 7*, 157–193. https://doi.org/10.1016/B978-0-08-099408-6.00005-0
Luoma, S. N., & Rainbow, P. S. (2005). Why is metal bioaccumulation so variable? Biodynamics as a unifying concept. *Environmental Science and Technology, 39*(7), 1921–1931. https://doi.org/10.1021/es048947e
Lv, X., Tao, J., Chen, B., & Zhu, X. (2016). Roles of temperature and flow velocity on the mobility of nano-sized titanium dioxide in natural waters. *Science of The Total Environment, 565*, 849–856. https://doi.org/10.1016/j.scitotenv.2016.03.001
Ma, R., Levard, C., Michel, F. M., Brown, G. E., & Lowry, G. V. (2013). Sulfidation mechanism for zinc oxide nanoparticles and the effect of sulfidation on their solubility. *Environmental Science and Technology, 47*(6), 2527–2534. https://doi.org/10.1021/es3035347
Merikanto, J., Spracklen, D. V., Mann, G. W., Pickering, S. J., & Carslaw, K. S. (2009). Impact of nucleation on global CCN. *Atmospheric Chemistry and Physics, 9*(21), 8601–8616. https://doi.org/10.5194/acp-9-8601-2009

Miao, L., Wang, C., Hou, J., Wang, P., Ao, Y., Li, Y., Lv, B., Yang, Y., You, G., & Xu, Y. (2016). Effect of alginate on the aggregation kinetics of copper oxide nanoparticles (CuO NPs): Bridging interaction and hetero-aggregation induced by $Ca2^{+}$. *Environmental Science and Pollution Research, 23*(12), 11611–11619. https://doi.org/10.1007/s11356-016-6358-1

Miao, A. J., Zhang, X. Y., Luo, Z., Chen, C. S., Chin, W. C., Santschi, P. H., & Quigg, A. (2010). Zinc oxide-engineered nanoparticles: Dissolution and toxicity to marine phytoplankton. *Environmental Toxicology and Chemistry, 29*(12), 2814–2822. https://doi.org/10.1002/etc.340

Montaño, M. D., Lowry, G. V., Von Der Kammer, F., Blue, J., & Ranville, J. F. (2014). Current status and future direction for examining engineered nanoparticles in natural systems. *Environmental Chemistry, 11*(4), 351–366. https://doi.org/10.1071/EN14037

Morones, J. R., Elechiguerra, J. L., Camacho, A., Holt, K., Kouri, J. B., Ramírez, J. T., & Yacaman, M. J. (2005). The bactericidal effect of silver nanoparticles. *Nanotechnology, 16*(10), 2346–2353. https://doi.org/10.1088/0957-4484/16/10/059

National Research Council. (2012). *A research strategy for environmental, health, and safety aspects of engineered nanomaterials.*

Nason, J. A., McDowell, S. A., & Callahan, T. W. (2012). Effects of natural organic matter type and concentration on the aggregation of citrate-stabilized gold nanoparticles. *Journal of Environmental Monitoring, 14*(7), 1885–1892. https://doi.org/10.1039/c2em00005a

Navarro, E., Baun, A., Behra, R., Hartmann, N. B., Filser, J., Miao, A. J., Quigg, A., Santschi, P. H., & Sigg, L. (2008). Environmental behavior and ecotoxicity of engineered nanoparticles to algae, plants, and fungi. *Ecotoxicology, 17*(5), 372–386. https://doi.org/10.1007/s10646-008-0214-0

Navarro, E., Piccapietra, F., Wagner, B., Marconi, F., Kaegi, R., Odzak, N., Sigg, L., & Behra, R. (2008). Toxicity of silver nanoparticles to Chlamydomonas reinhardtii. *Environmental Science and Technology, 42*(23), 8959–8964. https://doi.org/10.1021/es801785m

Oberdörster, E. (2004). Manufactured nanomaterials (fullerenes, C60) induce oxidative stress in the brain of juvenile largemouth bass. *Environmental Health Perspectives, 112*(10), 1058–1062. https://doi.org/10.1289/ehp.7021

Parekh, B., Debies, T., Knight, P., Santhanam, K. S. V., & Takacs, G. A. (2006). Surface functionalization of multiwalled carbon nanotubes with UV and vacuum UV photo-oxidation. *Journal of Adhesion Science and Technology, 20*(16), 1833–1846. https://doi.org/10.1163/156856106779116641

Peretyazhko, T. S., Zhang, Q., & Colvin, V. L. (2014). Size-controlled dissolution of silver nanoparticles at neutral and acidic pH conditions: Kinetics and size changes. *Environmental Science and Technology, 48*(20), 11954–11961. https://doi.org/10.1021/es5023202

Petosa, A. R., Jaisi, D. P., Quevedo, I. R., Elimelech, M., & Tufenkji, N. (2010). Aggregation and deposition of engineered nanomaterials in aquatic environments: Role of physicochemical interactions. *Environmental Science and Technology, 44*(17), 6532–6549. https://doi.org/10.1021/es100598h

Pulido-Reyes, G., Rodea-Palomares, I., Das, S., Sakthivel, T. S., Leganes, F., Rosal, R., Seal, S., & Fernández-Piñas, F. (2015). Untangling the biological effects of cerium oxide nanoparticles: The role of surface valence states. *Scientific Reports, 5*(1). https://doi.org/10.1038/srep15613

Rathnayake, S., Unrine, J. M., Judy, J., Miller, A. F., Rao, W., & Bertsch, P. M. (2014). Multitechnique investigation of the ph dependence of phosphate induced transformations of ZnO nanoparticles. *Environmental Science and Technology, 48*(9), 4757–4764. https://doi.org/10.1021/es404544w

Reinhardt, K., & Winkler, H. (2002). Cerium mischmetal, cerium alloys, and cerium compounds. In *Ullmann's Encyclopedia of Industrial Chemistry* (pp. 285–300). Weinheim, Germany: Wiley-VCH.

Scown, T. M., Santos, E. M., Johnston, B. D., Gaiser, B., Baalousha, M., Mitov, S., Lead, J. R., Stone, V., Fernandes, T. F., Jepson, M., van Aerle, R., & Tyler, C. R. (2010). Effects of aqueous exposure to silver nanoparticles of different sizes in rainbow trout. *Toxicological Sciences, 115*(2), 521–534. https://doi.org/10.1093/toxsci/kfq076

Selck, H., Handy, R. D., Fernandes, T. F., Klaine, S. J., & Petersen, E. J. (2016). Nanomaterials in the aquatic environment: A European union-United States perspective on the status of ecotoxicity testing, research priorities, and challenges ahead. *Environmental Toxicology and Chemistry, 35*(5), 1055–1067. https://doi.org/10.1002/etc.3385

Shevlin, D., O'Brien, N., & Cummins, E. (2018). Silver engineered nanoparticles in freshwater systems – likely fate and behaviour through natural attenuation processes. *Science of the Total Environment, 621*, 1033–1046. https://doi.org/10.1016/j.scitotenv.2017.10.123

Shrivastava, M., Srivastav, A., Gandhi, S., Rao, S., Roychoudhury, A., Kumar, A., Singhal, R. K., Jha, S. K., & Singh, S. D. (2019). Monitoring of engineered nanoparticles in soil-plant system: A review. *Environmental Nanotechnology, Monitoring & Management, 11*, 100218. https://doi.org/10.1016/j.enmm.2019.100218

Sikder, M., Lead, J. R., Chandler, G. T., & Baalousha, M. (2018). A rapid approach for measuring silver nanoparticle concentration and dissolution in seawater by UV–Vis. *Science of the Total Environment, 618*, 597–607. https://doi.org/10.1016/j.scitotenv.2017.04.055

Skompska, M., Vorotyntsev, M. A., Rajchowska, A., & Levin, O. V. (2010). Mixed solutions of silver cation and chloride anion in acetonitrile: Voltammetric and EQCM study. *Physical Chemistry Chemical Physics, 12*(35), 10525–10535. https://doi.org/10.1039/c003332d

Smith, C. J., Shaw, B. J., & Handy, R. D. (2007). Toxicity of single walled carbon nanotubes to rainbow trout, (*Oncorhynchus mykiss*): Respiratory toxicity, organ pathologies, and other physiological effects. *Aquatic Toxicology, 82*(2), 94–109. https://doi.org/10.1016/j.aquatox.2007.02.003

Smith, B., Wepasnick, K., Schrote, K. E., Cho, H. H., Ball, W. P., & Fairbrother, D. H. (2009). Influence of surface oxides on the colloidal stability of multi-walled carbon nanotubes: A structure-property relationship. *Langmuir, 25*(17), 9767–9776. https://doi.org/10.1021/la901128k

Sun, T. Y., Gottschalk, F., Hungerbühler, K., & Nowack, B. (2014). Comprehensive probabilistic modelling of environmental emissions of engineered nanomaterials. *Environmental Pollution, 185*, 69–76. https://doi.org/10.1016/j.envpol.2013.10.004

Surette, M. C., & Nason, J. A. (2016). Effects of surface coating character and interactions with natural organic matter on the colloidal stability of gold nanoparticles. *Environmental Science: Nano, 3*(5), 1144–1152. https://doi.org/10.1039/c6en00180g

Tolaymat, T., El Badawy, A., Genaidy, A., Abdelraheem, W., & Sequeira, R. (2017). Analysis of metallic and metal oxide nanomaterial environmental emissions. *Journal of Cleaner Production, 143*, 401–412. https://doi.org/10.1016/j.jclepro.2016.12.094

Vo, D. Q., Kim, E. J., & Kim, S. (2009). Surface modification of hydrophobic nanocrystals using short-chain carboxylic acids. *Journal of Colloid and Interface Science, 337*(1), 75–80. https://doi.org/10.1016/j.jcis.2009.04.078

Wagner, S., Gondikas, A., Neubauer, E., Hofmann, T., & Von Der Kammer, F. (2014). Spot the difference: Engineered and natural nanoparticles in the environment-release, behavior, and fate. *Angewandte Chemie - International Edition, 53*(46), 12398–12419. https://doi.org/10.1002/anie.201405050

Warren, P. B. (2000). A theory of void formation in charge-stabilized colloidal suspensions at low ionic strength. *Journal of Chemical Physics, 112*(10), 4683–4698. https://doi.org/10.1063/1.481024

Wu, A., Jing, L., Wang, J., Qu, Y., Xie, Y., Jiang, B., Tian, C., & Fu, H. (2015). ZnO-dotted porous ZnS cluster microspheres for high efficient, Pt-free photocatalytic hydrogen evolution. *Scientific Reports, 5*. https://doi.org/10.1038/srep08858

Wu, Y., Zhou, Q., Li, H., Liu, W., Wang, T., & Jiang, G. (2010). Effects of silver nanoparticles on the development and histopathology biomarkers of Japanese medaka (*Oryzias latipes*) using the partial-life test. *Aquatic Toxicology, 100*(2), 160–167. https://doi.org/10.1016/j.aquatox.2009.11.014

Xia, T., Kovochich, M., Liong, M., Mädler, L., Gilbert, B., Shi, H., Yeh, J. I., Zink, J. I., & Nel, A. E. (2008). Comparison of the mechanism of toxicity of zinc oxide and cerium oxide nanoparticles based on dissolution and oxidative stress properties. *ACS Nano, 2*(10), 2121–2134. https://doi.org/10.1021/nn800511k

Yin, Y., Liu, J., & Jiang, G. (2012). Sunlight-induced reduction of ionic Ag and Au to metallic nanoparticles by dissolved organic matter. *ACS Nano, 6*(9), 7910–7919. https://doi.org/10.1021/nn302293r

Yin, Yongguang, Yang, X., Zhou, X., Wang, W., Yu, S., Liu, J., & Jiang, G. (2015). Water chemistry controlled aggregation and photo-transformation of silver nanoparticles in environmental waters. *Journal of Environmental Sciences, 34*, 116–125. https://doi.org/10.1016/j.jes.2015.04.005

Yue, Y., Li, X., Sigg, L., Suter, M. J. F., Pillai, S., Behra, R., & Schirmer, K. (2017). Interaction of silver nanoparticles with algae and fish cells: A side by side comparison. *Journal of Nanobiotechnology, 15*(1). https://doi.org/10.1186/s12951-017-0254-9

Zhang, X., Du, A. J., Lee, P., Sun, D. D., & Leckie, J. O. (2008). TiO_2 nanowire membrane for concurrent filtration and photocatalytic oxidation of humic acid in water. *Journal of Membrane Science, 313*(1–2), 44–51. https://doi.org/10.1016/j.memsci.2007.12.045

Zhang, J., & Nosaka, Y. (2014). Mechanism of the OH radical generation in photocatalysis with TiO_2 of different crystalline types. *The Journal of Physical Chemistry C, 118*(20), 10824–10832. https://doi.org/10.1021/jp501214m

CHAPTER

6

Biophysicochemical transformations of ENMs in air

Tao Xu[1], *Fan Zheng*[2], *and Yong Zhao*[3]

[1] Center for Nanotechnology and Nanotoxicology, Harvard T. H. Chan School of Public Health, Harvard University, Boston, MA, United States; [2] Composites and Polymer Engineering Laboratory, South Dakota School of Mines & Technology, Rapid City, SD, United States; [3] W. M. Keck Center for Advanced Microscopy & Microanalysis, University of Delaware, Newark, DE, United States

6.1 Introduction

New engineered nanomaterials (ENMs) are continuously designed to meet the market demand in a wide array of fields such as automotive, electronics, chemical and material industries, agriculture, energy, as well as health and pharmaceuticals (Goswami et al., 2017; Rabajczyk et al., 2020). Meanwhile, the production of ENMs can bring significant unexplored risk factors for humans and the environment, which require substantial attention (Garner & Keller, 2014; Manke et al., 2013; Sajid et al., 2015). The physicochemical characteristics of anthropogenic and natural nanomaterials (NMs) might differ those corresponding to bulk materials in numerous ways, such as aspect ratio, reactivity, chemical composition, and toxicity (Montaño et al., 2014; Xia et al., 2009). Possible potential impacts of anthropogenic and naturally emitted nanomaterials on ecosystems are still under investigation, with regulatory concerns in air quality control (Abbott & Maynard, 2010; Farré et al., 2011; Kuhlbusch et al., 2014; Meesters et al., 2013).

https://doi.org/10.1016/B978-0-323-90558-9.00010-3

6.1.1 Types of ENMs

There is a large number of existing ENM derivatives, which are designed and synthesized with enhanced mechanical, catalytic, optical, and/or electrical (conductivity) properties (Klaine et al., 2012). They can be described based on the synthesis precursors, processes, and/or final products.

6.1.1.1 Metallic nanoparticles

Metallic NMs such as Ag, Au, Fe, Cu, Pt, Pd, Ni, and Co are widely used as antimicrobial, optics, catalytic, electronics, sensing, and doping agents. The available preparatory techniques for these NMs include chemical reduction, irradiation, electrochemical reaction, and photochemical reduction (Goswami et al., 2017). Many of those NMs, such as nanowires of Au, Cu, Si, and Co, are frequently used as conductors and semiconductors. Nanocrystal quantum dots with enhanced photoluminescence (Farré et al., 2011) have been widely used in biomedical fields, including in-vivo biomedical imaging, live-cell imaging, cell organelles labeling (e.g., neoplastic cells, peroxisomes, and DNA), and cell receptors imaging (Beaurepaire et al., 2004; Colton et al., 2004; Dubertret et al., 2002; Gao et al., 2004).

6.1.1.2 Metal oxide nanoparticles

Metal oxide NMs, including ZnO, FeO, SiO_2, CeO_2, and TiO_2, possess distinct photocatalytic properties, and have been frequently used as components in rubber additives, catalytic converters, photovoltaic cells, sensors, and environmental remediation (Ghosh et al., 2011; Mittal et al., 2013; Quinn et al., 2005). The common techniques for producing those NMs include chemical reaction, as well as hydrothermal and solvothermal reduction (Goswami et al., 2017).

6.1.1.3 Polymeric nanoparticles

Polymeric NMs are commonly used in medical fields (e.g., drug delivery) (Farré et al., 2011), and they are designed and synthesized by modifying their size, surface charge, morphology, and composition depending on their predetermined application (Auffan et al., 2009; Farré et al., 2011). Polymerization can be achieved via cascade synthesis, microemulsion, surfactant-free emulsion, interfacial emulsion, sonochemical reduction, chemical reduction, oil in water emulsion, and sol-gel synthesis (Iwamoto et al., 2009; Pádrová et al., 2016; R. Yang et al., 2012; Zhang et al., 2010). On the other hand, the synthesis of polymer-based NMs requires large amounts of organic solvents, which bring possible effects for humans and the environment (Goswami et al., 2017; Mehnert & Mäder, 2012).

6.1.1.4 Carbonaceous nanoparticles

This group includes carbonaceous nanomaterials and nanosheets. Among these, nanotubes with distinct cylindrical structures are roughly classified as single-, double-, or multiwalled CNTs (Dai, 2002; Niyogi et al., 2002), and have been widely used in electronics, optics, and materials science (Bianco & Prato, 2003). Graphene nanofoil is another carbonaceous NM of interest, which is composed of a series of superimposed layers of hexagonal networks of C atoms bound by three neighboring layers in a planar network; graphene has been widely used in anticorrosive electrodes and conductors in electronic engineering (Petersen & Henry, 2012; Vlassiouk et al., 2011). Fullerenes, as one of the major members of this group with extensive sp^2 carbon atoms arranged in fused benzene rings, possess exceptional electrical, chemical, and physical properties (Mauter & Elimelech, 2008), and have been widely applied in electrooptical devices and drug delivery systems (Bosi et al., 2003; Castro et al., 2017, 2018).

6.1.2 Source of ENMs in air

6.1.2.1 Natural source

A variety of natural sources can release NMs into the atmosphere. They include volcanic eruptions, dust storms, and cosmic produced particulates, among others. Tons of ash with temperatures over 1400°C have been released into the atmosphere after hot gas flow during volcanic eruptions. Those ashes have complex structures consisting of solid and liquid particulate matter (Symonds et al., 1988, 2018), which further form nano- and micro-sized particle clusters through chemical reactions or electrostatic forces of attraction as the temperature drops. Deserts are another major permanent source of NMs by air lifting into the atmosphere. The long-range migration of both mineral dust and anthropogenic pollutants on the continents has been the focus of recent related investigations, which have revealed that about 50% of aerosols in the troposphere are minerals originating from deserts (Symonds et al., 1988). Furthermore, cosmic dust is another major permanent source of NMs in the atmosphere. About 40,000 tons of cosmic dust are estimated to reach the Earth's surface every year, with most grains having a mass between 10^{-16} and 10^{-4} kg (Genge, 2008; Genge et al., 2017), from which the major fraction creates the interplanetary dust cloud surrounding the Sun, the planetoids between Mars and Jupiter, and the comets (Zook, 2001, pp. 75–92).

6.1.2.2 Anthropogenic sources

Currently, around 63%–91% of the produced ENMs are disposed in landfills, while 0.4%–7% of total global production is released into water. Air emissions represent the smallest flow of ENMs to the environment, with an estimation of less than 1.5%. However, considering the possible impact of these emissions at the local scale and the concerns about the potential health effects, air emissions of ENMs might develop the major concern with respect to environmental impacts and human health (John et al., 2017).

A variety of ENM manufacturing processes show variable emission potential. The major air emission could derive from the process step of NMs, and also from the subsequent phases such as drying, recovering from the reactor, further handling and transport, diffusion from stacks, during use, and accidental release. As an illustration, Table 6.1 lists a

TABLE 6.1 ENM production processes and their potential emissivity: H = potentially high emissivity during the production phase, L = Low emissivity in normal conditions during the production phase.

Process type	Process	Production conditions	Emissivity
Physical processes	Electrospinning	Closed	H
	Gas phase processes based on heating and fast cooling	Controlled atmosphere	L
	Laser pyrolysis	Controlled atmosphere	L
	Spray drying	Atmospheric conditions	H
Chemical processes	Liquid phase processes: precipitation, sol gel, exfoliation, sonochemistry	ENM produced as liquid suspension; in case of drying, it can be emissive	H
	Chemical vapor deposition	Controlled atmosphere	L
	Flame synthesis	Atmospheric conditions	H
	Hot plasma synthesis	Controlled atmosphere	L
Mechanical processes	High energy milling	Atmospheric conditions	H
	Severe plastic deformation	Atmospheric conditions	H
	Shock wave processes	Atmospheric conditions	H

From John, A. C., Küpper, M., Manders-Groot, A. M. M., Debray, B., Lacome, J. M. & Kuhlbusch, T. A. J. (2017). Emissions and possible environmental implication of engineered nanomaterials (ENMs) in the atmosphere. Atmosphere, *8(5). https://doi.org/10.3390/atmos8050084.*

series of ENM production processes (ISO/TS 80004-8; John et al., 2017) and indicates whether the process is potentially highly emissive (Vignes, 2013). This classification is qualitative and mainly based on expert judgment considering the intrinsic degree of tightness of the reactor, in which the main chemical synthesis reaction occurs. It is considered here that synthesis reaction that requires vacuum or controlled atmosphere is basically less emissive during the synthesis phase than the processes performed under atmospheric conditions. (Engineered Nanomaterials; Geraci et al., 2015; Gridelet et al., 2015; Report & Project; Virji & Stefaniak, 2014). The emission potential is also dependent on the form of the ENMs, i.e., whether it is produced as free powder, suspension in a liquid, or as solid material. In general, powders have a higher emission potential than suspensions and undivided solid forms.

6.2 Chemical, physical, and biological transformations of ENMs in air

Once released to the environment, ENMs can exhibit modifications in their behavior due to various factors. Several environmental processes including aggregation, stability, dissolution, deposition, photo transformation, sulfidation, and macromolecular transformations would involve and significantly affect the fate of ENMs in the air (Goswami et al., 2017).

6.2.1 Photochemical processes: photocatalytic degradation, oxidation, and reduction

Reduction and oxidation are coupled processes in natural systems and involve the transfer of electrons from and to chemical moieties. Sunlight-catalyzed redox reactions (i.e., photooxidation and photoreduction) are very important transformation processes affecting ENM coatings, surface properties, oxidation state, the release of reactive oxygen species (ROS), toxicity, and persistence (Fig. 6.1). Many ENMs including TiO_2 and CNTs might be innately photoactive, potentially producing ROS when exposed to sunlight (Lowry et al., 2012). NMs such as TiO_2, ZnO, and SiO_2 can be used to catalyze photolysis reactions with the generation of ROS through UV or sunlight irradiation (Hoffmann et al., 2007). The generation of ROS is also promoted by the interaction of Zn NMs with living cells (Xia et al., 2006). On the other hand, some ENMS may be oxidized or reduced by sunlight, changing their redox state, charge, and properties, and therefore potential toxicity. In some cases, oxidation may result in the accumulation of a relatively insoluble oxide surface coating on the NMs. When this occurs, the surface is passivated, subsequent oxidation diminishes, and metal-oxide phases with a high capacity for binding ions from solution

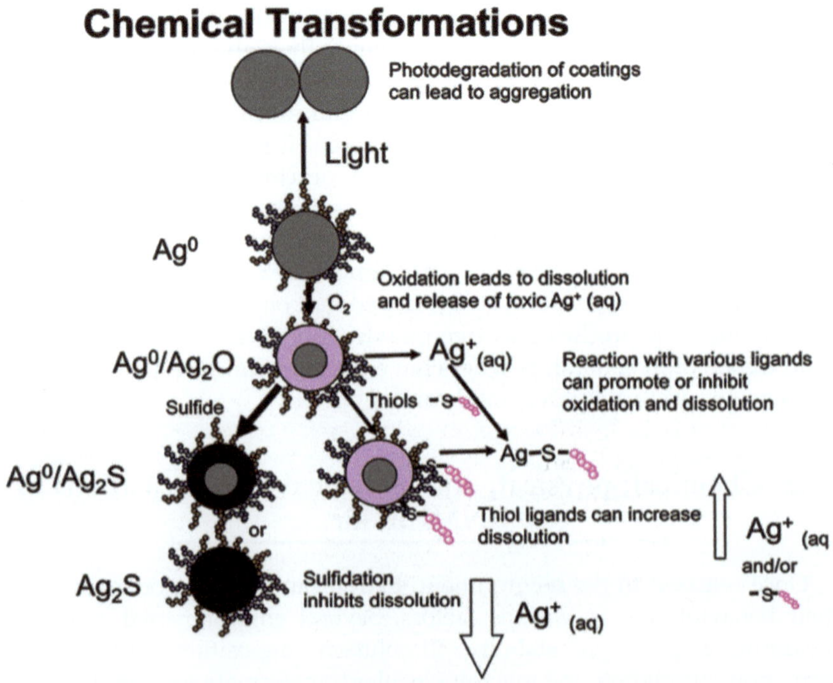

FIGURE 6.1 Representative chemical transformations of metal nanomaterials and the potential impacts on their behavior and effects in the environment. AgNPs are used to exemplify the types of transformations that may occur. *From Lowry, G. V., Gregory, K. B., Apte, S. C. & Lead, J. R. (2012). Transformations of nanomaterials in the environment.* Environmental Science & Technology, *46(13), 6893–6899. https://doi.org/10.1021/es300839e Copyright (2012) American Chemical Society.*

are formed (Lowry et al., 2012). In the case of Ag NMs, oxidation of Ag(0) to Ag(I) is accompanied by the dissolution and release of bactericidal Ag^+ (Lok et al., 2007). Furthermore, ENMs from Class B soft metal cations (-including Ag, Zn, and Cu) have a strong affinity with inorganic and organic sulfide ligands, making them highly reactive with sulfur-containing biomacromolecules and inorganic sulfur in sediments, soils, and air (Lowry et al., 2012). The formation of a relatively insoluble metal-sulfide shell on the NMs surface can modify the surface charge, dissolution, and stability against aggregation.

6.2.2 Interactions with other substances present in the environment: speciation/complexation

Macromolecular interaction is one of the most important transformation mechanisms of NMs in the environment. Adsorption of macromolecules, as well as organic and inorganic ligands on NM surfaces, can significantly affect their surface chemistry and behavior in biological and environmental systems (Fig. 6.2). For example, adsorption of polymer coatings on NMs generally decreases their attachment to silica surfaces, resulting in greater environmental mobility and, potentially, less effective removal during drinking water treatment (Phenrat et al., 2010). Adsorption of organic ligands, metal cations, and/or oxo-anions can occur on either the surface of core NMs or within the organic macromolecular coating of particles. Organic ligands with thiol groups may affect NM dissolution, surface charge, and stability against aggregation (Deonarine et al., 2011; Fauconnier et al., 1997). Organics present in the atmosphere can also condense onto the airborne ENMs altering their surface chemistry and behavior in the air (Lowry et al., 2012). Furthermore, most macromolecules contained in natural organic matter (NOM) are biopolymers and degradable organic products including humic and fulvic acid (Metsämuuronen et al., 2014). These molecules are able to

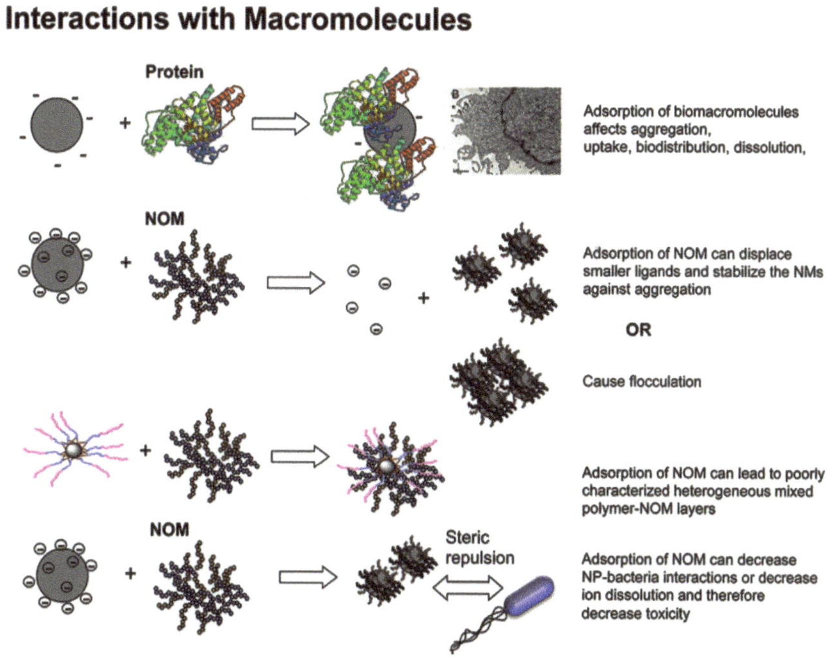

FIGURE 6.2 Effects of nanomaterial interactions with macromolecules such as proteins and natural organic matter. Adsorbed macromolecules can affect aggregation, nanoparticle-biointeractions, biouptake, and fate, transport, and effects in the environment. *From Lowry, G. V., Gregory, K. B., Apte, S. C. & Lead, J. R. (2012). Transformations of nanomaterials in the environment.* Environmental Science & Technology, *46(13), 6893–6899. https://doi.org/10.1021/es300839e Copyright (2012) American Chemical Society.*

substantially modify the properties and behaviors of the associated ENMs (Nel et al., 2009). For example, the –COOH and –OH groups in NOMs may interact with ENMs (e.g., TiO_2) to form insoluble complexes via electrostatic and steric forces, and further change the surface charge, solubility, and stability of ENMs (Chen & Elimelech, 2006; King & Jarvie, 2012; Sedlmair et al., 2012; Van Hoecke et al., 2011; Yang & Xing, 2009).

6.2.3 Biological processes: transformation mediated by living organisms

Biological transformations of ENMs are inevitable when they interact with living tissues and environmental media (e.g., soil, water, and air) (Fig. 6.3). Redox reactions are fundamental to growth processes in all biological systems. These reactions take place in the cytoplasm, cell wall, cell membrane, and extracellularly via redox-labile enzymes and cytochromes, or through ancillary intracellular ROS production such as hydroxyl radicals and H_2O_2. For example, the redox reactions between bacteria and NMs, such as nanoscale iron oxide, are well understood (Lowry et al., 2012; Richter et al., 2012). It was recently found that *Geobacter* and *Shewanella* spp. are able to produce nanoscale silver particles by

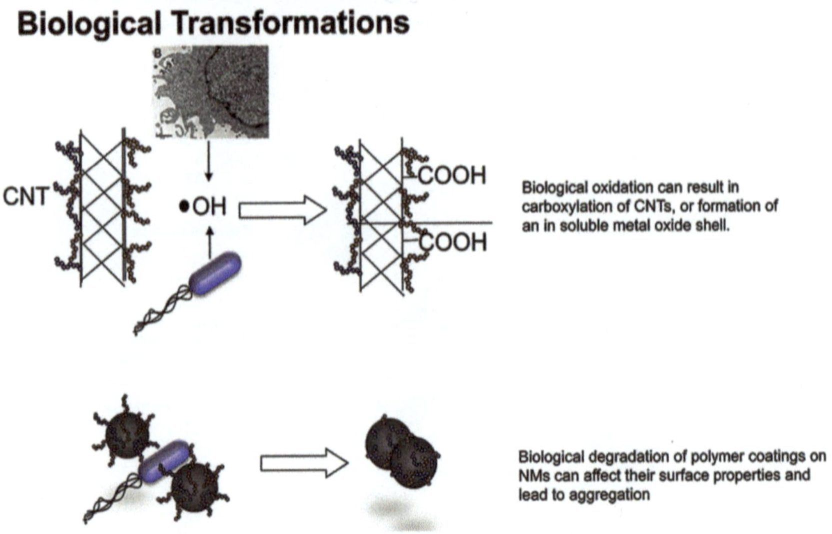

FIGURE 6.3 Biologically mediated transformations of nanomaterials and their coatings, and the subsequent impact on fate, transport, and effects. *From Lowry, G. V., Gregory, K. B., Apte, S. C. & Lead, J. R. (2012). Transformations of nanomaterials in the environment.* Environmental Science & Technology, *46(13), 6893–6899. https://doi.org/10.1021/es300839e Copyright (2012) American Chemical Society.*

the reduction of Ag^+ from solution (Law et al., 2008). Biologically mediated transformations of both underlying NM core and outer coatings are feasible, and those transformations can affect the behavior of the ENMs including surface charge, aggregation state, and reactivity, which can further affect the transport, bioavailability, and toxicity of ENMs. The oxidation and carboxylation of CNTs by OH radicals produced from the horseradish peroxidase enzyme have been well demonstrated (Allen et al., 2008). This oxidation increases the surface charge of CNTs and the stability against aggregation, but decreases hydrophobicity; this biological oxidation and surface functionalization may also affect the toxicity of CNTs (Kang et al., 2008). Furthermore, biotransformation of polymer coatings has been applied on many ENMs in biomedical applications. Covalently bound poly(ethylene glycol) (PEG) coatings on ENMs were demonstrated to be bioavailable to microorganisms isolated from an urban stream (Kirschling et al., 2011), while the PEG coating caused the ENMs to aggregate. Biological transformations of ENMs, especially carbon-based ones and their organic coatings may eventually cause the aggregations and reduce their concentrations in the environment, further affecting their fate and transport. However, it is still under investigation whether those processes occur at certain rates, which are high enough to be important (Lowry et al., 2012). Additionally, there is an infinite number of biomacromolecules in living cells and in the environment. The presence of biomacromolecules can affect the dissolution rate, bioaccumulation, ROS production, toxicity, and may further determine the fate of NMs in the organism, for example excretion or accumulation in specific organs. In living cells, adsorption of a wide range of proteins on NMs, also known as protein coronas (Cedervall et al., 2007; Lundqvist et al., 2008), are dynamic in nature. The proteins and other macromolecules adsorbed on NMs partially determine the properties (e.g., size, electrophoretic mobility, and surface composition), as well as the subsequent environmental behavior and biological response of NMs. The protein coronas may also offer a way for the entry of ENMs into cells through the process of receptor-mediated endocytosis (Lynch & Dawson, 2008).

6.2.4 Physical processes: aggregation, sedimentation, and dissolution

Over time, aggregation of NMs into clusters is almost unavoidable, unless they contain engineered or incidental coatings that inhibit aggregation (Fig. 6.4). This process in the natural environment generally proceeds via two major pathways, i.e., homoaggregation (aggregation of similar particles) and heteroaggregation (aggregation of dissimilar particles). Both processes can immobilize ENMs in either soil or airborne particles, while in most of the cases the much greater concentration of

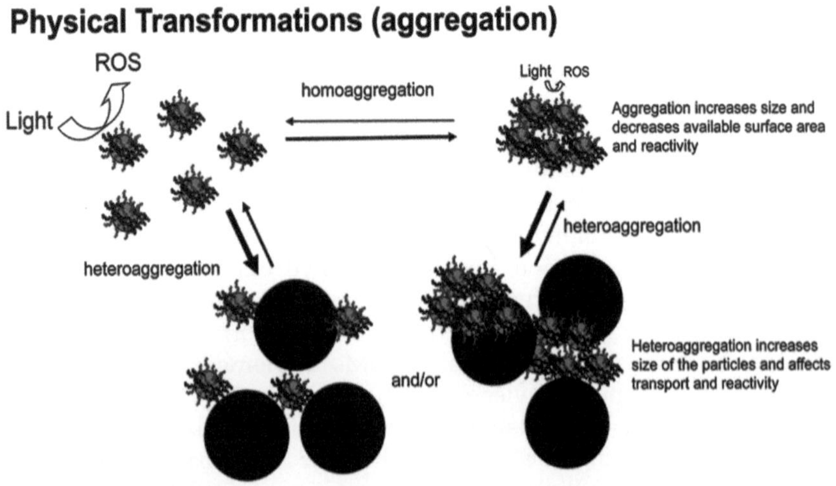

FIGURE 6.4 Effects of physical transformations including aggregation and heteroaggregation on the reactivity and transport of nanomaterials. *From Lowry, G. V., Gregory, K. B., Apte, S. C. & Lead, J. R. (2012). Transformations of nanomaterials in the environment.* Environmental Science & Technology, *46(13), 6893–6899. https://doi.org/10.1021/es300839e Copyright (2012) American Chemical Society.*

environmental particles results in heteroaggregation (Goswami et al., 2017). The phenomenon of ENM aggregation in natural systems could be explained by the intermolecular and electrostatic forces (Liang, Bradford, Simunek, Heggen, et al., 2013; Liang, Bradford, Simunek, Vereecken, & Klumpp, 2013). In addition, the combined effects of several other factors including pH, spatial configuration, ionic strength, and ionic balance may also greatly influence aggregation and distribution patterns of ENMs (shown in Table 6.2). Aggregation of NMs reduces the specific surface area and increases the aggregate size, which in turn affects their transport in porous media, sedimentation, reactivity, uptake by organisms, and toxicity. For example, 30–70 nm in diameter Fe(0) NMs rapidly aggregate in water to form microsized aggregates (Phenrat et al., 2007) that greatly decrease their mobility in the subsurface and the pathways of exposure to sensitive receptors. Heteroaggregation between NMs and comparatively larger particles (e.g., clay) may change NM behavior, and the NM–clay heteroaggregates eventually move like a clay particle than the associated NMs (Hotze et al., 2010). Aggregation may serve to increase the persistence of the NMs if it decreases the rate of dissolution or degradation. The size of a NM may also affect its bioavailability to organisms when heteroaggregates become too large for direct transport across the cell wall and membrane. Conversely, heteroaggregation with soft biogenic particles might increase NM

TABLE 6.2 The effects of different factors on the aggregation of ENMs.

Types of NMs	Factors	Impacts
Ag-PVP	Surface charge, Al content	Homo- and heteroaggregation of negatively charged Ag NPs and retention in soil matrix
CuO	SOM, hydraulic conductivity	Reduction of pore size; reduced hydraulic conductivity
Fullerene	Clay content; NOM; ionic strength	Increased ionic strength and replacement of NaCl and $CaCl_2$ helps in deposition
Fullerene	Humic acid, hydrophobic attraction	Forms complexes due to strong hydrophobic nature
SiO_2-PEG	Pore size, hydraulic conductivity	Reduction in swelling index of montmorillonite; change in wettability of natural surface; kaolinite and or metal oxide impurities in sand
TiO_2	pH; humic acid	Forms ligands with humic acids especially with COOH groups; changes rock wettability

From Goswami, L., Kim, K. H., Deep, A., Das, P., Bhattacharya, S. S., Kumar, S. & Adelodun, A. A. (2017). Engineered nano particles: Nature, behavior, and effect on the environment. Journal of Environmental Management*, 196, 297–315. https://doi.org/10.1016/j.jenvman.2017.01.011*

bioavailability such as the uptake by filter feeders that preferentially remove larger particles. The aggregation kinetics of ENMs can also be affected by the spatial distribution of functional groups despite differences in surface oxidation (Dwivedi et al., 2015). For example, the aggregation of polyvinyl pyrrolidone (PVP) coated Ag NMs was inhibited by steric hindrance from a brushlike coating on the outer surface. Moreover, pH regulates the transformation of ENMs into structurally and morphologically heterogeneous products. For example, graphene oxide (GO) NMs in both natural and synthetic water had a high hydrodynamic diameter and electrophoretic mobility within the pH range of 2–4 (Chowdhury et al., 2013; Goswami et al., 2017). Also, the enhanced aggregation of iron oxide NMs at pH values close to the point of zero charge (PZC) has been observed (Baalousha, 2009).

6.3 Interaction of airborne ENMs with plant surface

Plants represent the major interface between the ecological system and environment, and they provide a large surface area for ENMs exposure through aboveground biomass such as leaves (Dietz & Herth, 2011; Singh et al., 2017). A portion of ENMs released into the environment (either natural atmospheric or anthropogenic source) can reach plants via wind dispersion (Uzu et al., 2009). These ENMs can be absorbed through the leaves (Corredor et al., 2009; Larue, Castillo-Michel, Sobanska, Cécillon, et al., 2014), and interact with plants as well (Hong et al., 2014; Xiong et al., 2019). Since plants may act as a potential pathway for ENMs transport and a route for bioaccumulation into the food chain, it is essential to understand the interactions between ENMs and plants, such as uptake, accumulation, translocation, and biotransformation (Rico et al., 2011; Xiong et al., 2021). Additionally, a fundamental understanding of the interactions between ENMs and plants is critical for developing nano-enabled agriculture (Rai et al., 2018). In recent years, the scientific community has paid increasing attention to study the interactions between airborne ENMs and plant leaves (Khan et al., 2019). To better understand such interaction, several key questions need to be addressed: (1) Which plant part is the preferred way of entrance? (2) What properties of ENMs affect their interaction with the plant? (3) How do airborne ENMs interact with plant surface? In this section, the aforementioned questions are explained and discussed.

6.3.1 Uptake, translocation, and biotransformation of ENMs in plant leaves

6.3.1.1 Uptake pathway

The exact mode in which foliar applied ENMs penetrate the leaves is not yet clearly understood. There are two possible entrance routes, stomata and cuticle, as shown in Fig. 6.5 (Avellan et al., 2019). The stomatal pathway likely occurs via a passive loading, while the cuticular entrance might take place through small hydrophilic pores or cuticle diffusion, direct disruption, or both.

Stomatal pathway is the most investigated route. Stomatal uptake of ENMs mainly depends on the stomata size (ranging from 10 to 100 μm), density on plant leaves, and aperture cycles (Raliya et al., 2016). Eichert et al. investigated the size exclusion limits of the stomatal foliar uptake pathway and found that only smaller particles occasionally entered the interior of the leaf through stomatal pores after 2–9 days, whereas penetration of larger particles was never detected (Eichert et al., 2008). In another study, aerosolized NPs were directly applied to leaf surfaces and only a certain fraction of NPs (<100 nm) entered the leaf following the stomatal pathway (Wang et al., 2013).

Cuticle is a waxy and porous layer and has small hydrophilic cuticular pores in the nanometer range (Wang et al., 2016). The primary function of the cuticle is to control transpiration, prevent water loss, and control

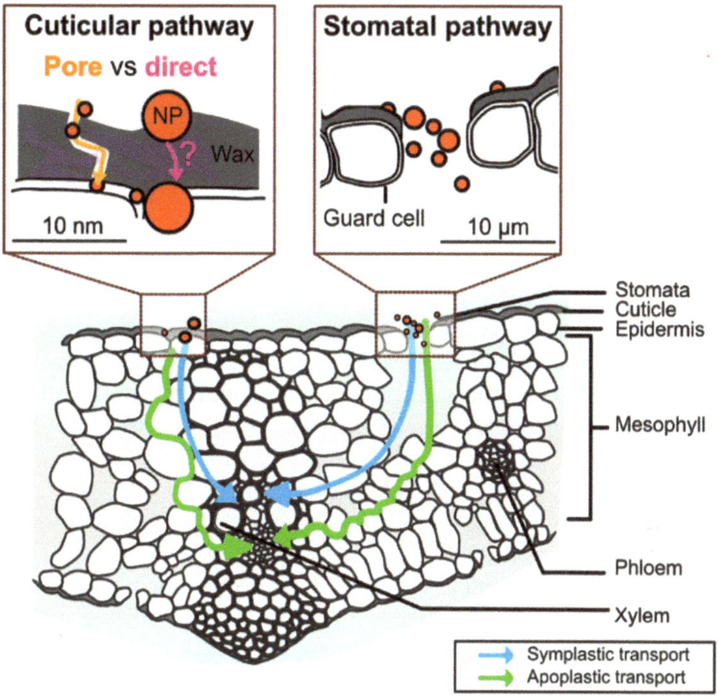

FIGURE 6.5 Schematic showing a wheat leaf cross-section along with the routes for ENMs uptake following foliar exposure. *From Avellan, A., Yun, J., Zhang, Y., Spielman-Sun, E., Unrine, J. M., Thieme, J., Li, J., Lombi, E., Bland, G. & Lowry, G. V. (2019). Nanoparticle size and coating chemistry control foliar uptake pathways, translocation, and leaf-to-rhizosphere transport in wheat.* ACS Nano, 13*(5), 5291–5305. https://doi.org/10.1021/acsnano.8b09781.*

interactions with harmful or beneficial environmental factors. The cuticular pathway may take place through small hydrophilic pores, and ENMs larger than the cuticular pores can also penetrate leaves by diffusion and/or direct disruption (Avellan et al., 2019). It should be noted that ENMs can be trapped by the cuticular wax and then diffuse in the leaf, since the wax lipids may quickly adsorb on the large surface of ENMs (Birbaum et al., 2010; Schreck et al., 2012). Compared to stomata, cuticle has significantly larger area on the leaves, which makes cuticular uptake potentially a more effective pathway for ENMs uptake than stomatal pathway.

6.3.1.2 Translocation pathway

After entering leaves, ENMs can move upward and downward through xylem or phloem tissues (Hauck et al., 2008; Pullagurala et al., 2018). ENMs can localize in leaf cells in the epidermis (e.g., guard cells), extracellular space, or organelles (e.g., chloroplasts) (Hu et al., 2020). As shown in Fig. 6.5, ENMs cross the leaf epidermal barrier either via stomatal and/or cuticular pathways, then translocate through two possible pathways: (1) apoplastic pathway through the extracellular space and in between cell walls, and (2) symplastic pathway through the leaf mesophyll cells (i.e., translocate between cells through the cytosol).

6.3.1.3 Biotransformation of ENMs in plants

The uptake of Ag NPs by foliar exposure and their possible biotransformation were studied (Larue, Castillo-Michel, Sobanska, Cécillon, et al., 2014). The authors found that Ag-NPs can be internalized in the leaf tissue and partially or totally oxidized. Furthermore, bioaccumulated Ag-NPs underwent dissolution and Ag^+ was bounded to either organic molecules such as thiol-containing molecules (complexation) or Cl ions (AgCl) to form secondary species.

6.3.2 Key properties of ENMs that affect their interaction with plants

The physical and chemical properties of ENMs such as size, surface charge, and surface coating play a critical role in determining the uptake, translocation, and biotransformation (Chen, 2018). In this section, different properties of ENMs and their influence on the interactions with plants are discussed.

6.3.2.1 Size

The plant leaf surface is formed by a cuticle containing nanoscale hydrophilic pores and micron-scale stomatal pores. ENMs with a size larger than the stomata usually cannot enter leaves through the stomata pathway. For the cuticular entrance, there are two possible mechanisms. ENMs can go through small hydrophilic cuticular pores within the cuticles, and these pores usually have a size-exclusion limit in the nanometer range (Wang et al., 2016). Furthermore, ENMs larger than the cuticular pores can also cross the cuticle and penetrate into leaves via diffusion and/or direct disruption. One study reported that 4–100 nm NPs were able to cross the cuticle by disruption of the waxy layer (Larue, Castillo-Michel, Sobanska, Trcera, et al., 2014). Another study reported that fluorescent accumulation from tagged NPs (>50 nm) was observed in the epidermis layer underneath the cuticle in areas without stomata (Nadiminti et al., 2013).

The pathways for cellular translocation (apoplastic vs. symplastic; Fig. 6.5) may also be affected by the size and/or the shape of ENMs (Raliya et al., 2016). Therefore, the size of ENMs is an important property that could affect foliar uptake and translocation in the plant (Etxeberria et al., 2016).

6.3.2.2 *Surface coating*

Foliar uptake of ENMs is limited by a number of barriers, being the first one the leaf cuticle. ENMs and their solutes are usually taken up by plant leaves through two pathways: aqueous pores of the cuticle and stomata for hydrophilic compounds, and diffusion through the cuticle for lipophilic ones (Larue, Castillo-Michel, Sobanska, Cécillon, et al., 2014). Different leaf features, for example stomata, trichomes, and cuticle, present a wide range of surface wettability, with stomatal guard cells being most hydrophilic, and leaf cuticle being the most hydrophobic (Popp et al., 2005).

ENMs are usually surface coated with different materials, which results in different surface wettability. The chemical properties of coatings play a key role in the interaction of ENMs with plant leaves, their entry path, and the subsequent impacts. In one study, ZnO-NPs suspensions with coatings of different hydrophobicity were applied to wheat and sunflower leaves (Read et al., 2020), and the surface coating did not affect Zn absorption but enhanced particle adhesion to leaves. In another study, Au NPs with different surface coatings (citrate vs. polyvinylpyrrolidone, PVP) were foliarly applied to wheat (Avellan et al., 2019). The relatively more hydrophobic PVP coating enhanced the interaction of Au NPs with the leaf surface. The surface coatings also reduced the ability of Au NPs to cross the cuticle, enter the mesophyll, and move through the plant vasculature.

6.3.2.3 *Surface charge*

Surface charge (ζ potential) also plays an important role in ENMs translocation across the plant plasma membrane and organelle lipid bilayers. Positively or negatively charged ENMs with ζ potential magnitudes higher than 20 or 30 mV were uptaken by plant cell or chloroplast membranes, respectively, whereas more neutral ENMs did not penetrate plant lipid bilayers (Lew et al., 2018; Wong et al., 2016). Moreover, a larger magnitude of ζ potential is essential for translocation across lipid membranes when the size of ENMs decreases. Hu et al. studied the effects of ζ potential and hydrodynamic size of hydrophilic nanoparticles on delivery efficiency and pathways to specific leaf cells and organelles (Hu et al., 2020). The results showed that hydrophilic nanoparticles (<20 nm for cotton and 11 nm for maize), and positively charged (>15 mV), exhibited the highest foliar delivery efficiencies into guard cells (100%), extracellular space (90.3%), and chloroplasts (55.8%).

6.3.2.4 *Surface chemistry*

The adhesion of ENMs to the leaf surface strongly depended on the functional groups of ENMs' surface. ENMs capped with various functional groups usually interacted with leaf surface through hydrogen bond, electrostatic attraction, covalent bond, and/or electrostatic repulsion. To increase the adhesion to leaf, Yu et al. developed three types of functionalized abamectin poly(lactic acid) (Abam-PLA) NPs (CH_3CO-PLA-NS, HOOC-PLA- NS, and H_2N-PLA-NS) as the nanocarrier of pesticides with different adhesion abilities to cucumber leaf (Yu et al., 2017). The Abam-PLA nanoparticles exhibited favorable deposition on the surface of the leaf, while the adhesion to the leaf surface followed the order: H_2N-PLA-NS > CH_3CO-PLA-NS > HOOC-PLA-NS.

The surface chemistry of ENMs may also affect their translocation in leaves. After crossing the cuticle, ENMs encounter epidermis and mesophyll tissues (Fig. 6.5). The surface chemistry interact with these tissues and structures that affect ENMs translocation to the leaf vasculature. Furthermore, the internal leaf cells might resent a wide range of functional groups that could interact with the ENMs surfaces (Lew et al., 2018).

6.3.3 Key properties of plant surface that affect their interaction with ENMs

The surface properties of plant leaf such as wax content, surface wettability, stomatal density, and type of trichomes, may significantly affect the interaction with ENMs (Li et al., 2019; Winkler & Zotz, 2010). For example, the interactions between ENMs and plant leaves were reported to be inevitably affected by cuticular wax (Hu et al., 2017). In addition, plant leaves with high wax content such as citrus, generally uptake less amounts of ENMs through the leaf surface than species with a less wax content (e.g., tomato leaves) (Du et al., 2015). Kranjc et al. found that plants with low surface free energy (SFE) were easier to adhere and retain ENMs than plants with high SFE (Kranjc et al., 2018). Li et al. found that trichomes played a critical role in Zn absorption in sunflower but not in soybean and tomato (Li et al., 2018), while Winkler et al. revealed a correlation between the foliar uptake of rubidium (Rb) and trichome density in epiphytic bromeliads (Winkler and Zotz, 2010).

6.3.4 Bioavailability of airborne ENMs to plant surface

Bioavailability of airborne ENMs to plants occurs in a restricted manner because the supply channels are available only at the surfaces of the leaves. The probability of penetration depends on several factors such as size/shape, surface charge, surface coating, surface chemistry, and their concentration in air (Judy et al., 2012). It is also important to note that the chemical, physical, and biological transformations of airborne ENMs in the air can possibly change their physical and chemical properties, thus further affecting the bioavailability of ENMs to plant leaf surface. However, most of current studies are focused on the interaction of pristine ENMs with plant leaves, rather than transformed ENMs. Studies on the bioavailability and uptake of ENMs transformed in the air are urgently needed.

6.4 Impact of airborne ENMs on plant

6.4.1 Negative impact: atmospheric deposition of ENMs on plant surface

Atmospheric deposition and uptake of contaminative ENMs (metal-enriched) by plants, particularly vegetables and crops grown near industrial or urban areas, are the major contributor to the pollutant load of plants. Such contaminative ENMs may not only cause phytotoxicity, but also pose potential risks for human exposure through the ingestion of contaminated vegetables and crops (Natasha et al., 2019).

6.4.1.1 Phytotoxicity of atmospheric deposited contaminative ENMs

Atmospheric ENMs are generally harmful for plant growth since they may cause alterations in transpiration rates, thermal balance, photosynthesis, and metabolism (Wang et al., 2013). For example, arsenic (As) NPs can be absorbed via foliar pathways (0.50–0.73 mg/kg in leaves) by spinach (*Spinacia oleracea*) and translocated toward root tissues (0.35–0.68 mg/kg), which results in a decrease in plant dry biomass (up to 84%) and pigment content (up to 38%) (Natasha et al., 2019). Ceria (Ce) NPs foliar applied to cucumber seedling leaves (3.1, 6.2, and 12.5 mg per plant) decreased the net photosynthesis rate and transpiration rate by 22% and 11%, respectively (Hong et al., 2016). After foliar exposure of TiO_2 NPs on grapevine (*Vitis vinifera* L.), the net CO_2 assimilation significantly decreased and stomatal conductance increased, indicating metabolic (nonstomatal) inhibition of the photosynthesis (Teszlák et al., 2018). High foliar Cu uptake occurred after exposure of CuO-NPs for 15 days to leafy edible vegetables (i.e., lettuce and cabbage), resulting in the reduction of plant weight, net photosynthesis level, and water content, along with necrotic areas near deformed stomata containing CuO-NPs (Xiong et al., 2017).

The phytotoxicity of ENMs is usually dose-dependent. For example, low concentrations of CeO_2 NPs had positive effect on marigold, but higher concentrations resulted in oxidative stress (Jahani et al., 2019). Foliar application of Ce NPs can alter the stomatal density and length, and photosynthesis of bean seedlings at relatively high doses, which induce membrane damage (Salehi et al., 2018). Ce NPs and Ce^{3+} ions were repeatedly applied to common bean plants by foliar pathway, and both treatments triggered dose-dependent oxidative damage in the leaves, while Ce^{3+} ions were significantly more toxic than equimolar Ce NPs (Xie et al., 2019).

6.4.1.2 Food chain contamination and possible human exposure

It has been reported that atmospheric deposition at industrial locations contributes more to plant foliar Pb concentrations than soil pollution (Schreck et al., 2014). Concentrations of Pb, Cd, Cu, and Zn in vegetables sold at the roadside market were elevated, possibly because of atmospheric deposition of industrial or automobile exhausts (Al Jassir et al., 2005). In addition, a substantial airborne contribution to the Pb concentration in *Aster subulatus* plant leaves was reported (Hu et al., 2011).

After foliar uptake, accumulation and translocation of ENMs occur within the plant. For example, foliar application of PbO NPs leads to a significant accumulation of Pb in spinach leaves (42.25 μg/g), with limited translocation toward root tissues (4.46 μg/g) (Natasha et al., 2020). Higher concentrations of Ce and Cu were detected in the cucumber seedling roots after foliar exposure (Hong et al., 2016). Ce was detected in all tissues of the CeO_2 NPs treated plants, indicating uptake through the leaves and translocation to the other plant parts including stems, roots, and flowers (Hong et al., 2014). After foliar exposure of various concentrations of Fe_2O_3 (0, 60, and 180 μg per plant) on wheat seedlings for 21 days, the concentrations of Fe in leaves, stems, and roots were 1100, 280, and 160 mg/kg, respectively (Lu et al., 2020). In another study, lettuce plants were exposed to pristine TiO_2-NPs and aged paint leachate containing TiO_2-NPs and microparticles (TiO_2-MPs). All the tested materials (NPs and MPs) were internalized in lettuce leaves, and detected in all types of tissues (Larue, Castillo-Michel, Sobanska, Trcera, et al., 2014). It is highly possible that such ENMs in air can be eventually transferred to the food chain, posing risks to food safety and human health due to the ingestion of metal-enriched vegetables and crops. Therefore, atmospheric levels of metal-enriched ENMs need to be monitored to avoid their food chain contamination and possible human exposure.

6.4.2 Positive impact: agrichemical delivery through foliar

Nanotechnology in agriculture has been widely studied in recent years and has become a hot research subject (Kah et al., 2018, 2019; White & Gardea-Torresdey, 2018). ENMs containing macronutrients and micronutrients can be used as nano-enabled agrichemicals, and foliar delivery of such nano products provides a promising approach to enhance nutrient uptake and reduce environmental impacts of agrichemicals compared to those conventionally applied through the soil route.

6.4.2.1 *Enhancing plant development*

Foliar-applied ENMs are more convenient and efficient for plant response at the field scale (Fernández & Eichert, 2009). The application of ENMs containing microelements for foliar uptake can enhance plant growth and fruit/grains development. ZnO NPs were foliar applied to different plants and showed beneficial effects. For example, ZnO-NPs treatment induced generally better sunflower physiological responses under field conditions (Kolenčík et al., 2020). The grains of ZnO NP-treated plant (foxtail millet) had significantly higher oil and total nitrogen content and significantly lower crop water stress index. (Marek et al., 2019). Compared to the conventional Zn salts, ZnO NPs exhibited more positive impact on coffee growth and physiology, possibly because of their increased ability to be uptaken by the leaf (Rossi et al., 2019). In another study, foliar spraying with ZnO NPs and SiO_2 NPs enhanced the chlorophyll content, chlorophyll fluorescence parameters, and the fresh/dry weights of leaves and roots of cucumber (Li et al., 2021).

Foliar uptake of other ENMs such as Cu, TiO_2, Fe_3O_4 NPs was reported to improve the plant development. For example, Cu NPs induced the production of fruits with greater firmness, vitamin C, lycopene, and the ABTS antioxidant capacity (Lopez-Vargas et al., 2018). Proper doses of MoS_2 and TiO_2 NPs boosted plant production by increasing plant photosynthetic rate without inducing excessive stress or toxicity (Song et al., 2020). Foliar applied Fe_3O_4 NPs significantly increased the fresh and dry weights of shoot and root compared to the control (Rostamizadeh et al., 2021).

Interestingly, not only metal oxide NPs, but also biopolymer-based NPs were used to treat plants. For example, foliar application of chitosan (Ch) NPs significantly increased the growth, yield, and mineral content when compared to the chitosan and untreated control (Sathiyabama & Manikandan, 2021). Chitosan NPs also induced several defense-related enzymes in leaves of finger millet plants by increasing the innate immune response.

6.4.2.2 *Alleviating the adverse effects of abiotic stresses*

Foliar uptake of ENMs was reported to mitigate the adverse effects of abiotic stresses such as toxic trace elements, soil salinity, drought, and chilling. The contamination of large areas of arable land with toxic trace elements such as cadmium (Cd) and Pb is a severe concern worldwide, and environmentally feasible amendments are necessary to minimize the accumulation of toxic trace elements in plants. Foliar delivery of ENMs is reported to alleviate the adverse impacts of toxic trace elements on different plants. For example, foliar spray of ZnO NPs significantly reduced the Cd content in rice or maize shoot and roots (Ali et al., 2019;

Rizwan et al., 2019), as well as wheat grain (Adrees et al., 2021). Foliar application of TiO_2 NPs decreased shoot Cd content from 15.2% to 17.8% and showed a greater effect on mitigating Cd-induced toxicity by increasing superoxide dismutase and glutathione S-transferase activities and upregulating several metabolic pathways (Lian et al., 2020). Foliar uptake of nano-Fe_3O_4 alleviated Cd toxicity by reducing Cd accumulation and increasing nutrient intake (Zadeh et al., 2019). In another study, the foliar-applied Si NPs mitigated the adverse impacts of Pb on coriander plants, resulting in the minimization of Pb concentration in plants and enhancements in the plant defense system (Fatemi et al., 2021). These studies provided a promising strategy for the application of ENMs in crop safety production in toxic trace element contaminated soils.

Water shortage and soil salinity are major challenges for sustaining global food security. Applying nutrients in the nano form is a novel and promising strategy for crops to alleviate the adverse effects of such stresses. The foliar uptake of ZnO NPs, under drought stress, improved growth characteristics and increased fruit yield of eggplant by up to 22.6%, compared to fully irrigated plants and nonapplied ZnO NPs (Semida et al., 2021). Foliar application of fullerenol nanoparticles (FNPs) can modify the intracellular water metabolism in drought-exposed plants, and FNPs acted as intracellular binders of water, which created an additional water reserve, enabling adaptation to drought stress (Borisev et al., 2016). Under soil salinity stress, foliar application of Se NPs enhanced salinity tolerance in strawberries improving its growth and yield, due to the ability to protect photosynthetic pigments (Zahedi et al., 2019).

Chilling is one of the major abiotic stresses in marginal tropical regions where its incidence occurs with seasonal changes, negatively affecting crop yield. ENMs such as SiO_2, ZnO, Se, and graphene were foliar applied on sugarcane leaves as a mitigation strategy against chilling stress. It was reported that the NP treatments reduced the negative effects of chilling by regulating photosynthesis (Elsheery et al., 2020). Among the NPs, SiO_2 demonstrated higher amelioration effects, and they can be utilized alone or in combination with other NPs to mitigate chilling stress in sugarcane.

6.4.2.3 Mitigating the adverse effects of biotic stresses

The development of ENMs for disease management is essential for agriculture and food security. Foliar application of ENMs has been recently employed as a new strategy to protect plants from pathogens (e.g., bacteria and fungus) for mitigating the adverse effects of biotic stresses. For example, two forms of nanoscale Cu (CuO NPs and $Cu_3(PO_4)_2 \cdot 3H_2O$ nanosheets) were foliar applied to tomatoes in the presence of the pathogen *Fusarium oxysporum* f. sp. *lycopersici* (Shen et al., 2020). In the greenhouse, weekly foliar applications of the nanosheets and

NPs increased seedling biomass by 90.9% and 93.3%, respectively, compared to diseased and ionic Cu controls. In the field, $Cu_3(PO_4)_2 \cdot 3H_2O$ nanosheets decreased disease progress by 26.0% and significantly improved fruit yield by over 45.5% per plant, relative to the other treatments in infested soil. Anusuya et al. found that foliar application of beta-D-glucan NPs (0.1%, w/v) protected turmeric plants against rot disease by reinforcing the plant's defense mechanism, which provoked a marked increase in the activity of defense enzymes at different age levels (Anusuya & Sathiyabama, 2015).

6.5 Concluding remarks

In this chapter, the potential transformations of ENMs in the air were discussed, including physical, chemical, and biological reactions of ENMs with atmospheric components such as humidity, vapors, gases, colloids, and light, among others. Subsequently, the interactions of airborne ENMs with plant surface including the uptake pathway, translocation pathway, and biotransformation of ENMs in plants were explained. Finally, the negative and positive impacts of airborne ENMs on the plants were discussed. Note that the research field of ENMs in the air has a large gap regarding the phytotoxicity and fate of ENMs after leaf exposure. The uptake, bioaccumulation, biotransformation, and risks of ENMs for food crops are not well understood. Thus far, very few ENMs and plant species have been studied, and mainly at the early growth stages of the plants. A limited number of studies depicted the biotransformation of ENMs in vegetable and other crops, and the possible transmission of the ENMs to the next generation of plants exposed to ENMs is unknown (Rico et al., 2011). In order to move forward, it is essential to systematically assess the ENM–plant interactions and develop models that predict the translocation, distribution, and biotransformation of ENMs in plants based on their chemical and physical properties.

References

Abbott, L. C., & Maynard, A. D. (2010). Exposure assessment approaches for engineered nanomaterials. *Risk Analysis, 30*(11), 1634–1644. https://doi.org/10.1111/j.1539-6924.2010.01446.x

Adrees, M., Khan, Z. S., Hafeez, M., Rizwan, M., Hussain, K., Asrar, M., Alyemeni, M. N., Wijaya, L., & Ali, S. (2021). Foliar exposure of zinc oxide nanoparticles improved the growth of wheat (*Triticum aestivum* L.) and decreased cadmium concentration in grains under simultaneous Cd and water deficient stress. *Ecotoxicology and Environmental Safety, 208*. https://doi.org/10.1016/j.ecoenv.2020.111627

Al Jassir, M. S., Shaker, A., & Khaliq, M. A. (2005). Deposition of heavy metals on green leafy vegetables sold on roadsides of Riyadh City, Saudi Arabia. *Bulletin of Environmental*

Contamination and Toxicology, 75(5), 1020–1027. https://doi.org/10.1007/s00128-005-0851-4

Ali, S., Rizwan, M., Noureen, S., Anwar, S., Ali, B., Naveed, M., Abd_Allah, E. F., Alqarawi, A. A., & Ahmad, P. (2019). Combined use of biochar and zinc oxide nanoparticle foliar spray improved the plant growth and decreased the cadmium accumulation in rice (*Oryza sativa* L.) plant. *Environmental Science and Pollution Research, 26*(11), 11288–11299. https://doi.org/10.1007/s11356-019-04554-y

Allen, B. L., Kichambare, P. D., Gou, P., Vlasova, I. I., Kapralov, A. A., Konduru, N., Kagan, V. E., & Star, A. (2008). Biodegradation of single-walled carbon nanotubes through enzymatic catalysis. *Nano Letters, 8*(11), 3899–3903. https://doi.org/10.1021/nl802315h

Anusuya, S., & Sathiyabama, M. (2015). Foliar application of β-d-glucan nanoparticles to control rhizome rot disease of turmeric. *International Journal of Biological Macromolecules, 72*, 1205–1212. https://doi.org/10.1016/j.ijbiomac.2014.10.043

Auffan, M., Rose, J., Bottero, J. Y., Lowry, G. V., Jolivet, J. P., & Wiesner, M. R. (2009). Towards a definition of inorganic nanoparticles from an environmental, health and safety perspective. *Nature Nanotechnology, 4*(10), 634–641. https://doi.org/10.1038/nnano.2009.242

Avellan, A., Yun, J., Zhang, Y., Spielman-Sun, E., Unrine, J. M., Thieme, J., Li, J., Lombi, E., Bland, G., & Lowry, G. V. (2019). Nanoparticle size and coating chemistry control foliar uptake pathways, translocation, and leaf-to-rhizosphere transport in wheat. *ACS Nano, 13*(5), 5291–5305. https://doi.org/10.1021/acsnano.8b09781

Baalousha, M. (2009). Aggregation and disaggregation of iron oxide nanoparticles: Influence of particle concentration, pH and natural organic matter. *Science of the Total Environment, 407*(6), 2093–2101. https://doi.org/10.1016/j.scitotenv.2008.11.022

Beaurepaire, E., Buissette, V., Sauviat, M. P., Giaume, D., Lahlil, K., Mercuri, A., Casanova, D., Huignard, A., Martin, J. L., Gacoin, T., Boilot, J. P., & Alexandrou, A. (2004). Functionalized fluorescent oxide nanoparticles: Artificial toxins for sodium channel targeting and imaging at the single-molecule level. *Nano Letters, 4*(11), 2079–2083. https://doi.org/10.1021/nl049105g

Bianco, A., & Prato, M. (2003). Can carbon nanotubes be considered useful tools for biological applications? *Advanced Materials, 15*(20), 1765–1768. https://doi.org/10.1002/adma.200301646

Birbaum, K., Brogioli, R., Schellenberg, M., Martinoia, E., Stark, W. J., Günther, D., & Limbach, L. K. (2010). No evidence for cerium dioxide nanoparticle translocation in maize plants. *Environmental Science and Technology, 44*(22), 8718–8723. https://doi.org/10.1021/es101685f

Borisev, M., Borisev, I., Zupunski, M., Arsenov, D., Pajevic, S., Curcic, Z., Vasin, J., & Djordjevic, A. (2016). Drought impact is alleviated in sugar beets (beta vulgaris L.) by foliar application of fullerenol nanoparticles. *PLoS One, 11*.

Bosi, S., Da Ros, T., Spalluto, G., & Prato, M. (2003). Fullerene derivatives: An attractive tool for biological applications. *European Journal of Medicinal Chemistry, 38*(11–12), 913–923. https://doi.org/10.1016/j.ejmech.2003.09.005

Castro, E., Garcia, A. H., Zavala, G., & Echegoyen, L. (2017). Fullerenes in biology and medicine. *Journal of Materials Chemistry B, 5*(32), 6523–6535. https://doi.org/10.1039/C7TB00855D

Castro, E., Murillo, J., Fernandez-Delgado, O., & Echegoyen, L. (2018). Progress in fullerene-based hybrid perovskite solar cells. *Journal of Materials Chemistry C, 6*(11), 2635–2651. https://doi.org/10.1039/C7TC04302C

Cedervall, T., Lynch, I., Lindman, S., Berggard, T., Thulin, E., Nilsson, H., Dawson, K. A., & Linse, S. (2007). Understanding the nanoparticle-protein corona using methods to quantify exchange rates and affinities of proteins for nanoparticles. *Proceedings of the National Academy of Sciences, 104*(7), 2050–2055. https://doi.org/10.1073/pnas.0608582104

Chen, H. (2018). Metal based nanoparticles in agricultural system: Behavior, transport, and interaction with plants. *Chemical Speciation and Bioavailability, 30*(1), 123–134. https://doi.org/10.1080/09542299.2018.1520050

Chen, K. L., & Elimelech, M. (2006). Aggregation and deposition kinetics of fullerene (C60) nanoparticles. *Langmuir, 22*(26), 10994–11001. https://doi.org/10.1021/la062072v

Chowdhury, I., Duch, M. C., Mansukhani, N. D., Hersam, M. C., & Bouchard, D. (2013). Colloidal properties and stability of graphene oxide nanomaterials in the aquatic environment. *Environmental Science and Technology, 47*(12), 6288–6296. https://doi.org/10.1021/es400483k

Colton, H. M., Falls, J. G., Ni, H., Kwanyuen, P., Creech, D., McNeil, E., Casey, W. M., Hamilton, G., & Cariello, N. F. (2004). Visualization and quantitation of peroxisomes using fluorescent nanocrystals: Treatment of rats and monkeys with fibrates and detection in the liver. *Toxicological Sciences, 80*(1), 183–192. https://doi.org/10.1093/toxsci/kfh144

Corredor, E., Testillano, P. S., Coronado, M. J., González-Melendi, P., Fernández-Pacheco, R., Marquina, C., Ibarra, M. R., De La Fuente, J. M., Rubiales, D., Pérez-De-Luque, A., & Risuẽo, M. C. (2009). Nanoparticle penetration and transport in living pumpkin plants: In situ subcellular identification. *BMC Plant Biology, 9*. https://doi.org/10.1186/1471-2229-9-45

Dai, H. (2002). Carbon nanotubes: Synthesis, integration, and properties. *Accounts of Chemical Research, 35*(12), 1035–1044. https://doi.org/10.1021/ar0101640

Deonarine, A., Lau, B. L. T., Aiken, G. R., Ryan, J. N., & Hsu-Kim, H. (2011). Effects of humic substances on precipitation and aggregation of zinc sulfide nanoparticles. *Environmental Science and Technology, 45*(8), 3217–3223. https://doi.org/10.1021/es1029798

Dietz, K. J., & Herth, S. (2011). Plant nanotoxicology. *Trends in Plant Science, 16*(11), 582–589. https://doi.org/10.1016/j.tplants.2011.08.003

Du, Y., Kopittke, P. M., Noller, B. N., James, S. A., Harris, H. H., Xu, Z. P., Li, P., Mulligan, D. R., & Huang, L. (2015). In situ analysis of foliar zinc absorption and short-distance movement in fresh and hydrated leaves of tomato and citrus Using synchrotron-based X-ray fluorescence microscopy. *Annals of Botany, 115*(1), 41–53. https://doi.org/10.1093/aob/mcu212

Dubertret, B., Skourides, P., Norris, D. J., Noireaux, V., Brivanlou, A. H., & Libchaber, A. (2002). In vivo imaging of quantum dots encapsulated in phospholipid micelles. *Science, 298*(5599), 1759–1762. https://doi.org/10.1126/science.1077194

Dwivedi, A. D., Dubey, S. P., Sillanpää, M., Kwon, Y. N., Lee, C., & Varma, R. S. (2015). Fate of engineered nanoparticles: Implications in the environment. *Coordination Chemistry Reviews, 287*, 64–78. https://doi.org/10.1016/j.ccr.2014.12.014

Marek, K., et al. (2019). Effect of foliar spray application of zinc oxide nanoparticles on quantitative, nutritional, and physiological parameters of foxtail millet (*Setaria italica* L.) under field conditions. *Nanomaterials, 1559*. https://doi.org/10.3390/nano9111559

Eichert, T., Kurtz, A., Steiner, U., & Goldbach, H. E. (2008). Size exclusion limits and lateral heterogeneity of the stomatal foliar uptake pathway for aqueous solutes and water-suspended nanoparticles. *Physiologia Plantarum, 134*(1), 151–160. https://doi.org/10.1111/j.1399-3054.2008.01135.x

Elsheery, N. I., Sunoj, V. S. J., Wen, Y., Zhu, J. J., Muralidharan, G., & Cao, K. F. (2020). Foliar application of nanoparticles mitigates the chilling effect on photosynthesis and photoprotection in sugarcane. *Plant Physiology and Biochemistry, 149*, 50–60. https://doi.org/10.1016/j.plaphy.2020.01.035

Engineered nanomaterials: Evidence on the effectiveness of workplace controls to prevent exposure (Vol. 75). (n.d.).

Etxeberria, E., Gonzalez, P., Bhattacharya, P., Sharma, P., & Ke, P. C. (2016). Determining the size exclusion for nanoparticles in citrus leaves. *HortScience, 51*(6), 732–737. https://doi.org/10.21273/hortsci.51.6.732

Farré, M., Sanchís, J., & Barceló, D. (2011). Analysis and assessment of the occurrence, the fate and the behavior of nanomaterials in the environment. *TrAC - Trends in Analytical Chemistry, 30*(3), 517–527. https://doi.org/10.1016/j.trac.2010.11.014

Fatemi, H., Esmaiel Pour, B., & Rizwan, M. (2021). Foliar application of silicon nanoparticles affected the growth, vitamin C, flavonoid, and antioxidant enzyme activities of coriander (Coriandrum sativum L.) plants grown in lead (Pb)-spiked soil. *Environmental Science and Pollution Research, 28*(2), 1417–1425. https://doi.org/10.1007/s11356-020-10549-x

Fauconnier, N., Pons, J. N., Roger, J., & Bee, A. (1997). Thiolation of maghemite nanoparticles by dimercaptosuccinic acid. *Journal of Colloid and Interface Science, 194*(2), 427–433. https://doi.org/10.1006/jcis.1997.5125

Fernández, V., & Eichert, T. (2009). Uptake of hydrophilic solutes through plant leaves: Current state of knowledge and perspectives of foliar fertilization. *Critical Reviews in Plant Sciences, 28*(1–2), 36–68. https://doi.org/10.1080/07352680902743069

Gao, X., Cui, Y., Levenson, R. M., Chung, L. W. K., & Nie, S. (2004). In vivo cancer targeting and imaging with semiconductor quantum dots. *Nature Biotechnology, 22*(8), 969–976. https://doi.org/10.1038/nbt994

Garner, K. L., & Keller, A. A. (2014). Emerging patterns for engineered nanomaterials in the environment: A review of fate and toxicity studies. *Journal of Nanoparticle Research, 16*(8). https://doi.org/10.1007/s11051-014-2503-2

Genge, M. J. (2008). Koronis asteroid dust within antarctic ice. *Geology, 36*(9), 687–690. https://doi.org/10.1130/G24493A.1

Genge, M. J., Larsen, J., van Ginneken, M., & Suttle, M. D. (2017). An urban collection of modern-day large micrometeorites: Evidence for variations in the extraterrestrial dust flux through the quaternary. *Geology, 45*(2), 119–122. https://doi.org/10.1130/G38352.1

Geraci, C., Heidel, D., Sayes, C., Hodson, L., Schulte, P., Eastlake, A., & Brenner, S. (2015). Perspectives on the design of safer nanomaterials and manufacturing processes. *Journal of Nanoparticle Research, 17*(9). https://doi.org/10.1007/s11051-015-3152-9

Ghosh, S., Patil, S., Ahire, M., Kitture, R., Jabgunde, A., Kale, S., Pardesi, K., Bellare, J., Dhavale, D. D., & Chopade, B. A. (2011). Synthesis of gold nanoanisotrops using dioscorea bulbifera tuber extract. *Journal of Nanomaterials, 2011*. https://doi.org/10.1155/2011/354793

Goswami, L., Kim, K. H., Deep, A., Das, P., Bhattacharya, S. S., Kumar, S., & Adelodun, A. A. (2017). Engineered nano particles: Nature, behavior, and effect on the environment. *Journal of Environmental Management, 196*, 297–315. https://doi.org/10.1016/j.jenvman.2017.01.011

Gridelet, L., Delbecq, P., Hervé, L., Boissolle, P., Fleury, D., Kowal, S., & Fayet, G. (2015). Proposal of a new risk assessment method for the handling of powders and nanomaterials. *Industrial Health, 53*(1), 56–68. https://doi.org/10.2486/indhealth.2014-0046

Hauck, T. S., Ghazani, A. A., & Chan, W. C. W. (2008). Assessing the effect of surface chemistry on gold nanorod uptake, toxicity, and gene expression in mammalian cells. *Small, 4*(1), 153–159. https://doi.org/10.1002/smll.200700217

Hoffmann, M., Holtze, E., & Wiesner, M. R. (2007). Reactive oxygen species generation on nanoparticulate material. In *Environmental nanotechnology. Applications and impacts of nanomaterials*.

Hong, J., Peralta-Videa, J. R., Rico, C., Sahi, S., Viveros, M. N., Bartonjo, J., Zhao, L., & Gardea-Torresdey, J. L. (2014). Evidence of translocation and physiological impacts of foliar applied CeO_2 nanoparticles on cucumber (Cucumis sativus) plants. *Environmental Science and Technology, 48*(8), 4376–4385. https://doi.org/10.1021/es404931g

Hong, J., Wang, L., Sun, Y., Zhao, L., Niu, G., Tan, W., Rico, C. M., Peralta-Videa, J. R., & Gardea-Torresdey, J. L. (2016). Foliar applied nanoscale and microscale CeO_2 and CuO

alter cucumber (Cucumis sativus) fruit quality. *Science of the Total Environment, 563–564*, 904–911. https://doi.org/10.1016/j.scitotenv.2015.08.029
Hotze, E. M., Phenrat, T., & Lowry, G. V. (2010). Nanoparticle aggregation: Challenges to understanding transport and reactivity in the environment. *Journal of Environmental Quality, 39*(6), 1909–1924. https://doi.org/10.2134/jeq2009.0462
Hu, J., Guo, H. Y., Li, J. L., Wang, Y. Q., Xiao, L., & Xing, B. S. (2017). Interaction of gamma-Fe_2O_3 nanoparticles with citrus maxima leaves and the corresponding physiological effects via foliar application. *Journal of Nanobiotechnology, 15*.
Hu, P., An, J., Faulkner, M. M., Wu, H., Li, Z., Tian, X., & Giraldo, J. P. (2020). Nanoparticle charge and size control foliar delivery efficiency to plant cells and organelles. *ACS Nano, 14*(7), 7970–7986. https://doi.org/10.1021/acsnano.9b09178
Hu, X., Zhang, Y., Luo, J., Xie, M., Wang, T., & Lian, H. (2011). Accumulation and quantitative estimates of airborne lead for a wild plant (Aster subulatus). *Chemosphere, 82*(10), 1351–1357. https://doi.org/10.1016/j.chemosphere.2010.11.079
ISO/TS 80004-8: 2013. Nanotechnologies — Vocabulary — Part 8: Nanomanufacturing processes, classification index: T 16-101-8. (n.d.).
Iwamoto, T., Kitamoto, Y., & Toshima, N. (2009). Anomalous magnetic behavior in FePt nanoparticles chemically synthesized with polymer protective agent. *Physica B: Condensed Matter, 404*(14–15), 2080–2085. https://doi.org/10.1016/j.physb.2009.03.048
Jahani, S., Saadatmand, S., Mahmoodzadeh, H., & Khavari-Nejad, R. A. (2019). Effect of foliar application of cerium oxide nanoparticles on growth, photosynthetic pigments, electrolyte leakage, compatible osmolytes and antioxidant enzymes activities of Calendula officinalis L. *Biologia, 74*(9), 1063–1075. https://doi.org/10.2478/s11756-019-00239-6
John, A. C., Küpper, M., Manders-Groot, A. M. M., Debray, B., Lacome, J. M., & Kuhlbusch, T. A. J. (2017). Emissions and possible environmental implication of engineered nanomaterials (ENMs) in the atmosphere. *Atmosphere, 8*(5). https://doi.org/10.3390/atmos8050084
Judy, J. D., Unrine, J. M., Rao, W., Wirick, S., & Bertsch, P. M. (2012). Bioavailability of gold nanomaterials to plants: Importance of particle size and surface coating. *Environmental Science and Technology, 46*(15), 8467–8474. https://doi.org/10.1021/es3019397
Kah, M., Kookana, R. S., Gogos, A., & Bucheli, T. D. (2018). A critical evaluation of nanopesticides and nanofertilizers against their conventional analogues. *Nature Nanotechnology, 13*(8), 677–684. https://doi.org/10.1038/s41565-018-0131-1
Kah, M., Tufenkji, N., & White, J. C. (2019). Nano-enabled strategies to enhance crop nutrition and protection. *Nature Nanotechnology, 14*(6), 532–540. https://doi.org/10.1038/s41565-019-0439-5
Khan, M. R., Adam, V., Rizvi, T. F., Zhang, B. H., Ahamad, F., Josko, I., Zhu, Y., Yang, M. Y., & Mao, C. B. (2019). *Small, 15*(37).
Kang, S., Mauter, M. S., & Elimelech, M. (2008). Physicochemical determinants of multiwalled carbon nanotube bacterial cytotoxicity. *Environmental Science and Technology, 42*(19), 7528–7534. https://doi.org/10.1021/es8010173
King, S. M., & Jarvie, H. P. (2012). Exploring how organic matter controls structural transformations in natural aquatic nanocolloidal dispersions. *Environmental Science and Technology, 46*(13), 6959–6967. https://doi.org/10.1021/es2034087
Kirschling, T. L., Golas, P. L., Unrine, J. M., Matyjaszewski, K., Gregory, K. B., Lowry, G. V., & Tilton, R. D. (2011). Microbial bioavailability of covalently bound polymer coatings on model engineered nanomaterials. *Environmental Science and Technology, 45*(12), 5253–5259. https://doi.org/10.1021/es200770z
Klaine, S. J., Koelmans, A. A., Horne, N., Carley, S., Handy, R. D., Kapustka, L., Nowack, B., & von der Kammer, F. (2012). Paradigms to assess the environmental impact of manufactured nanomaterials. *Environmental Toxicology and Chemistry, 31*(1), 3–14. https://doi.org/10.1002/etc.733

Kolenčík, M., Ernst, D., Urík, M., Ďurišová, Ľ., Bujdoš, M., Šebesta, M., Dobročka, E., Kšiňan, S., Illa, R., Qian, Y., Feng, H., Černý, I., Holišová, V., & Kratošová, G. (2020). Foliar application of low concentrations of titanium dioxide and zinc oxide nanoparticles to the common sunflower under field conditions. *Nanomaterials, 1619*. https://doi.org/10.3390/nano10081619

Kranjc, E., Mazej, D., Regvar, M., Drobne, D., & Remškar, M. (2018). Foliar surface free energy affects platinum nanoparticle adhesion, uptake, and translocation from leaves to roots in arugula and escarole. *Environmental Science: Nano, 5*(2), 520–532. https://doi.org/10.1039/c7en00887b

Kuhlbusch, T. A. J., Quincey, P., Fuller, G. W., Kelly, F., Mudway, I., Viana, M., Querol, X., Alastuey, A., Katsouyanni, K., Weijers, E., Borowiak, A., Gehrig, R., Hueglin, C., Bruckmann, P., Favez, O., Sciare, J., Hoffmann, B., EspenYttri, K., Torseth, K., ... Quass, U. (2014). New directions: The future of European urban air quality monitoring. *Atmospheric Environment, 87*, 258–260. https://doi.org/10.1016/j.atmosenv.2014.01.012

Larue, C., Castillo-Michel, H., Sobanska, S., Cécillon, L., Bureau, S., Barthès, V., Ouerdane, L., Carrière, M., & Sarret, G. (2014). Foliar exposure of the crop Lactuca sativa to silver nanoparticles: Evidence for internalization and changes in Ag speciation. *Journal of Hazardous Materials, 264*, 98–106. https://doi.org/10.1016/j.jhazmat.2013.10.053

Larue, C., Castillo-Michel, H., Sobanska, S., Trcera, N., Sorieul, S., Cécillon, L., Ouerdane, L., Legros, S., & Sarret, G. (2014). Fate of pristine TiO_2 nanoparticles and aged paint-containing TiO_2 nanoparticles in lettuce crop after foliar exposure. *Journal of Hazardous Materials, 273*, 17–26. https://doi.org/10.1016/j.jhazmat.2014.03.014

Law, N., Ansari, S., Livens, F. R., Renshaw, J. C., & Lloyd, J. R. (2008). Formation of nanoscale elemental silver particles via enzymatic reduction by Geobacter sulfurreducens. *Applied and Environmental Microbiology, 74*(22), 7090–7093. https://doi.org/10.1128/AEM.01069-08

Lew, T. T. S., Wong, M. H., Kwak, S. Y., Sinclair, R., Koman, V. B., & Strano, M. S. (2018). Rational design principles for the transport and subcellular distribution of nanomaterials into plant protoplasts. *Small, 14*(44). https://doi.org/10.1002/smll.201802086

Li, C., Wang, P., Lombi, E., Cheng, M., Tang, C., Howard, D. L., Menzies, N. W., & Kopittke, P. M. (2018). Absorption of foliar-applied Zn fertilizers by trichomes in soybean and tomato. *Journal of Experimental Botany, 69*(10), 2717–2729. https://doi.org/10.1093/jxb/ery085

Li, C., Wang, P., Van Der Ent, A., Cheng, M., Jiang, H., Read, T. L., Lombi, E., Tang, C., De Jonge, M. D., Menzies, N. W., & Kopittke, P. M. (2019). Absorption of foliar-applied zn in sunflower (helianthus annuus): Importance of the cuticle, stomata and trichomes. *Annals of Botany, 123*(1), 57–68. https://doi.org/10.1093/aob/mcy135

Li, S., Liu, J., Wang, Y., Gao, Y., Zhang, Z., Xu, J., & Xing, G. (2021). Comparative physiological and metabolomic analyses revealed that foliar spraying with zinc oxide and silica nanoparticles modulates metabolite profiles in cucumber (*Cucumis sativus* L.). *Food and Energy Security, 10*(1). https://doi.org/10.1002/fes3.269

Lian, J., Zhao, L., Wu, J., Xiong, H., Bao, Y., Zeb, A., Tang, J., & Liu, W. (2020). Foliar spray of TiO_2 nanoparticles prevails over root application in reducing Cd accumulation and mitigating Cd-induced phytotoxicity in maize (*Zea mays* L.). *Chemosphere, 239*. https://doi.org/10.1016/j.chemosphere.2019.124794

Liang, Y., Bradford, S. A., Simunek, J., Heggen, M., Vereecken, H., & Klumpp, E. (2013). Retention and remobilization of stabilized silver nanoparticles in an undisturbed loamy sand soil. *Environmental Science and Technology, 47*(21), 12229–12237. https://doi.org/10.1021/es402046u

Liang, Y., Bradford, S. A., Simunek, J., Vereecken, H., & Klumpp, E. (2013). Sensitivity of the transport and retention of stabilized silver nanoparticles to physicochemical factors. *Water Research, 47*(7), 2572–2582. https://doi.org/10.1016/j.watres.2013.02.025

Lok, C. N., Ho, C. M., Chen, R., He, Q. Y., Yu, W. Y., Sun, H., Tam, P. K. H., Chiu, J. F., & Che, C. M. (2007). Silver nanoparticles: Partial oxidation and antibacterial activities. *Journal of Biological Inorganic Chemistry, 12*(4), 527–534. https://doi.org/10.1007/s00775-007-0208-z

Lopez-Vargas, E. R., Ortega-Ortiz, H., Cadenas-Pliego, G., Romenus, K. D., Fuente, M. C., Benavides-Mendoza, A., & Juarez-Maldonado, A. (2018). *Foliar application of copper nanoparticles increases the fruit quality and the content of bioactive compounds in tomatoes* (Vol. 8).

Lowry, G. V., Gregory, K. B., Apte, S. C., & Lead, J. R. (2012). Transformations of nanomaterials in the environment. *Environmental Science & Technology, 46*(13), 6893–6899. https://doi.org/10.1021/es300839e

Lu, K., Shen, D., Liu, X., Dong, S., Jing, X., Wu, W., Tong, Y., Gao, S., & Mao, L. (2020). Uptake of iron oxide nanoparticles inhibits the photosynthesis of the wheat after foliar exposure. *Chemosphere, 259*. https://doi.org/10.1016/j.chemosphere.2020.127445

Lundqvist, M., Stigler, J., Elia, G., Lynch, I., Cedervall, T., & Dawson, K. A. (2008). Nanoparticle size and surface properties determine the protein corona with possible implications for biological impacts. *Proceedings of the National Academy of Sciences, 105*(38), 14265–14270. https://doi.org/10.1073/pnas.0805135105

Lynch, I., & Dawson, K. A. (2008). Protein-nanoparticle interactions. *Nano Today, 3*(1–2), 40–47. https://doi.org/10.1016/S1748-0132(08)70014-8

Manke, A., Wang, L., & Rojanasakul, Y. (2013). Mechanisms of nanoparticle-induced oxidative stress and toxicity. *BioMed Research International, 2013*. https://doi.org/10.1155/2013/942916

Mauter, M. S., & Elimelech, M. (2008). Environmental applications of carbon-based nanomaterials. *Environmental Science and Technology, 42*(16), 5843–5859. https://doi.org/10.1021/es8006904

Meesters, J. A., Veltman, K., Hendriks, A. J., & van de Meent, D. (2013). Environmental exposure assessment of engineered nanoparticles: Why REACH needs adjustment. *Integrated Environmental Assessment and Management, 9*(3), e15–e26. https://doi.org/10.1002/ieam.1446

Mehnert, W., & Mäder, K. (2012). Solid lipid nanoparticles: Production, characterization and applications. *Advanced Drug Delivery Reviews, 64*, 83–101. https://doi.org/10.1016/j.addr.2012.09.021

Metsämuuronen, S., Sillanpää, M., Bhatnagar, A., & Mänttäri, M. (2014). Natural organic matter removal from drinking water by membrane technology. *Separation and Purification Reviews, 43*(1), 1–61. https://doi.org/10.1080/15422119.2012.712080

Mittal, A. K., Chisti, Y., & Banerjee, U. C. (2013). Synthesis of metallic nanoparticles using plant extracts. *Biotechnology Advances, 31*(2), 346–356. https://doi.org/10.1016/j.biotechadv.2013.01.003

Montaño, M. D., Lowry, G. V., Von Der Kammer, F., Blue, J., & Ranville, J. F. (2014). Current status and future direction for examining engineered nanoparticles in natural systems. *Environmental Chemistry, 11*(4), 351–366. https://doi.org/10.1071/EN14037

Nadiminti, P. P., Dong, Y. D., Sayer, C., Hay, P., Rookes, J. E., Boyd, B. J., & Cahill, D. M. (2013). Nanostructured liquid crystalline particles as an alternative delivery vehicle for plant agrochemicals. *ACS Applied Materials and Interfaces, 5*(5), 1818–1826. https://doi.org/10.1021/am303208t

Natasha, Shahid, M., Dumat, C., Khalid, S., Rabbani, F., Farooq, A. B. U., Amjad, M., Abbas, G., & Niazi, N. K. (2019). Foliar uptake of arsenic nanoparticles by spinach: An assessment of physiological and human health risk implications. *Environmental Science and Pollution Research, 26*(20), 20121–20131. https://doi.org/10.1007/s11356-018-3867-0

Natasha, Shahid, M., Farooq, A. B. U., Rabbani, F., Khalid, S., & Dumat, C. (2020). Risk assessment and biophysiochemical responses of spinach to foliar application of lead oxide nanoparticles: A multivariate analysis. *Chemosphere, 245*. https://doi.org/10.1016/j.chemosphere.2019.125605

Nel, A. E., Mädler, L., Velegol, D., Xia, T., Hoek, E. M. V., Somasundaran, P., Klaessig, F., Castranova, V., & Thompson, M. (2009). Understanding biophysicochemical interactions at the nano-bio interface. *Nature Materials, 8*(7), 543–557. https://doi.org/10.1038/nmat2442

Niyogi, S., Hamon, M. A., Hu, H., Zhao, B., Bhowmik, P., Sen, R., Itkis, M. E., & Haddon, R. C. (2002). Chemistry of single-walled carbon nanotubes. *Accounts of Chemical Research, 35*(12), 1105–1113. https://doi.org/10.1021/ar010155r

Pádrová, K., Maťátková, O., Šiková, M., Füzik, T., Masák, J., Čejková, A., & Jirků, V. (2016). Mitigation of Fe0 nanoparticles toxicity to Trichosporon cutaneum by humic substances. *New Biotechnology, 33*(1), 144–152. https://doi.org/10.1016/j.nbt.2015.09.007

Petersen, E. J., & Henry, T. B. (2012). Methodological considerations for testing the ecotoxicity of carbon nanotubes and fullerenes: Review. *Environmental Toxicology and Chemistry, 31*(1), 60–72. https://doi.org/10.1002/etc.710

Phenrat, T., Saleh, N., Sirk, K., Tilton, R. D., & Lowry, G. V. (2007). Aggregation and sedimentation of aqueous nanoscale zerovalent iron dispersions. *Environmental Science and Technology, 41*(1), 284–290. https://doi.org/10.1021/es061349a

Phenrat, T., Song, J. E., Cisneros, C. M., Schoenfelder, D. P., Tilton, R. D., & Lowry, G. V. (2010). Estimating attachment of nano- and submicrometer-particles coated with organic macromolecules in porous media: Development of an empirical model. *Environmental Science and Technology, 44*(12), 4531–4538. https://doi.org/10.1021/es903959c

Popp, C., Burghardt, M., Friedmann, A., & Riederer, M. (2005). Characterization of hydrophilic and lipophilic pathways of Hedera helix L. cuticular membranes: Permeation of water and uncharged organic compounds. *Journal of Experimental Botany, 56*(421), 2797–2806. https://doi.org/10.1093/jxb/eri272

Project Report, M. (n.d.). Managing risks of nanomaterials; summary Report on processes and activities related to NM lifecycle within small and large industries. In Grant Agreement Number 263215.

Pullagurala, V. L. R., Rawat, S., Adisa, I. O., Hernandez-Viezcas, J. A., Peralta-Videa, J. R., & Gardea-Torresdey, J. L. (2018). Plant uptake and translocation of contaminants of emerging concern in soil. *Science of the Total Environment, 636*, 1585–1596. https://doi.org/10.1016/j.scitotenv.2018.04.375

Quinn, J., Geiger, C., Clausen, C., Brooks, K., Coon, C., O'Hara, S., Krug, T., Major, D., Yoon, W. S., Gavaskar, A., & Holdsworth, T. (2005). Field demonstration of DNAPL dehalogenation using emulsified zero-valent iron. *Environmental Science and Technology, 39*(5), 1309–1318. https://doi.org/10.1021/es0490018

Rabajczyk, A., Zielecka, M., Porowski, R., & Hopke, P. K. (2020). Metal nanoparticles in the air: State of the art and future perspectives. *Environmental Science: Nano, 7*(11), 3233–3254. https://doi.org/10.1039/d0en00536c

Rai, P. K., Kumar, V., Lee, S. S., Raza, N., Kim, K. H., Ok, Y. S., & Tsang, D. C. W. (2018). Nanoparticle-plant interaction: Implications in energy, environment, and agriculture. *Environment International, 119*, 1–19. https://doi.org/10.1016/j.envint.2018.06.012

Raliya, R., Franke, C., Chavalmane, S., Nair, R., Reed, N., & Biswas, P. (2016). Quantitative understanding of nanoparticle uptake in watermelon plants. *Frontiers in Plant Science, 7*(2016). https://doi.org/10.3389/fpls.2016.01288

Read, T. L., Doolette, C. L., Li, C., Schjoerring, J. K., Kopittke, P. M., Donner, E., & Lombi, E. (2020). Optimising the foliar uptake of zinc oxide nanoparticles: Do leaf surface properties and particle coating affect absorption? *Physiologia Plantarum, 170*(3), 384–397. https://doi.org/10.1111/ppl.13167

Richter, K., Schicklberger, M., & Gescher, J. (2012). Dissimilatory reduction of extracellular electron acceptors in anaerobic respiration. *Applied and Environmental Microbiology, 78*(4), 913–921. https://doi.org/10.1128/AEM.06803-11

Rico, C. M., Majumdar, S., Duarte-Gardea, M., Peralta-Videa, J. R., & Gardea-Torresdey, J. L. (2011). Interaction of nanoparticles with edible plants and their possible implications in the food chain. *Journal of Agricultural and Food Chemistry, 59*(8), 3485–3498. https://doi.org/10.1021/jf104517j

Rizwan, M., Ali, S., Zia ur Rehman, M., Adrees, M., Arshad, M., Qayyum, M. F., Ali, L., Hussain, A., Chatha, S. A. S., & Imran, M. (2019). Alleviation of cadmium accumulation in maize (Zea mays L.) by foliar spray of zinc oxide nanoparticles and biochar to contaminated soil. *Environmental Pollution, 248*, 358–367. https://doi.org/10.1016/j.envpol.2019.02.031

Rossi, L., Fedenia, L. N., Sharifan, H., Ma, X., & Lombardini, L. (2019). Effects of foliar application of zinc sulfate and zinc nanoparticles in coffee (Coffea arabica L.) plants. *Plant Physiology and Biochemistry, 135*, 160–166. https://doi.org/10.1016/j.plaphy.2018.12.005

Rostamizadeh, E., Iranbakhsh, A., Majd, A., Arbabian, S., & Mehregan, I. (2021). Physiological and molecular responses of wheat following the foliar application of iron oxide nanoparticles. *International Journal of Nano Dimension, 12*(2), 128–134.

Sajid, M., Ilyas, M., Basheer, C., Tariq, M., Daud, M., Baig, N., & Shehzad, F. (2015). Impact of nanoparticles on human and environment: Review of toxicity factors, exposures, control strategies, and future prospects. *Environmental Science and Pollution Research, 22*(6), 4122–4143. https://doi.org/10.1007/s11356-014-3994-1

Salehi, H., Chehregani, A., Lucini, L., Majd, A., & Gholami, M. (2018). Morphological, proteomic and metabolomic insight into the effect of cerium dioxide nanoparticles to phaseolus vulgaris L. under soil or foliar application. *Science of the Total Environment, 616–617*, 1540–1551. https://doi.org/10.1016/j.scitotenv.2017.10.159

Sathiyabama, M., & Manikandan, A. (2021). Foliar application of chitosan nanoparticle improves yield, mineral content and boost innate immunity in finger millet plants. *Carbohydrate Polymers, 258*. https://doi.org/10.1016/j.carbpol.2021.117691

Schreck, E., Dappe, V., Sarret, G., Sobanska, S., Nowak, D., Nowak, J., Stefaniak, E. A., Magnin, V., Ranieri, V., & Dumat, C. (2014). Foliar or root exposures to smelter particles: Consequences for lead compartmentalization and speciation in plant leaves. *Science of the Total Environment, 476–477*, 667–676. https://doi.org/10.1016/j.scitotenv.2013.12.089

Schreck, E., Foucault, Y., Sarret, G., Sobanska, S., Cécillon, L., Castrec-Rouelle, M., Uzu, G., & Dumat, C. (2012). Metal and metalloid foliar uptake by various plant species exposed to atmospheric industrial fallout: Mechanisms involved for lead. *Science of the Total Environment, 427–428*, 253–262. https://doi.org/10.1016/j.scitotenv.2012.03.051

Sedlmair, J., Gleber, S. C., Wirick, S., Guttmann, P., & Thieme, J. (2012). Interaction between carbon nanotubes and soil colloids studied with X-ray spectromicroscopy. *Chemical Geology, 329*, 32–41. https://doi.org/10.1016/j.chemgeo.2011.08.009

Semida, W. M., Abdelkhalik, A., Mohamed, G. F., Abd El-Mageed, T. A., Abd El-Mageed, S. A., Rady, M. M., & Ali, E. F. (2021). Foliar application of zinc oxide nanoparticles promotes drought stress tolerance in eggplant (*Solanum melongena* l.). *Plants, 10*(2), 1–18. https://doi.org/10.3390/plants10020421

Shen, Y., Borgatta, J., Ma, C., Elmer, W., Hamers, R. J., & White, J. C. (2020). Copper nanomaterial morphology and composition control foliar transfer through the cuticle and mediate resistance to root fungal disease in tomato (*Solanum lycopersicum*). *Journal of Agricultural and Food Chemistry, 68*(41), 11327–11338. https://doi.org/10.1021/acs.jafc.0c04546

Singh, S., Vishwakarma, K., Singh, S., Sharma, S., Dubey, N. K., Singh, V. K., Liu, S., Tripathi, D. K., & Chauhan, D. K. (2017). Understanding the plant and nanoparticle interface at transcriptomic and proteomic level: A concentric overview. *Plant Gene, 11*, 265–272. https://doi.org/10.1016/j.plgene.2017.03.006

Song, C., Huang, M., White, J. C., Zhang, X., Wang, W., Sarpong, C. K., Jamali, Z. H., Zhang, H., Zhao, L., & Wang, Y. (2020). Metabolic profile and physiological response of cucumber foliar exposed to engineered MoS_2 and TiO_2 nanoparticles. *NanoImpact, 20*. https://doi.org/10.1016/j.impact.2020.100271

Symonds, R. B., Rose, W. I., Bluth, G. J. S., & Gerlach, T. M. (2018). Volcanic-gas studies: Methods, results, and applications. In *Volatiles in magmas* (Vol. 30, pp. 1–66). De Gruyter Mouton. https://www.degruyter.com/view/product/502068.

Symonds, R. B., Rose, W. I., & Reed, M. H. (1988). Contribution of C1- and F-bearing gases to the atmosphere by volcanoes. *Nature, 334*(6181), 415–418. https://doi.org/10.1038/334415a0

Teszlák, P., Kocsis, M., Scarpellini, A., Jakab, G., & Kőrösi, L. (2018). Foliar exposure of grapevine (*Vitis vinifera* L.) to TiO_2 nanoparticles under field conditions: Photosynthetic response and flavonol profile. *Photosynthetica, 56*(4), 1378–1386. https://doi.org/10.1007/s11099-018-0832-6

Uzu, G., Sobanska, S., Aliouane, Y., Pradere, P., & Dumat, C. (2009). Study of lead phytoavailability for atmospheric industrial micronic and sub-micronic particles in relation with lead speciation. *Environmental Pollution, 157*(4), 1178–1185. https://doi.org/10.1016/j.envpol.2008.09.053

Van Hoecke, K., De Schamphelaere, K. A. C., Ramirez-Garcia, S., Van der Meeren, P., Smagghe, G., & Janssen, C. R. (2011). Influence of alumina coating on characteristics and effects of SiO_2 nanoparticles in algal growth inhibition assays at various pH and organic matter contents. *Environment International, 37*(6), 1118–1125. https://doi.org/10.1016/j.envint.2011.02.009

Vignes, A. (2013). In *Panorama des Principaux Procédés de Production, Utilisation, Transformation de Nanomatériaux, Rapport DRA-14-133557-00736A; INERIS: Verneuil-En-Halatte* (Vol. 82).

Virji, M. A., & Stefaniak, A. B. (2014). A review of engineered nanomaterial manufacturing processes and associated exposures. *Comprehensive Materials Processing, 8*, 103–125. https://doi.org/10.1016/B978-0-08-096532-1.00811-6. Elsevier Ltd.

Vlassiouk, I., Regmi, M., Fulvio, P., Dai, S., Datskos, P., Eres, G., & Smirnov, S. (2011). Role of hydrogen in chemical vapor deposition growth of large single-crystal graphene. *ACS Nano, 5*(7), 6069–6076. https://doi.org/10.1021/nn201978y

Wang, P., Lombi, E., Zhao, F. J., & Kopittke, P. M. (2016). Nanotechnology: A new opportunity in plant sciences. *Trends in Plant Science, 21*(8), 699–712. https://doi.org/10.1016/j.tplants.2016.04.005

Wang, W. N., Tarafdar, J. C., & Biswas, P. (2013). Nanoparticle synthesis and delivery by an aerosol route for watermelon plant foliar uptake. *Journal of Nanoparticle Research, 15*(1). https://doi.org/10.1007/s11051-013-1417-8

White, J. C., & Gardea-Torresdey, J. (2018). Achieving food security through the very small. *Nature Nanotechnology, 13*(8), 627–629. https://doi.org/10.1038/s41565-018-0223-y

Winkler, U., & Zotz, G. (2010). "And then there were three": Highly efficient uptake of potassium by foliar trichomes of epiphytic bromeliads. *Annals of Botany, 106*(3), 421–427. https://doi.org/10.1093/aob/mcq120

Wong, M. H., Misra, R. P., Giraldo, J. P., Kwak, S. Y., Son, Y., Landry, M. P., Swan, J. W., Blankschtein, D., & Strano, M. S. (2016). Lipid exchange envelope penetration (LEEP) of nanoparticles for plant engineering: A universal localization mechanism. *Nano Letters, 16*(2), 1161–1172. https://doi.org/10.1021/acs.nanolett.5b04467

Xia, T., Kovochich, M., Brant, J., Hotze, M., Sempf, J., Oberley, T., Sioutas, C., Yeh, J. I., Wiesner, M. R., & Nel, A. E. (2006). Comparison of the abilities of ambient and manufactured nanoparticles to induce cellular toxicity according to an oxidative stress paradigm. *Nano Letters, 6*(8), 1794–1807. https://doi.org/10.1021/nl061025k

Xia, T., Li, N., & Nel, A. E. (2009). Potential health impact of nanoparticles. *Annual Review of Public Health, 30*, 137–150. https://doi.org/10.1146/annurev.publhealth.031308.100155

Xie, C., Ma, Y., Yang, J., Zhang, B., Luo, W., Feng, S., Zhang, J., Wang, G., He, X., & Zhang, Z. (2019). Effects of foliar applications of ceria nanoparticles and $CeCl_3$ on common bean (*Phaseolus vulgaris*). *Environmental Pollution, 250*, 530–536. https://doi.org/10.1016/j.envpol.2019.04.042

Xiong, T., Dumat, C., Dappe, V., Vezin, H., Schreck, E., Shahid, M., Pierart, A., & Sobanska, S. (2017). Copper oxide nanoparticle foliar uptake, phytotoxicity, and consequences for sustainable urban agriculture. *Environmental Science and Technology, 51*(9), 5242–5251. https://doi.org/10.1021/acs.est.6b05546

Xiong, T., Zhang, T., Dumat, C., Sobanska, S., Dappe, V., Shahid, M., Xian, Y., Li, X., & Li, S. (2019). Airborne foliar transfer of particular metals in Lactuca sativa L.: Translocation, phytotoxicity, and bioaccessibility. *Environmental Science and Pollution Research, 26*(20), 20064–20078. https://doi.org/10.1007/s11356-018-3084-x

Xiong, T., Zhang, T., Xian, Y., Kang, Z., Zhang, S., Dumat, C., Shahid, M., & Li, S. (2021). Foliar uptake, biotransformation, and impact of CuO nanoparticles in Lactuca sativa L. var. ramosa Hort. *Environmental Geochemistry and Health, 43*(1), 423–439. https://doi.org/10.1007/s10653-020-00734-9

Yang, K., & Xing, B. (2009). Sorption of phenanthrene by humic acid-coated nanosized TiO_2 and ZnO. *Environmental Science and Technology, 43*(6), 1845–1851. https://doi.org/10.1021/es802880m

Yang, R., Dong, G., Liu, Y., Sun, J., & Xu, Y. (2012). Preparation and characterization of the structure of PSt-b-PAA/Co composite material. *Journal of Applied Polymer Science, 124*(2), 1290–1294. https://doi.org/10.1002/app.34803

Yu, M., Yao, J., Liang, J., Zeng, Z., Cui, B., Zhao, X., Sun, C., Wang, Y., Liu, G., & Cui, H. (2017). Development of functionalized abamectin poly(lactic acid) nanoparticles with regulatable adhesion to enhance foliar retention. *RSC Advances, 7*(19), 11271–11280. https://doi.org/10.1039/c6ra27345a

Zadeh, R. R., Arvin, S. M. J., Jamei, R., Mozaffari, H., & Reza Nejhad, F. (2019). Response of tomato plants to interaction effects of magnetic (Fe_3O_4) nanoparticles and cadmium stress. *Journal of Plant Interactions, 14*(1), 474–481. https://doi.org/10.1080/17429145.2019.1626922

Zahedi, S. M., Abdelrahman, M., Hosseini, M. S., Hoveizeh, N. F., & Tran, L. S. P. (2019). Alleviation of the effect of salinity on growth and yield of strawberry by foliar spray of selenium-nanoparticles. *Environmental Pollution, 253*, 246–258. https://doi.org/10.1016/j.envpol.2019.04.078

Zhang, M. Y., Wang, Y., Zhao, D. Y., & Pan, G. (2010). Immobilization of arsenic in soils by stabilized nanoscale zero-valent iron, iron sulfide (FeS), and magnetite (Fe_3O_4) particles. *Chinese Science Bulletin, 55*(4), 365–372. https://doi.org/10.1007/s11434-009-0703-4

Zook, H. A. (2001). *Spacecraft measurements of the cosmic dust flux*. Springer Science and Business Media LLC. https://doi.org/10.1007/978-1-4419-8694-8_5

SECTION III

Molecular interaction of ENMs and plants: Reactions and mechanisms

CHAPTER

7

Biophysicochemical transformation of ENMs at root level

Alma Hortensia Serafin-Muñoz, and Berenice Noriega-Luna

Universidad de Guanajuato, Division de Ingenierias, Campus Guanajuato, Guanajuato, Mexico

7.1 The root system structure

Roots are important for plant support as well as for the absorption of nutrients, minerals, and water (Aiken & Smucker, 1996); plant survival depends on proper root function. Root development, architecture, and morphology depend on environmental factors such as nutrient availability, humidity, and temperature (Ristova & Busch, 2014). However, root evolution is not clearly understood; in addition, the genetic regulation of root trait development is only known in *Arabidopsis thaliana*. Knowledge about *Arabidopsis* root development derives from classical genetic experiments and the use of modern molecular biology and genomics techniques. However, there is still very little knowledge about the molecular basis and phenotypic variations because plants are sessile organisms that rely on genetic and epigenetic mechanisms to adapt to the environment in which they develop (Eichten et al., 2014; Falke et al., 2013).

7.1.1 Embryonic root formation in Arabidopsis

The Arabidopsis zygote divides asymmetrically to produce an apical cell, vertically to produce a larger basal cell, and horizontall to produce a suspensor that connects the embryo to maternal tissue (Laux et al., 2004). The embryo provides stem cells for the root's vascular, underground, and epidermal tissues (Scheres et al., 1994). The quiescent center (QC) cells,

Physicochemical Interactions of Engineered Nanoparticles and Plants
https://doi.org/10.1016/B978-0-323-90558-9.00008-5

which divide infrequently, are necessary for stem cell niche specification and maintenance of early stem cells (Van Den Berg et al., 1997). Stem cells continually undergo asymmetric divisions to produce daughter cells that move out of the stem cell niche and differentiate (Benfey & Scheres, 2000; Dolan et al., 1993). It is important to note that the hormone auxin regulates the formation of embryonic roots.

7.1.2 The root structure of *Arabidopsis*

The root is a radial and symmetrical organ formed by concentric rows of different cell types: (outside-in) the epidermis (Epi), the cortex (Cor), the endodermis (Endo), the pericycle, and the stele, which is in the center (Fig. 7.1A). Along the longitudinal axis, the primary root is composed of

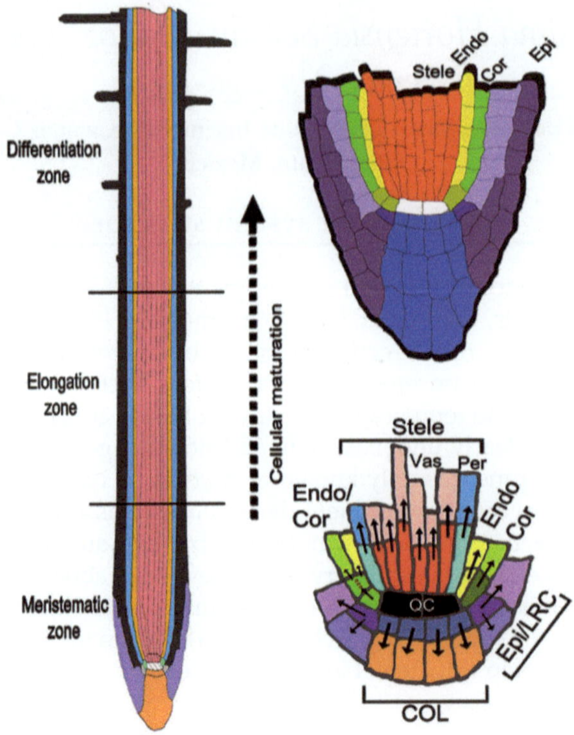

FIGURE 7.1 (A) *Arabidopsis* root meristem tissues: red, stele (pericycle and vasculature); yellow, endodermis (Endo); green, cortex (Cor); purple, epidermis (Epi). (B) stem cells in the *Arabidopsis* root. Left side: young *Arabidopsis* roots with an endodermal-cortical (Endo/Cor) stem cell; right side: organization in older roots with separate Endo and Cor stem cells. Daughter stem cells are shown with lighter shading: red, stele; yellow, Endo/Cor; yellow, Endo; green, Cor; purple, epidermal/lateral root cap stem cells (Epi/LRC); lilac, Epi; dark purple, LRC; orange, COL; black, QC. (C) the longitudinal organization of zones in the *Arabidopsis* root meristem; differentiation zone; elongation zone; meristem zone.

three zones: at the tip of the root is the root apical meristem (RAM), formed by the stem cell niche comprising the group of cells of lower mitotic activity, the (QC), which is surrounded by four types of stem or initial cells: epidermis/lateral root cap, cortex/endodermis, vascular/pericycle, and columella (Fig. 7.1B). These stem cells divide asymmetrically to give rise to self-renewing stem cells and daughter cells that, in turn, divide several times to form epidermis and lateral root cap cells, cortex and endodermal cells, and stele and columella cells (Dolan et al., 1993). The meristematic zone contains the proliferation domain, where cells reach the highest proliferation rate, and the transition domain, where cells begin to elongate but do not undergo elongation and continue to proliferate. After the RAM, cells enter the Elongation Zone (EZ), and finally, acquire their final characteristics in the Maturation Zone (MZ) (Fig. 7.1C).

7.1.3 The function of the stele

The vascular system, which consists of xylem and phloem, allows efficient transport of water, mineral nutrients, sugars, hormones, and other signaling molecules. The procambium, the meristematic tissue from which the primary vascular tissues derive, contains the xylem and phloem precursor cells distributed in a species-specific manner. The vasculature in *Arabidopsis* roots takes the form of a central cylinder or stele containing xylem and phloem interspersed with undifferentiated procambial cells and a surrounding layer of pericycle. These tissues derive from the initial pericycle/vasculature assembly proximal to the QC (Dolan et al., 1993; Ohashi-Ito et al., 2014). The vasculature grows longitudinally during primary growth and radially during secondary growth (Zhang et al., 2011). The outer tissues of the root have radial symmetry, and the stele is bilaterally symmetric. A mature *Arabidopsis* root cross-section reveals a central axis xylem cell midshaft with protoxylem at the poles and the metaxylem in the center. The phloem consists of sieve elements and cells arising from the initial phloem/procambial cells (Baum et al., 2002; Brady et al., 2007). The two types of xylem, protoxylem and metaxylem, differ in structure and function. Protoxylem cells have spiral cell wall thickenings, are smaller, differentiate early, and are usually destroyed as the plant matures (Fig. 7.2). In contrast, metaxylem cells have pitted cell wall thickenings, differentiate later, are more prominent, and form the water-conducting vessels of the plant.

The hormone auxin is necessary for establishing the central stele and breaking the root's initial radial symmetry with the formation of a xylem axis. This process is strengthened by the antagonistic action of cytokinin, which is also necessary for cell proliferation in the procambium. A high auxin level in the xylem axis activates a specific cytokinin signaling

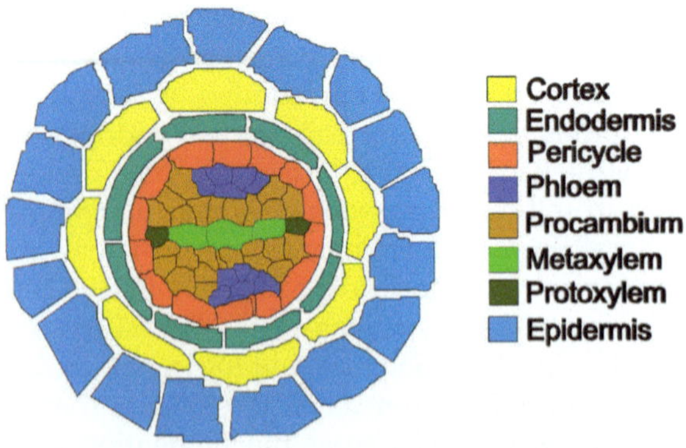

FIGURE 7.2 Organization of cell types within the stele. Cross-section of the root tip.

component, AHP6, which inhibits cytokinin signaling. In *Arabidopsis*, auxin and cytokinin's mutually inhibitory action defines the root vascular pattern. If either of these components is altered, the patterning of the xylem and procambium in the stele will be changed; in particular, the protoxylem is sensitive to auxin and cytokinin perturbations (Bishopp et al., 2011).

7.2 NP interactions with plant roots

Two main routes of NPs uptake by plant roots have been reported: (a) apoplastic and (b) symplastic. In the apoplastic route, nanoparticles traverse the apoplast through extracellular spaces, longitudinal channels, middle lamellae, intercellular spaces, and xylem (Rawat et al., 2017; Tripathi et al., 2017). Some NPs form complexes with membrane transporter proteins or root exudates and translocate them into the plant system (Avellan et al., 2017; Martínez-Fernández et al., 2016). NPs penetrate plant cells through cytoplasmic membrane proteins and ion channels or endocytosis, forming vesicles around the transport component (Ma et al., 2010). When plant roots take up NPs, they penetrate the cell wall formed by complex structures (Tripathi et al., 2017), including a network of cellulose microfibrils cross-linked with hemicellulose and lignin embedded in pectin. Hypothetically, to pass through the cell wall, NPs have to move passively through the cell wall pores; therefore, the uptake of NPs by plants is highly size-selective. NPs with sizes larger than 20 nm have limited access to the root epidermal cell wall; hardly NPs with sizes larger than 20 nm penetrate through the cell wall because the pore size is in the range of 2–20 nm (Burello & Worth, 2011). The xylem is a crucial plant tissue for the distribution and translocation of NPs (Aslani et al., 2014).

Possible interactions of NPs with plant roots include (A) adsorption to the root surface and (B) penetration of the cell wall and cell membrane (Lin et al., 2009). Subsequently, the NPs move into the stele via the symplastic route to be translocated to the leaves (Martínez-Fernández et al., 2016). The transport of NPs via the symplastic route occurs through organelles, transporter proteins, transmural channels, endocytosis, or through pores generated in the membrane (Rico et al., 2011; Tripathi et al., 2017). Once NPs enter cells, interaction with organelles can occur and modify metabolic processes, producing oxidative stress and genetic alterations (Fig. 7.3). Some of these processes are further described in this chapter.

7.3 NP interactions with water in plant roots

Some metallic NPs reduce root hydraulic conductivity and consequently decrease plant water content, turgor potential, and growth (Martínez-Fernández et al., 2011); however, it is unknown how this process occurs. Martínez-Fernández et al. (2016) found a reduction in hydraulic conductivity in roots of *H. annuus* L. plants treated with nFe_2O_3 in hydroponic culture. Still, no change was found in the internal water status of plants grown in soil amended with the same NPs. The reduction in dry biomass of plants exposed to nFeOx was related to the aggregation of NPs on the root surface, affecting water entry, implying a decrease in plant growth (Martínez-Fernández et al., 2016; Trujillo-Reyes et al., 2013).

Asli and Neumann (2009) reported reduced water transport capacity in primary roots of maize (Zea mays L.) plants exposed to $nTiO_2$ (30 nm). They also found that $nTiO_2$ inhibits transpiration and reduces hydraulic conductivity in the root apices of maize, causing contraction of the cell wall pores, which inhibited shoot and root growth. Interaction of NPs with microorganisms related to roots is also important. This topic is discussed in Chapter 8 of this book.

7.4 Plant roots and soil

7.4.1 Plant roots change the physicochemical properties of the soil

Plant roots secrete a large number of exudates that include inorganic ions, small molecular weight organic substances (phenols, aldehydes, amino acids, and organic acids), and high molecular weight organic substances (mycelium and cellulose), forming a microenvironment around the root called the "rhizosphere" (Bais et al., 2006). Root tissues and associated root exudates provide a physical refuge for many

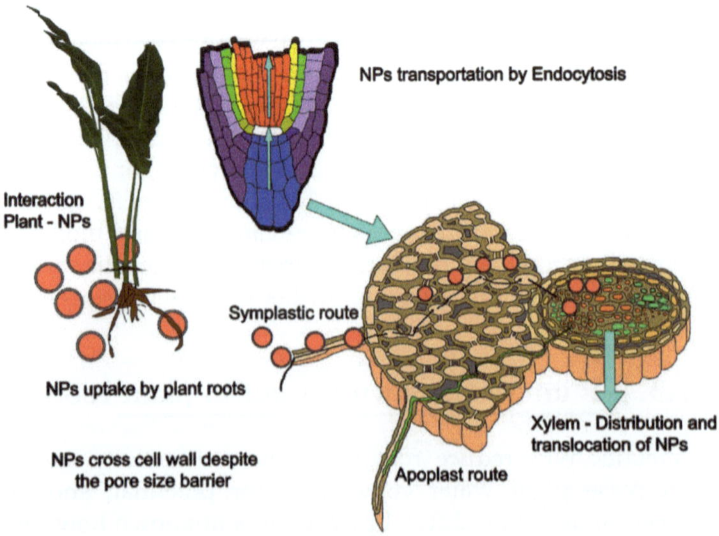

FIGURE 7.3 Interaction Plant-NPs. Apoplastic and symplastic routes.

microorganisms. Roots and soil particles are frequently in contact with root hairs, mucilage, and microbes that form a zone of multiple plant-soil interactions. Mucilage is a polysaccharide that protects root cells during their penetration into the soil, it can be of bacterial and plant origin, and can bind soil particles upon drying (Czarnes et al., 2000; Materechera et al., 1992). When the particles are rewetted, the binding with plant mucilage is through sugars, while the binding with bacterial mucilage is mediated by proteins (Bais et al., 2001, 2006).

Roots utilize the cracks and fissures of rocks and, through repeated cycles of wetting and drying and chemical modifications in the rhizosphere, increase soil volume. Roots growing in rock fissures adapt morphologically; their outer bark becomes flattened, while the internal water-conducting tissues remain cylindrical (Zwieniecki & Newton, 1995). It has been shown that roots and microflora favor the chemical weathering of minerals to generate soil materials. Hinsinger and Jaillard (1993) studied the vermiculitization of phlogopite mica induced by ryegrass and rapeseed roots under laboratory conditions. Some of the chemical processes involved in these studies are the release of protons and organic anions by roots, e.g., depletion of cations at sufficiently low concentrations destabilizes crystal lattices. Other studies focus on chemical transformations. For example, in soil seed rape, the irreversible transformation of phlogopite to vermiculite was caused by severe root-induced acidification of the rhizosphere, leading to the acid dissolution of the phlogopite network (Hinsinger & Jaillard, 1993).

The development of water-stable aggregates is an essential process in the genesis of soils because it largely determines soil characteristics, such as aeration, infiltration, and erodibility. Herein, plant roots play a crucial role through the release of carbon compounds (Young & Crawford, 2004) and directly through wetting and drying phenomena, accumulation of inorganic chemicals on the root surface that acts as cementing agents, the release of organic compounds that promote particle aggregation, and the role of undecayed senescent roots acting as steel rods in reinforced concrete. Traoré et al. (2000) investigated different organic materials released by roots involved in soil aggregation. The experiment consisted of mixing luvisol with corn root mucilage, glucose, polygalacturonic acid, and a "model" soluble exudate composed of a mixture of glucose, amino acids, and organic acids. In that experiment, the addition of amendments was more significant than the concentration of soluble exudates typically found in soils, and significant effects on soil structure were observed. The additions increased the stability of the aggregates in water, although the effect of glucose was lower relative to the other amendments used. The proportion of water-stable aggregates increased when mucilage and model exudates were added but decreased in the polygalacturonic acid treatment. Other studies have shown that cycles of soil wetting by rain and drying by roots affect soil aggregation (Materechera et al., 1992). Also, Czarnes et al. (2000) examined the interaction of dextran, xanthan, and polygalacturonic acid exudates and soil wetting and drying. In general, it was observed that soil wetting and drying increased sorptivity and decreased repellency, except in soils treated with polygalacturonic acid. Thus, polygalacturonic acid stabilized the rhizosphere soil structure by increasing the strength of interparticle bonds and reducing the rate of wetting.

7.4.2 Formation of metallic complexes with root exudate components

Low molecular weight organic acids (LMWOAs), such as oxalic acid, citric acid, and malic acid present in root exudates play an important role in heavy metal chelation, nutrient dissolution, and acidification in the rhizosphere, which impacts heavy metal adsorption capacity by soil particles and heavy metal uptake by plants (Lu et al., 2013). On the other hand, mucilages are highly hydrated materials that improve the contact between roots and soil particles and facilitate exchange processes between the root surface and adjacent soil under dry conditions (Nambiar, 1976). Apical mucilages consist of a central cellulosic core coated with polysaccharides containing glucose, fucose, and galacturonic acid (Wright & Northcote, 1974). Mench et al. (1985) proved that corn root mucilages bind heavy metals, particularly with the carboxyl groups of uronic acids located in the polysaccharide chains. Proteins or glycoproteins could also be involved in metal binding. In addition, the presence of uronic acids in root exudates allows the formation of high molecular weight gels in cation-rich media, where cations act as bridges between consecutive uronic chains (Cortez, 1982). Metals probably bind in two ways to root mucilages, depending on their binding sites: (a) ions may be integrated into the gel acting as a bridge, and (b) ions may be bound by more labile forms to isolated uronic units. The introduction of metal ions to mucilages occurs through an exchange of the central cations (Ca^{2+} and Mg^{2+}) which have a lower affinity for pectin binding sites than heavy metals (Makridou, 1977; Morel et al., 1986). Likewise, in another study (Morel et al., 1986), it was observed that lead is the most efficient metal in the desorption of major cations from mucilage. In addition, other root components, clays, and microbial products, among others, combine with mucilages to form a mixed product called mucigel (Jenny & Grossenbacher, 1963), which also appears to be a site of heavy metal accumulation. Such accumulation is of considerable importance since heavy metals can alter the biodegradation of mucilages.

Low molecular weight exudates display strong, weak, and very weak acidity. Strong acidity is related to the presence of simple organic acids (citric acid, malic acid) and amino acids (glutamic acid, lysine, glycine); nitrate and phosphate residues also contribute to strong acidity (Mench et al., 1988). Weak acidity was attributed to the high content of simple compounds in the soluble exudates (Mosoni et al., 1979) and the presence of uronic acids (Sillén & Martell, n.d.). Hydroxyl or OH-phenolic groups are related to very weak acidity (Bizri et al., 1984); low concentrations of phenolic compounds were also associated with very weak acidity of exudates. The study analyzed the formation of complexes with copper (Cu), lead (Pb), and zinc (Zn), and soluble root exudates of low molecular weight, finding that the stability of the complexes formed followed the order Cu > Pb > Zn at pH 5.0. However, the order of affinity found in this work was different from that obtained with exudates of higher molecular weight (Pb > Cu) (Morel et al., 1986). The main difference seems to derive from the composition of the exudates, since high molecular weight exudates are rich in polysaccharides with uronic acid units that provide some of the characteristics of pectins, such as complexing properties (Pb > Cu > Zn) (Makridou, 1977; Morel et al., 1986).

Different aluminum (Al)-tolerant plant species can secrete organic acids from the root apex in response to stress caused by Al exposure and form stable and innocuous complexes with Al^{3+} to detoxify the rhizosphere from this element. The organic acids secreted in response to the presence of Al differ according to plant species. For example, wheat and *A. thaliana* secrete malate and citrate, maize and soybean secrete mainly citrate, and *Fagopyrum esculentum* (buckwheat) secretes oxalate (Hoekenga et al., 2003; Jia et al., 2005; Pellet et al., 1996; Ryan et al., 2009).

Citrate was the most effective detoxifying compound due to its strong Al^{3+} chelating capacity (Kochian et al., 2004).

7.4.3 Bioavailability of metal oxide nanoparticles in the soil-root system

Root exudates can induce the transformation of NPs in the rhizosphere, altering their bioavailability. Stegemeier et al. (2015) found that alfalfa (*Medicago sativa* L.) root exudates dissolved Ag_2S NPs, making them more available to plants. Huang et al. (2017) demonstrated that the association of organic acids from plant roots with Cu NPs significantly decreased Cu uptake and bioaccumulation in plants. Other studies have suggested that the aggregation, sedimentation, and dissolution processes of CuO NPs in the rhizosphere were related to citric acid, the most abundant low molecular weight organic acid in rice root exudates (Peng et al., 2019).

Some reports on the biotransformation of ZnO NPS indicate these NPS are not internalized in plants but are present as Zn(II) species. For example, in soybean plants exposed to ZnO NPs, Zn was found as Zn citrate (Hernandez-Viezcas et al., 2013), whereas in sand-grown wheat, it was identified as Zn phosphate (Dimkpa et al., 2012), and as Zn citrate in soil-grown cowpea plants (Wang et al., 2013). Depending on the Zn^{2+}concentration, there may be, at least partially, Zn toxicity.

Peng et al. (2019) determined the bioavailability of metal oxide nanoparticles (MONP) in the soil-root system. They examined the accumulation and microdistribution of ZnO, CuO, and CeO_2 nanoparticles (NPs) (50, 100, and 500 mg/kg) in rice field soil. The results showed that exposure to 500 mg/kg CuO and CeO_2 NPs induced primary accumulation of Cu and Ce elements in rice roots, up to 235.48 mg Cu/kg and 164.84 mg Ce/kg, respectively, while the Zn concentration in shoots was up to 313.18 mg/kg. Therefore, the effect of NPs on plant growth is mainly related to the chemical species and solubility of the NPs. Using micro-XRF, these researchers showed that Zn was mainly localized in the root cortex, while Cu was accumulated primarily on the root exodermis and little Ce was distributed in the root (Peng et al., 2019). Thus, it is of great importance to assess the environmental risks of MONPs in soil and to ensure the safety of agricultural products.

Iron oxide NPs are also used as nanofertilizers. A study on the efficacy of Fe_2O_3 NPs as a fertilizer for *Arachis hypogaea* has revealed that Fe_2O_3 NPs and Fe_2O_3-EDTA effectively increased root length, plant height, and biomass by regulating phytohormones and antioxidant enzymes' activity (Nhan et al., 2016). In addition, Fe_2O_3 NPs were adsorbed onto the soil, increasing iron availability to plant roots. Another study showed that growth parameters of *Solanum lycopersicum* improved under the influence of Fe_2O_3 NPs (Shankramma et al., 2016). The NPs were poured into the root hairs and root tips, followed by the nodal and mid-zones of the plant. The authors proposed that biomineralization of the NPs occurred due to the presence of rich phytochemicals in the plants. Furthermore, exposure of Bt transgenic and nontransgenic cotton to Fe_2O_3 NPs ($1000\ mg\ L^{-1}$) showed the presence of dark spots (particles) located mainly in the endodermis and vascular cilium (Nhan et al., 2016). The uptake of Fe_2O_3 NPs and their aggregation in roots was evident; iron content in shoots and roots increased in a dose-dependent manner. It was speculated that bioaccumulation of Fe_2O_3 NPs in these milkweeds could represent a potential risk to the environment and human health.

The fate of these ENMs in the ecosystem and food chain integrity is of great concern (Fig. 7.4). The application of ENMs to some plant species has shown remarkable changes in the antioxidant enzymes' activity and the upregulation of heat-shock proteins. Various plant species have evolved antioxidant defense mechanisms to prevent oxidative damage and enhance plant resistance to toxicity caused by the metal or metal oxide ENMs; these mechanisms are yet to be fully explored and understood (Anjum et al., 2015). The ENM absorption and translocation to different plant parts depend on their bioavailability, concentration, solubility, and exposure time (Husen & Iqbal, 2019). The type of nanomaterial and plant species are also determining factors.

7.5 Nanoparticle internalization in root cells

NPs uptake by roots occurs when a particle penetrates the cell wall and plasma membranes of the epidermal layers. Subsequently, it enters the

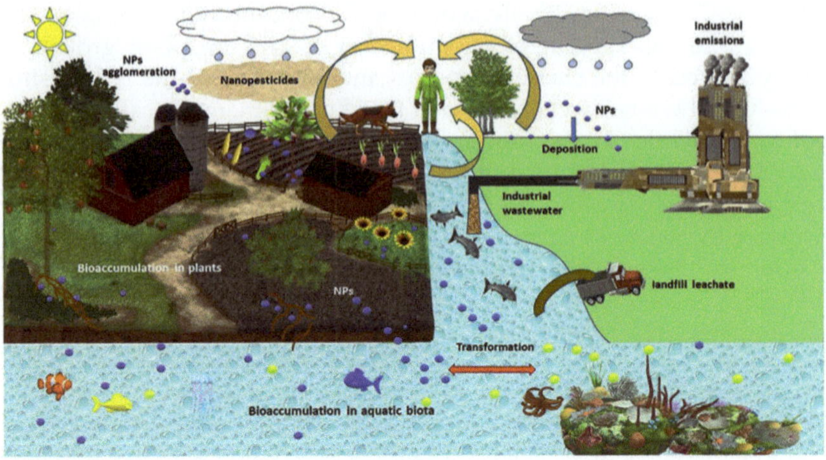

FIGURE 7.4 Interaction and fate of nanoparticles (NPs) in the ecosystem and the integrity of the food chain.

vascular tissues (xylem) and is translocated through the stems and leaves (Fig. 7.5). From its embryonic stage, the root allows the incorporation of ENMs into the plant (Behzadi et al., 2017). In root cells of *Phaseolus radiatus* and *Sorghum bicolor*, the uptake of Ag nanoparticles was confirmed (Lee et al., 2012). The root surface generally exhibits a negative charge that attracts positively charged NPs. For example, it has been demonstrated that CeO_2 NPs, with negative charges or neutral moieties on the surface, are more easily translocated than those with positive charges, which are more likely to remain adsorbed to the root surface (Asati et al., 2010). In the processes of uptake and transport, NPs must pass through a series of barriers from the surface to the xylem vessels, including the root surface cuticle, epidermis, cortex, and endodermis, being transported through the shoots via the xylem. Three routes have been described for this purpose: cell wall pores, apoplastic pathway, and symplastic pathway through plasmodesmata (Lv et al., 2019).

7.5.1 Cell wall pores

The plant cell wall acts as a semipermeable barrier that regulates the trafficking of materials across the membrane through various pores to the plant cells. In order to pass through these pores, the size, orientation, and properties of NPs play a critical role. NPs or their aggregates having a size smaller than the diameter of the cell wall pores easily penetrate the cell wall

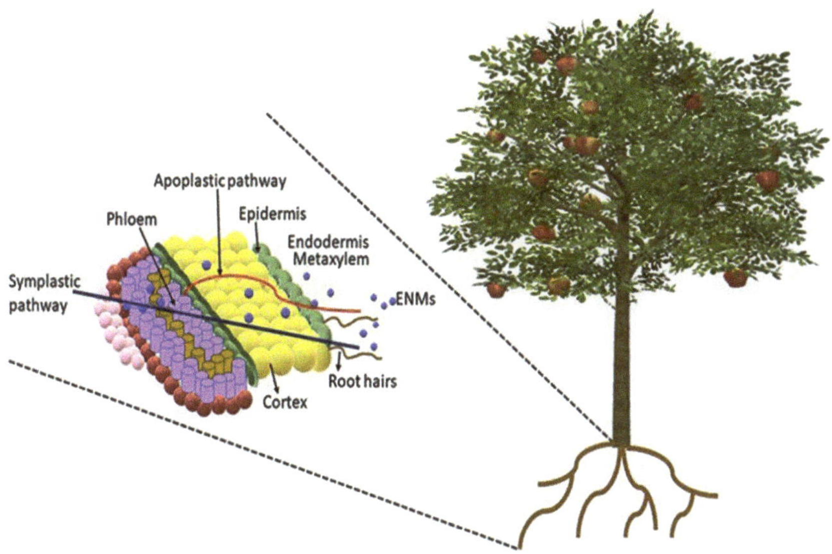

FIGURE 7.5 Uptake of ENMs by cell walls and plasma membranes of the epidermal layers of the roots.

and reach the plasma membrane (Navarro et al., 2008). NPs may enter plant cells by binding to carrier proteins and ion channels, by endocytosis, by binding to organic acids present in the culture media, or by forming complexes with the transporters of the plasma membrane (Kurepa et al., 2010). The cell walls, through which water molecules and other solutes must pass to enter the roots, are a porous network of polysaccharide fiber matrices. The pore size of plant walls is usually in the range of 3–8 nm, which is much smaller than that of many tested NPs (Carpita & Gibeaut, 1993). Navarro et al. (2008) hypothesized that cell walls, which are typically 5–20 nm thick, function as natural sieves. Therefore, aggregates of NPs smaller than the largest pore size are expected to pass through and reach the plasma membrane, and aggregates of larger particles are not likely to enter plant cells. However, studies indicate that NPs can induce the formation of new large pores that allow the internalization of large NPs through cell walls (Gracia-Pinilla et al., 2008).

7.5.2 Apoplastic pathway

In wheat plants grown in hydroponics, the uptake of citric acid-coated Fe_2O_3 NPs (20 mg L^{-1}) took place through the apoplastic pathway in roots (Iannone et al., 2016). Fluorescence imaging and transmission electron microscopy revealed that magnetic Fe_2O_3 NPs penetrated the root epidermis of maize and migrated from the epidermis to endodermal cells via the apoplastic route (Li et al., 2016). Therefore, it was hypothesized that macromolecular exudates excreted by root cells could be responsible for the accumulation of NPs in the root epidermis.

Another related study analyzed the uptake and accumulation of anatase TiO_2 NPs (12 nm), demonstrating the uptake of these NPs by roots of wheat plants from the hydroponic medium. After seven days of exposure, TiO_2 NPs were observed in root parenchyma cells and in the vascular cylinder, indicating the transfer of NPs from the medium to wheat plant tissues (Larue et al., 2014). On the other hand, the uptake and transport of TiO_2 NPs were studied in two crops (*Phaseolus vulgaris* and *T. aestivum*), a wetland species (*Rumex crispus*), and a floating aquatic plant (*Elodea canadensis*). All rooted plants showed higher concentrations of Ti in the roots. Ti from these NPs was translocated from roots to shoots in *R. crispus*. It has been proposed that this species could increase the availability of metal NPs by producing siderophores or by altering rhizosphere pH (Jacob et al., 2013).

Quantum dots (QDs) are important metal NPs used in various fields (Bera et al., 2010). In *A. thaliana* exposed to CdSe/CdZnS QDs, intact QDs were not internalized by plants. Analysis of Cd and Se in roots and leaves showed that QDs were mainly present on the outer surfaces of roots and were not transported to plant leaves (Navarro et al., 2012; Sun et al., 2021).

Thus, the strong adsorption of QDs on root surfaces is possibly due to van der Waals interactions and the cross-linking of CO_2 clusters on plant roots and QDs. Uptake of QDs occurs when QDs are conjugated with nitrogenous organic compounds such as arginine, glycine, and chitosan (Silver & Ou, 2005).

7.5.3 Symplastic pathway

Several studies have confirmed that the symplastic pathway is the dominant pathway for NPs transport within plants (Lv et al., 2019). This pathway is facilitated by important transporter proteins, aquaporins, and interconnected channels, as well as by using the endocytosis process. Particulate forms of mineral nutrients can be immobilized and mobilized through the xylem and phloem. Both transport systems, the plasma membrane (apoplast) and the plasmodesic (symplasmic), play a crucial role along with water, which is the carrier not only for nutrients but also for nanoparticles (de la Rosa et al., 2021).

7.5.4 Root water transport

The water present in soil is absorbed by root structures, primary roots, root cap, lateral roots, and root hairs and is mobilized radially through the xylem tracheid. The movement of NPs is determined by the nature of these nanomaterials and by the biology of the plant (Martínez-Fernández et al., 2016; Zhao et al., 2021). The movement of water through the root apoplast is driven by pressure and osmotic gradient; it has been suggested that NPs travel through the epidermis, cortex, and ultimately, the entire plant (Aiken & Smucker, 1996). Maintenance of root water transport capacity is essential to meet water utilization for evaporation and leaf growth in plants. Reduced water supply in roots can lead to xylem stress, inhibition of leaf growth, stomatal closure, reduced transpiration, and ultimately plant wilting. Imbalance in water transport by roots can occur because of environmental stress. Any material present in external water sources must pass through the cell wall pores of the root epidermal layer to reach the root xylem by parallel transport of molecules through the symplastic and apoplastic pathways. Plant cell walls act as semipermeable membranes that allow the passage of water and small molecules, blocking the entrance of relatively large species (Asli & Neumann, 2009, 2010). Therefore, the presence of NPs in soil water sources can affect root hydraulic conductivity and, ultimately, transpiration. Asli and Neumann (2010) mentioned that the physical interaction between NPs and the root cell wall, the increased interfacial viscosity, and the subsequent decrease in pore diameter are considered reasons behind the inhibition of root hydraulic conductivity.

7.5.5 Mechanisms: passive diffusion and facilitated transport

After the first exposure of NPs to the root, two mechanisms occur: passive diffusion and facilitated transport (da Cruz et al., 2019; Rossi et al., 2017; Zhao, Peralta-Videa et al., 2011).

Passive diffusion—the passage of water across the cell wall is mediated by passive transport. The passive diffusion of NPs is usually associated with the transport of low molecular weight substances and endogenous metabolites across cell membranes (Santner et al., 2012). This pathway does not require energy investment and allows the free movement of molecules from higher to lower concentrations. In experiments performed by Zhao, Peng, et al. (2012), the existence of the CeO_2 was demonstrated in roots of corn plants grown in soil amended with these nanoparticles, and a passive uptake was suggested. The same study also revealed that alginic acid coating increased the transport of CeO_2 NPs to the root cortex. Rossi et al. (2017) investigated the effect of CeO_2 and Cd^{2+} NPs on their uptake and accumulation by soybean seedlings grown in soil. They determined that the presence of Cd increased Ce concentration in soybean roots and leaves. It was observed that the root barrier did not prevent the uptake of Cd. This also confirmed the presence of passive diffusion through plant cell walls and membranes. The joint presence of both species increased the production of root exudates, which influence the dissolution of NPs; due to this, Ce uptake increased. Under the same context, it is known that plant roots contain many ion transporters and proteins in their membranes, so they may act concomitantly during the transport of other metal ions.

Facilitated transport—facilitated transport is a process of transport of molecules or ions across a biological membrane by specific integral transmembrane proteins. In this regard, it has been reported that the 13 mineral nutrients (Fe, B, Cu, Mn, Mo, Si, Zn, Ni, Cl, Se, Na, Al, and Co) required in trace amounts are transported into the plant with the help of specific transport proteins (transporters-channels) present in roots (de la Rosa et al., 2021). In the same context, several studies have been carried out to explain the facilitated transport of NPs from roots to shoots. da Cruz et al. (2019) analyzed the uptake of Zn by *P. vulgaris* treated with ZnO NPs (40 and 300 nm) and $ZnSO_4$ (100 and 1000 mg Zn/L) for 48 h and described the physiological effects of these treatments. Their results indicated that 40 nm ZnO NPs were more easily dissolved by the roots than those with a size of 300 nm. According to the data, root-to-shoot Zn-translocation presented a decreasing gradient of concentration and rate of absorption. It was also observed that Zn movement occurred through the axial xylem system. Transcriptional analysis of genes encoding for metal carriers indicated a higher expression of tonoplast-localized transporters. These transporters regulate the content and volume of vacuoles in plant cells and are integrated by three major components: (a) two proton pumps, (b) the vacuolar H^+-ATPase (V-ATPase) and H^+-pyrophosphatase (V-PPase), and (c) aquaporins. Tonoplast transporters include metal ion transporters, vacuolar Ca^{2+}-ATPase, vacuolar Ca^{2+}/H^+ antiporter, vacuolar Na^+/H^+ antiporter, and Zn transporter (Pandey & Sanyal, 2020). No specific information has been provided to determine if these transporters are also involved in NPs' movement.

7.5.6 Toxicity of NPs and root pathways

The toxicity of NPs in the plant system depends on their dose, form, and size (Husen & Iqbal, 2019). NPs are able to infiltrate living plant tissues and migrate to different regions of the plant system (Corredor et al., 2009). For example, in the aqueous medium or soil matrix, they move through the symplastic or apoplastic region to penetrate the root epidermis, pass through the cortex, and finally translocate to the stem and leaves through the xylem and phloem, so they usually cause a wide impact on the plant system.

It is worth mentioning that plant growth inhibition may not derive directly from the chemical phytotoxicity of nanoparticles (Hochella et al., 2019). Studies have mentioned that toxicity may also derive from physical interactions between nanoparticles and plant cell transport pathways, the inhibition of apoplastic trafficking by blocking cell wall intercellular spaces or cell wall pores, or the inhibition of symplastic connections between cells by blocking nanometer-sized plasmodesmata (Ma et al., 2010; Phogat et al., 2016; Yan & Chen, 2019). The latest work showed that the diameter of maize root cell wall pores was reduced from 6.6 to 3.0 nm by pretreatment of nanoparticles (Ma et al., 2010). Another factor that needs to be considered in phytotoxicity studies is the solvent effect (Vijayakumar et al., 2021). The duration of nanoparticles in solution is very short without stabilizers and most commercial products contain certain stabilizers. Other researchers who investigated the uptake of ZnO NPs by ryegrass (*L. perenne*) noticed no upward translocation of ZnO NPs from roots to shoots (Lin & Xing, 2008). ZnO NPs primarily adhered to the root surface, and individual nanoparticles were also observed in the apoplast and protoplast spaces in root endodermis and stele.

Ma et al. (2010) observed that AgNPs as large as 40 nm can be taken up by the roots of *A. thaliana* and transported to the shoots, even though the majority of AgNPs adhered to the root cap. In addition, it was observed that AgNPs accumulate in an unexpected tissue (columella) in the roots. Martínez-Fernández et al. (2016) exposed *Helianthus annus* L. plants for 55 days to γ-Fe_2O_3 NPs concentrations of 50 and 100 mg/L in hydroponics. According to their results, blockage of IRT-1 membrane protein transporters in roots occurred. In addition, a change in the spatial distribution of these proteins and modification of complex systems, such as the water transport system, were also observed. Another study showed that uptake and transport of nanocrystals from roots to leaves occurs within a few days in *Phalaenopsis* spp. and *A. Thaliana* when watered with an aqueous colloidal solution of NaYF4:YbEr nanocrystals (Hischemöller et al., 2009). NaYF4:YbEr are upconversion fluorescence nanocrystals that emit visible light under near-infrared light. Orchid roots were soaked in the NaYF4:YbEr colloidal solution for a different time, and NaYF4:YbEr nanoparticles in plant tissues were visualized using a confocal laser scanning microscope. The images clearly demonstrated the route of the penetration of the nanocrystals into plant tissues: from velamen radicum (a special epidermis occurring only in aerial roots) to passage cells in about 10 minutes, until they finally reached vascular tissues in days. This work is probably the first study to show the uptake kinetics and illustrate NPs' potential penetration routes into plant tissues. However, whether this observation is true for subterranean roots is unknown.

7.6 Biotransformation of ENMs inside root structures

As it has been previously stated, ENMs can undergo different types of transformation with the help of plant components. It is well known that root exudates in the rhizosphere play a key role in determining the behavior and toxicity of heavy metals. Most ENMs are readily adsorbed and aggregated on the root surface (Phogat et al., 2016). This physical transformation will limit the uptake of NPs in roots and their subsequent translocation. On the other hand, the direct contact of NPs with root exudates increases the possibility of NP transformation. It should be noted that in the case of metallic NPs (Zahraee et al., 2020), dissolution is the most common transformation process affecting their behavior and fate in plants. Some studies have focused on the biotransformation of ZnO nanoparticles (Dimkpa et al., 2013). Synchrotron-based studies (XANES) have indicated no pristine ZnO is internalized in plants, instead it is mostly presented as transformed Zn(II) species such as Zn citrate in soybean plants (Hernandez-Viezcas et al., 2013), Zn phosphate in sand-grown wheat (Dimkpa et al., 2012), and Zn citrate, histidine, and phytate in soil-grown cowpea plants (Wang et al., 2013). Consequently, toxic effects will be induced by the released Zn^{2+} cations. Another study reported that ZnO NPs were transformed from Zn^{2+} to Zn nitrite or Zn acetate in germinating soybean roots, while CeO_2 NPs remained unchanged (López-Moreno et al., 2010). On the other hand, metal-based nanoparticles such as CuO, ZnO, and Ag can dissolve and release metal ions and chemically transform by reacting with inorganic (e.g., sulfur and phosphate) or organic substances that exist in the environment and in plants (Dimkpa et al., 2012; Zhang et al., 2018). ENMs can also undergo these transformations by interacting with root exudates, as ENMs directly interact with the plant root in most cases (Huang et al., 2017). These transformations will affect the final fate and toxicity of ENMs in plants. In the same context, organic acids excreted by plant roots play a critical role in the biotransformation of ENMs in plants by promoting dissolution. Interesting studies have addressed the study of the biotransformation of rare earth oxide NMs in root exudates. Another study found a large number of needlelike $LaPO_4$ clusters in the intercellular regions, as well as in the vacuole and cytoplasm of cucumber root germinated with La_2O_3 NP treatment for five days, indicating significant biotransformation of NPs in plants. An in vitro experiment suggested that organic acids excreted by the root play a critical role in the transformation process (Y. Ma et al., 2011). Organic acids promote the release of La^{3+} through the formation of La-carboxylate complexes that can be subsequently precipitated by phosphates that exist widely in plants. Zhao et al. (2021) showed that a concentration as low as 0.32 $mgL^{-1}$$Yb_2O_3$ ENMs for 14 days reduced cucumber plant biomass production. It was also observed that Yb^{3+} concentrations in the rhizosphere solution were much higher than those in the exposure solution. The organic acids promoted the dissolution of Yb_2O_3. These nanoparticles are very reactive and release Yb^{3+} ions more easily than those of their bulk counterpart, resulting in a higher concentration of Yb^{3+} in the rhizosphere solution, thus resulting in a higher toxicity. Other studies have observed that ENMs can not only transform outside the roots, but they can also first enter the roots and then transform in the plants. For example, in Yb^{3+} treated cucumber roots, flocculent $YbPO_4$ was only found in the intercellular regions; however, in Yb_2O_3 treatment, $YbPO_4$ was also found in the vacuole and cytoplasm, indicating that Yb_2O_3 NP uses the symplastic pathway, since it probably crosses the cell wall, enters the cytoplasm and vacuole, and transforms into $YbPO_4$. Reduction and oxidation are important reactions that commonly occur in soil-plant systems. As mentioned above, several ENMs contain metal elements with changing valence states that can undergo reduction, oxidation, and subsequent transformation upon interaction with biogenic redox agents in plants (Wang et al., 2012; Yin et al., 2011; Zhou et al., 2011). In this approach, studies have been conducted in detail on the environmental transformation of Ag. Yin et al. (2011) observed that metallic Ag NPs were oxidized as Ag_2S or Ag_2O in the roots of *Lolium multiflorum*.

Other studies have found that CuO NPs translocated from shoots to the root and were partially reduced and transformed into Cu_2S and Cu_2O during translocation (Wang et al., 2012). A similar transformation of CuO NPs was observed in sand-grown wheat (Dimkpa et al., 2012).

7.7 Conclusions

The nature of ENMs has a high impact on plants, which are essential elements of ecosystems, as well as on the global food supply and environmental health. Interactions of ENMS with roots include uptake, translocation, and accumulation, which may change plant metabolism. The available literature on ENM-plant interactions allows the following conclusions:

1) ENMs present in the soil accumulate in the root epidermis and adhere to surface tissues as individual particles or aggregates that are affected not only by ENMs properties but also by soil properties.
2) In terms of soil mechanics, nanosols readily dissolve and destroy or disrupt plant roots; moreover, they reduce leaf size, stem growth, and root proliferation.
3) ENMs have a very high specific surface area. Thus, they react very actively with other soil matrix particles and plant root surfaces.
4) The presence of nanoparticles, even in small quantities, can affect different soil properties including alterations in soil strength, permeability, hardness indices, and resistance.
5) It is important to explore whether ENMs interacting with plant roots compromise crop yields, soil properties, and even soil organisms and soil functional activities.

References

Aiken, R. M., & Smucker, A. J. M. (1996). Root system regulation of whole plant growth. *Annual Review of Phytopathology, 34,* 325–346. https://doi.org/10.1146/annurev.phyto.34.1.325

Anjum, N. A., Adam, V., Kizek, R., Duarte, A. C., Pereira, E., Iqbal, M., Lukatkin, A. S., & Ahmad, I. (2015). Nanoscale copper in the soil-plant system—toxicity and underlying potential mechanisms. *Environmental Research, 138,* 306–325. https://doi.org/10.1016/j.envres.2015.02.019

Asati, A., Santra, S., Kaittanis, C., & Perez, J. M. (2010). Surface-charge-dependent cell localization and cytotoxicity of cerium oxide nanoparticles. *ACS Nano, 4*(9), 5321–5331. https://doi.org/10.1021/nn100816s

Aslani, F., Bagheri, S., Muhd Julkapli, N., Juraimi, A. S., Hashemi, F. S. G., & Baghdadi, A. (2014). *Effects of engineered nanomaterials on plants growth: An overview.* Scientific World Journal. https://doi.org/10.1155/2014/641759, 2014.

Asli, S., & Neumann, P. M. (2009). Colloidal suspensions of clay or titanium dioxide nanoparticles can inhibit leaf growth and transpiration via physical effects on root water transport. *Plant, Cell and Environment, 32*(5), 577–584. https://doi.org/10.1111/j.1365-3040.2009.01952.x

Asli, S., & Neumann, P. M. (2010). Rhizosphere humic acid interacts with root cell walls to reduce hydraulic conductivity and plant development. *Plant and Soil, 336*(1), 313–322. https://doi.org/10.1007/s11104-010-0483-2

Avellan, A., Schwab, F., Masion, A., Chaurand, P., Borschneck, D., Vidal, V., Rose, J., Santaella, C., & Levard, C. (2017). Nanoparticle uptake in plants: Gold nanomaterial localized in roots of *Arabidopsis thaliana* by X-ray computed nanotomography and hyperspectral imaging. *Environmental Science & Technology, 51*(15), 8682–8691. https://doi.org/10.1021/acs.est.7b01133

Bais, H. P., Loyola-Vargas, V. M., Flores, H. E., & Vivanco, J. M. (2001). Root-specific metabolism: The biology and biochemistry of underground organs. In , *Vol. 37. In Vitro cellular and developmental biology* (pp. 730–741). Plant. https://doi.org/10.1007/s11627-001-0122-y, 6.

Bais, H. P., Weir, T. L., Perry, L. G., Gilroy, S., & Vivanco, J. M. (2006). The role of root exudates in rhizosphere interactions with plants and other organisms. *Annual Review of Plant Biology, 57*, 233–266. https://doi.org/10.1146/annurev.arplant.57.032905.105159

Baum, S. F., Dubrovsky, J. G., & Rost, T. L. (2002). Apical organization and maturation of the cortex and vascular cylinder in *Arabidopsis thaliana* (Brassicaceae) roots. *American Journal of Botany, 89*(6), 908–920. https://doi.org/10.3732/ajb.89.6.908

Behzadi, S., Serpooshan, V., Tao, W., Hamaly, M. A., Alkawareek, M. Y., Dreaden, E. C., Brown, D., Alkilany, A. M., Farokhzad, O. C., & Mahmoudi, M. (2017). Cellular uptake of nanoparticles: Journey inside the cell. *Chemical Society Reviews, 46*(14), 4218–4244. https://doi.org/10.1039/c6cs00636a

Benfey, P. N., & Scheres, B. (2000). Root development. *Current Biology, 10*(22), R813–R815. https://doi.org/10.1016/S0960-9822(00)00814-9

Bera, D., Qian, L., Tseng, T.-K., & Holloway, P. H. (2010). Quantum dots and their multimodal applications: A review. *Materials*, 2260–2345. https://doi.org/10.3390/ma3042260

Bishopp, A., Help, H., El-Showk, S., Weijers, D., Scheres, B., Friml, J., Benková, E., Mähönen, A. P., & Helariutta, Y. (2011). A mutually inhibitory interaction between auxin and cytokinin specifies vascular pattern in roots. *Current Biology, 21*(11), 917–926. https://doi.org/10.1016/j.cub.2011.04.017

Bizri, Y., Cromer, M., Scharff, J. P., Guillet, B., & Rouiller, J. (1984). Constantes de stabilité de complexes organo-minéraux. Interactions des ions plombeux avec les composés organiques hydrosolubles des eaux gravitaires de podzol. *Geochimica et Cosmochimica Acta, 48*(2), 227–234. https://doi.org/10.1016/0016-7037(84)90246-1

Brady, S. M., Orlando, D. A., Lee, J. Y., Wang, J. Y., Koch, J., Dinneny, J. R., Mace, D., Ohler, U., & Benfey, P. N. (2007). A high-resolution root spatiotemporal map reveals dominant expression patterns. *Science, 318*(5851), 801–806. https://doi.org/10.1126/science.1146265

Burello, E., & Worth, A. P. (2011). QSAR modeling of nanomaterials. *WIREs Nanomedicine and Nanobiotechnology, 3*(3), 298–306. https://doi.org/10.1002/wnan.137

Carpita, N. C., & Gibeaut, D. M. (1993). Structural models of primary cell walls in flowering plants: Consistency of molecular structure with the physical properties of the walls during growth. *Plant Journal, 3*(1), 1–30. https://doi.org/10.1111/j.1365-313X.1993.tb00007.x

Corredor, E., Testillano, P. S., Coronado, M.-J., González-Melendi, P., Fernández-Pacheco, R., Marquina, C., Ibarra, M. R., de la Fuente, J. M., Rubiales, D., Pérez-de-Luque, A., & Risueño, M.-C. (2009). Nanoparticle penetration and transport in living pumpkin plants: In situsubcellular identification. *BMC Plant Biology, 9*(1). https://doi.org/10.1186/1471-2229-9-45

Cortez, J. (1982). *Rôle des ions calcium dans la formation du mucigel de Zea mays.*

da Cruz, T. N. M., Savassa, S. M., Montanha, G. S., Ishida, J. K., de Almeida, E., Tsai, S. M., Lavres Junior, J., & Pereira de Carvalho, H. W. (2019). A new glance on root-to-shoot in vivo zinc transport and time-dependent physiological effects of ZnSO4 and ZnO nanoparticles on plants. *Scientific Reports, 9*(1). https://doi.org/10.1038/s41598-019-46796-3

Czarnes, S., Hallett, P. D., Bengough, A. G., & Young, I. M. (2000). Influence de mucilages racinaire et microbiens modeles sur la structure du sol et le transport d'eau. *European Journal of Soil Science, 51*(3), 435–443. https://doi.org/10.1046/j.1365-2389.2000.00327.x

Dimkpa, C. O., McLean, J. E., Britt, D. W., & Anderson, A. J. (2013). Antifungal activity of ZnO nanoparticles and their interactive effect with a biocontrol bacterium on growth antagonism of the plant pathogen Fusarium graminearum. *BioMetals, 26*(6), 913–924. https://doi.org/10.1007/s10534-013-9667-6

Dimkpa, C. O., McLean, J. E., Latta, D. E., Manangón, E., Britt, D. W., Johnson, W. P., Boyanov, M. I., & Anderson, A. J. (2012). CuO and ZnO nanoparticles: Phytotoxicity, metal speciation, and induction of oxidative stress in sand-grown wheat. *Journal of Nanoparticle Research, 14*(9). https://doi.org/10.1007/s11051-012-1125-9

Dolan, L., Janmaat, K., Willemsen, V., Linstead, P., Poethig, S., Roberts, K., & Scheres, B. (1993). Cellular organisation of the *Arabidopsis thaliana* root. *Development, 119*(1), 71–84.

Eichten, S. R., Schmitz, R. J., & Springer, N. M. (2014). Epigenetics: Beyond chromatin modifications and complex genetic regulation. *Plant Physiology, 165*(3), 933–947. https://doi.org/10.1104/pp.113.234211

Falke, K. C., Glander, S., He, F., Hu, J., De Meaux, J., & Schmitz, G. (2013). The spectrum of mutations controlling complex traits and the genetics of fitness in plants. *Current Opinion in Genetics and Development, 23*(6), 665–671. https://doi.org/10.1016/j.gde.2013.10.006

Gracia-Pinilla, M.Á., Pérez-Tijerina, E., Antúnez-García, J., Fernández-Navarro, C., Tlahuice-Flores, A., Mejía-Rosales, S., Montejano-Carrizales, J. M., & José-Yacamán, M. (2008). On the structure and properties of silver nanoparticles. *The Journal of Physical Chemistry C, 112*(35), 13492–13498. https://doi.org/10.1021/jp804085q

Hernandez-Viezcas, J. A., Castillo-Michel, H., Andrews, J. C., Cotte, M., Rico, C., Peralta-Videa, J. R., Ge, Y., Priester, J. H., Holden, P. A., & Gardea-Torresdey, J. L. (2013). In situ synchrotron X-ray fluorescence mapping and speciation of CeO 2 and ZnO nanoparticles in soil cultivated soybean (Glycine max). *ACS Nano, 7*(2), 1415–1423. https://doi.org/10.1021/nn305196q

Hinsinger, P., & Jaillard, B. (1993). Root-induced release of interlayer potassium and vermiculitization of phlogopite as related to potassium depletion in the rhizosphere of ryegrass. *Journal of Soil Science, 44*(3), 525–534. https://doi.org/10.1111/j.1365-2389.1993.tb00474.x

Hischemöller, A., Nordmann, J., Ptacek, P., Mummenhoff, K., & Haase, M. (2009). In-vivo imaging of the uptake of upconversion nanoparticles by plant roots. *Journal of Biomedical Nanotechnology, 5*(3), 278–284. https://doi.org/10.1166/jbn.2009.1032

Hochella, M. F., Mogk, D. W., Ranville, J., Allen, I. C., Luther, G. W., Marr, L. C., McGrail, B. P., Murayama, M., Qafoku, N. P., Rosso, K. M., Sahai, N., Schroeder, P. A., Vikesland, P., Westerhoff, P., & Yang, Y. (2019). Natural, incidental, and engineered nanomaterials and their impacts on the Earth system. *Science, 363*(6434). https://doi.org/10.1126/science.aau8299

Hoekenga, O. A., Vision, T. J., Shaff, J. E., Monforte, A. J., Lee, G. P., Howell, S. H., & Kochian, L. V. (2003). Identification and characterization of aluminum tolerance loci in Arabidopsis (Landsberg erecta x Columbia) by quantitative trait locus mapping. A physiologically simple but genetically complex trait. *Plant Physiology, 132*(2), 936–948. https://doi.org/10.1104/pp.103.023085

Huang, Y., Zhao, L., & Keller, A. A. (2017). Interactions, transformations, and bioavailability of nano-copper exposed to root exudates. *Environmental Science and Technology, 51*(17), 9774–9783. https://doi.org/10.1021/acs.est.7b02523

Husen, A., & Iqbal, M. (2019). *Nanomaterials and plant potential.*

Iannone, M. F., Groppa, M. D., de Sousa, M. E., Fernández van Raap, M. B., & Benavides, M. P. (2016). Impact of magnetite iron oxide nanoparticles on wheat (*Triticum aestivum* L.) development: Evaluation of oxidative damage. *Environmental and Experimental Botany, 131*, 77–88. https://doi.org/10.1016/j.envexpbot.2016.07.004

Jacob, D. L., Borchardt, J. D., Navaratnam, L., Otte, M. L., & Bezbaruah, A. N. (2013). Uptake and translocation of Ti from nanoparticles in crops and wetland plants. *International Journal of Phytoremediation, 15*(2), 142–153. https://doi.org/10.1080/15226514.2012.683209

Jenny, H., & Grossenbacher, K. (1963). Root-soil boundary zones as seen in the electron microscope. *Soil Science Society of America Journal, 273*. https://doi.org/10.2136/sssaj1963.03615995002700030018x

Jia, G., Wang, H., Yan, L., Wang, X., Pei, R., Yan, T., Zhao, Y., & Guo, X. (2005). Cytotoxicity of carbon nanomaterials: Single-wall nanotube, multi-wall nanotube, and fullerene. *Environmental Science and Technology, 39*(5), 1378–1383. https://doi.org/10.1021/es0487291

Kochian, L. V., Hoekenga, O. A., & Piñeros, M. A. (2004). How do crop plants tolerate acid soils? Mechanisms of aluminum tolerance and phosphorous efficiency. *Annual Review of Plant Biology, 55*, 459–493. https://doi.org/10.1146/annurev.arplant.55.031903.141655

Kurepa, J., Paunesku, T., Vogt, S., Arora, H., Rabatic, B. M., Lu, J., Wanzer, M. B., Woloschak, G. E., & Smalle, J. A. (2010). Uptake and distribution of ultrasmall anatase TiO2 alizarin red s nanoconjugates in arabidopsis thaliana. *Nano Letters, 10*(7), 2296–2302. https://doi.org/10.1021/nl903518f

Larue, C., Castillo-Michel, H., Sobanska, S., Trcera, N., Sorieul, S., Cécillon, L., Ouerdane, L., Legros, S., & Sarret, G. (2014). Fate of pristine TiO2 nanoparticles and aged paint-containing TiO2 nanoparticles in lettuce crop after foliar exposure. *Journal of Hazardous Materials, 273*, 17–26. https://doi.org/10.1016/j.jhazmat.2014.03.014

Laux, T., Würschum, T., & Breuninger, H. (2004). Genetic regulation of embryonic pattern formation. *Plant Cell, 16*, S190–S202. https://doi.org/10.1105/tpc.016014

Lee, W. M., Kwak, J. I., & An, Y. J. (2012). Effect of silver nanoparticles in crop plants Phaseolus radiatus and Sorghum bicolor: Media effect on phytotoxicity. *Chemosphere, 86*(5), 491–499. https://doi.org/10.1016/j.chemosphere.2011.10.013

Li, J., Hu, J., Ma, C., Wang, Y., Wu, C., Huang, J., & Xing, B. (2016). Uptake, translocation and physiological effects of magnetic iron oxide (γ-Fe2O3) nanoparticles in corn (Zea mays L.). *Chemosphere, 159*, 326–334.

Lu, L. L., Tian, S. K., Yang, X. E., Peng, H. Y., & Li, T. Q. (2013). Improved cadmium uptake and accumulation in the hyperaccumulator Sedum alfredii: the impact of citric acid and tartaric acid. *Journal of Zhejiang University Science B, 14*(2), 106–114. https://doi.org/10.1631/jzus.B1200211

Lv, J., Christie, P., & Zhang, S. (2019). Uptake, translocation, and transformation of metal-based nanoparticles in plants: recent advances and methodological challenges. *Environmental Science: Nano, 6*(1), 41–59.

Lin, D., & Xing, B. (2008). Root uptake and phytotoxicity of ZnO nanoparticles. *Environmental Science and Technology, 42*(15), 5580–5585. https://doi.org/10.1021/es800422x

Lin, X.q., Zhu, D.f., Chen, H.z., & Zhang, Y.p. (2009). Effects of plant density and nitrogen application rate on grain yield and nitrogen uptake of super hybrid rice. *Rice Science, 16*(2), 138–142. https://doi.org/10.1016/S1672-6308(08)60070-0

López-Moreno, M. L., De La Rosa, G., Hernández-Viezcas, J. A., Castillo-Michel, H., Botez, C. E., Peralta-Videa, J. R., & Gardea-Torresdey, J. L. (2010). Evidence of the differential biotransformation and genotoxicity of ZnO and CeO2 nanoparticles on soybean (Glycine max) plants. *Environmental Science and Technology, 44*(19), 7315–7320. https://doi.org/10.1021/es903891g

Ma, X., Geisler-Lee, J., Deng, Y., & Kolmakov, A. (2010). Interactions between engineered nanoparticles (ENPs) and plants: phytotoxicity, uptake and accumulation. *Science of the Total Environment, 408*(16), 3053–3061. https://doi.org/10.1016/j.scitotenv.2010.03.031

Ma, Y., He, X., Zhang, P., Zhang, Z., Guo, Z., Tai, R., … Chai, Z. (2011). Phytotoxicity and biotransformation of La_2O_3 nanoparticles in a terrestrial plant cucumber (Cucumis sativus). *Nanotoxicology, 5*(4), 743–753. https://doi.org/10.3109/17435390.2010.545487

Makridou, C. (1977). *Complexation de quelques ions métalliques par les acides galacturonique et glucuronique.*

Ryan, P. R., Raman, H., Gupta, S., Horst, W. J., & Delhaize, E. (2009). A second mechanism for aluminum resistance in wheat relies on the constitutive efflux of citrate from roots. *Plant Physiology, 149*(1), 340–351. https://doi.org/10.1104/pp.108.129155

Martínez-Fernández, D., Barroso, D., & Komárek, M. (2016). Root water transport of Helianthus annuus L. under iron oxide nanoparticle exposure. *Environmental Science and Pollution Research, 23*(2), 1732–1741. https://doi.org/10.1007/s11356-015-5423-5

Martínez-Fernández, D., Walker, D. J., Romero-Espinar, P., Flores, P., & del Río, J. A. (2011). Physiological responses of Bituminaria bituminosa to heavy metals. *Journal of Plant Physiology, 168*(18), 2206–2211. https://doi.org/10.1016/j.jplph.2011.08.008

Materechera, S. A., Dexter, A. R., & Alston, A. M. (1992). Formation of aggregates by plant roots in homogenised soils. *Plant and Soil, 142*(1), 69–79. https://doi.org/10.1007/BF00010176

Mench, M., Morel, J. M., & Guckert, A. (1985). Liaison du cadmium avec la fraction macromoleculaire soluble des exsudats racinaires de mais (*Zea mays l.*). *Comptes Rendus des Seances de l'Academie des Sciences-Series III, 301*(7), 379–382.

Mench, M., Morel, J. L., Guckert, A., & Guillet, B. (1988). Metal binding with root exudates of low molecular weight. *Journal of Soil Science, 39*(4), 521–527. https://doi.org/10.1111/j.1365-2389.1988.tb01236.x

Morel, J. L., Mench, M., & Guckert, A. (1986). Measurement of Pb^{2+}, Cu^{2+} and Cd^{2+} binding with mucilage exudates from maize (Zea mays L.) roots. *Biology and Fertility of Soils, 2*(1), 29–34. https://doi.org/10.1007/BF00638958

Mosoni, L., Petit-Ramel, M. M., Domard, A., & Rinaudo, M. (1979). Potentiometric studies on interactions of α-L-glutamic acid oligomers with Cu(II) in aqueous solutions. *Biopolymers, 18*(9), 2341–2351. https://doi.org/10.1002/bip.1979.360180920

Nambiar, E. K. S. (1976). Uptake of Zn65 from dry soil by plants. *Plant and Soil, 44*(1), 267–271. https://doi.org/10.1007/BF00016978

Navarro, E., Baun, A., Behra, R., Hartmann, N. B., Filser, J., Miao, A. J., Quigg, A., Santschi, P. H., & Sigg, L. (2008). Environmental behavior and ecotoxicity of engineered nanoparticles to algae, plants, and fungi. *Ecotoxicology, 17*(5), 372–386. https://doi.org/10.1007/s10646-008-0214-0

Navarro, D. A., Bisson, M. A., & Aga, D. S. (2012). Investigating uptake of water-dispersible CdSe/ZnS quantum dot nanoparticles by *Arabidopsis thaliana* plants. *Journal of Hazardous Materials, 211–212*, 427–435. https://doi.org/10.1016/j.jhazmat.2011.12.012

Nhan, L., Ma, C., Rui, Y., Cao, W., Deng, Y., Liu, L., & Xing, B. (2016). The effects of Fe2O3 nanoparticles on physiology and insecticide activity in non-transgenic and Bt-transgenic cotton. *Frontiers in Plant Science, 6.*

Ohashi-Ito, K., Saegusa, M., Iwamoto, K., Oda, Y., Katayama, H., Kojima, M., Sakakibara, H., & Fukuda, H. (2014). A BHLH complex activates vascular cell division via cytokinin action in root apical meristem. *Current Biology, 24*(17), 2053–2058. https://doi.org/10.1016/j.cub.2014.07.050

Pandey, G. K., & Sanyal, S. K. (2020). Functional dissection of calcium homeostasis and transport machinery in plants. *Environmental Toxicology and Chemistry, 29*(5), 1146–1154. https://doi.org/10.1002/etc.146

Pellet, D. M., Papernik, L. A., & Kochian, L. V. (1996). Multiple aluminum-resistance mechanisms in wheat (roles of root apical phosphate and malate exudation). *Plant Physiology, 112*(2), 591–597. https://doi.org/10.1104/pp.112.2.591

Peng, C., Tong, H., Yuan, P., Sun, L., Jiang, L., & Shi, J. (2019). Aggregation, sedimentation, and dissolution of copper oxide nanoparticles: Influence of low-molecular-weight organic acids from root exudates. *Nanomaterials, 9*(6). https://doi.org/10.3390/nano9060841

Phogat, N., Khan, S. A., Shankar, S., Ansary, A. A., & Uddin, I. (2016). Fate of inorganic nanoparticles in agriculture. *Advanced Materials Letters, 7*(1), 3–12. https://doi.org/10.5185/amlett.2016.6048

Rawat, M., Nayan, R., Negi, B., Zaidi, M. G. H., & Arora, S. (2017). Physio-biochemical basis of iron-sulfide nanoparticle induced growth and seed yield enhancement in B. juncea. *Plant Physiology and Biochemistry, 118*, 274–284. https://doi.org/10.1016/j.plaphy.2017.06.021

Rico, C. M., Majumdar, S., Duarte-Gardea, M., Peralta-Videa, J. R., & Gardea-Torresdey, J. L. (2011). Interaction of nanoparticles with edible plants and their possible implications in the food chain. *Journal of Agricultural and Food Chemistry, 59*(8), 3485–3498. https://doi.org/10.1021/jf104517j

Ristova, D., & Busch, W. (2014). Natural variation of root traits: From development to nutrient uptake. *Plant Physiology, 166*(2), 518–527. https://doi.org/10.1104/pp.114.244749

de la Rosa, G., Vázquez-Núñez, E., Molina-Guerrero, C., Serafín-Muñoz, A. H., & Vera-Reyes, I. (2021). Interactions of nanomaterials and plants at the cellular level: Current knowledge and relevant gaps. *Nanotechnology for Environmental Engineering, 6*(1). https://doi.org/10.1007/s41204-020-00100-1

Rossi, L., Zhang, W., Schwab, A. P., & Ma, X. (2017). Uptake, accumulation, and in planta distribution of coexisting cerium oxide nanoparticles and cadmium in glycine max (L.) Merr. *Environmental Science and Technology, 51*(21), 12815–12824. https://doi.org/10.1021/acs.est.7b03363

Santner, J., Smolders, E., Wenzel, W. W., & Degryse, F. (2012). First observation of diffusion-limited plant root phosphorus uptake from nutrient solution. *Plant, Cell and Environment, 35*(9), 1558–1566. https://doi.org/10.1111/j.1365-3040.2012.02509.x

Scheres, B., Wolkenfelt, H., Willemsen, V., Terlouw, M., Lawson, E., Dean, C., & Weisbeek, P. (1994). Embryonic origin of the Arabidopsis primary root and root meristem initials. *Development, 120*(9), 2475–2487. https://doi.org/10.1242/dev.120.9.2475

Shankramma, K., Yallappa, S., Shivanna, M. B., & Manjanna, J. (2016). Fe2O3 magnetic nanoparticles to enhance S. lycopersicum (tomato) plant growth and their biomineralization. *Applied Nanoscience*, 983–990. https://doi.org/10.1007/s13204-015-0510-y

Sillén, L. G., & Martell, A. E. (n.d.). Stability constants of metal-ion complexes.

Silver, J., & Ou, W. (2005). Photoactivation of quantum dot fluorescence following endocytosis. *Nano Letters, 5*(7), 1445–1449. https://doi.org/10.1021/nl050808n

Stegemeier, J. P., Schwab, F., Colman, B. P., Webb, S. M., Newville, M., Lanzirotti, A., Winkler, C., Wiesner, M. R., & Lowry, G. V. (2015). Speciation matters: Bioavailability of silver and silver sulfide nanoparticles to alfalfa (Medicago sativa). *Environmental Science and Technology, 49*(14), 8451–8460. https://doi.org/10.1021/acs.est.5b01147

Sun, H., Wang, M., Lei, C., & Li, R. (2021). Cell wall: An important medium regulating the aggregation of quantum dots in maize (Zea mays L.) seedlings. *Journal of Hazardous Materials, 403*, 123960. https://doi.org/10.1016/j.jhazmat.2020.123960

Traoré, O., Groleau-Renaud, V., Plantureux, S., Tubeileh, A., & Boeuf-Tremblay, V. (2000). Effect of root mucilage and modelled root exudates on soil structure. *European Journal of Soil Science, 51*(4), 575–581. https://doi.org/10.1046/j.1365-2389.2000.00348.x

Tripathi, D. K., Shweta, Singh, S., Singh, S., Pandey, R., Singh, V. P., Sharma, N. C., Prasad, S. M., Dubey, N. K., & Chauhan, D. K. (2017). An overview on manufactured nanoparticles in plants: Uptake, translocation, accumulation and phytotoxicity. *Plant Physiology and Biochemistry, 110*, 2–12. https://doi.org/10.1016/j.plaphy.2016.07.030

Trujillo-Reyes, J., Vilchis-Nestor, A. R., Majumdar, S., Peralta-Videa, J. R., & Gardea-Torresdey, J. L. (2013). Citric acid modifies surface properties of commercial CeO2 nanoparticles reducing their toxicity and cerium uptake in radish (Raphanus sativus) seedlings. *Journal of Hazardous Materials, 263*, 677–684. https://doi.org/10.1016/j.jhazmat.2013.10.030

Van Den Berg, C., Willemsen, V., Hendriks, G., Weisbeek, P., & Scheres, B. (1997). Short-range control of cell differentiation in the Arabidopsis root meristem. *Nature, 390*(6657), 287–289. https://doi.org/10.1038/36856

Vijayakumar, S., Chen, J., Amarnath, M., Tungare, K., Bhori, M., Divya, M., González-Sánchez, Z. I., Durán-Lara, E. F., & Vaseeharan, B. (2021). Cytotoxicity, phytotoxicity, and photocatalytic assessment of biopolymer cellulose-mediated silver nanoparticles.

Colloids and Surfaces A: Physicochemical and Engineering Aspects, 628, 127270. https://doi.org/10.1016/j.colsurfa.2021.127270

Wang, P., Menzies, N. W., Lombi, E., McKenna, B. A., Johannessen, B., Glover, C. J., Kappen, P., & Kopittke, P. M. (2013). Fate of ZnO nanoparticles in soils and cowpea (Vigna unguiculata). *Environmental Science and Technology, 47*(23), 13822–13830. https://doi.org/10.1021/es403466p

Wang, Z., Xie, X., Zhao, J., Liu, X., Feng, W., White, J. C., & Xing, B. (2012). Xylem- and phloem-based transport of CuO nanoparticles in maize (Zea mays L.). *Environmental Science and Technology, 46*(8), 4434–4441. https://doi.org/10.1021/es204212z

Wright, K., & Northcote, D. H. (1974). The relationship of root cap slimes to pectins. *Biochemical Journal, 139*(3), 525–534. https://doi.org/10.1042/bj1390525

Yan, A., & Chen, Z. (2019). Impacts of silver nanoparticles on plants: A focus on the phytotoxicity and underlying mechanism. *International Journal of Molecular Sciences, 20*(5), 1003. https://doi.org/10.3390/ijms20051003

Yin, L., Cheng, Y., Espinasse, B., Colman, B. P., Auffan, M., Wiesner, M., Rose, J., Liu, J., & Bernhardt, E. S. (2011). More than the ions: The effects of silver nanoparticles on *Lolium multiflorum*. *Environmental Science and Technology, 45*(6), 2360–2367. https://doi.org/10.1021/es103995x

Young, I. M., & Crawford, J. W. (2004). Interactions and self-organization in the soil-microbe complex. *Science, 304*(5677), 1634–1637. https://doi.org/10.1126/science.1097394

Zahraee, S. M., Shiwakoti, N., & Stasinopoulos, P. (2020). A review on water-energy-greenhouse gas nexus of the bioenergy supply and production system. *Current Sustainable/Renewable Energy Reports, 7*(2), 28–39.

Zhang, J., Elo, J., & Helariutta, Y. (2011). Arabidopsis as a model for wood formation. *Current Opinion in Biotechnology, 22*(2), 293–299. https://doi.org/10.1016/j.copbio.2010.11.008

Zhang, W., Xiao, B., & Fang, T. (2018). Chemical transformation of silver nanoparticles in aquatic environments: Mechanism, morphology and toxicity. *Chemosphere, 191*, 324–334. https://doi.org/10.1016/j.chemosphere.2017.10.016

Zhao, L., Peng, B., Hernandez-Viezcas, J. A., Rico, C., Sun, Y., Peralta-Videa, J. R., Tang, X., Niu, G., Jin, L., Varela-Ramirez, A., Zhang, J. Y., & Gardea-Torresdey, J. L. (2012). Stress response and tolerance of Zea mays to CeO2 nanoparticles: Cross talk among H2O2, heat shock protein, and lipid peroxidation. *ACS Nano, 6*(11), 9615–9622. https://doi.org/10.1021/nn302975u

Zhao, L., Peralta-Videa, J. R., Hernandez-Viezcas, J. A., Hong, J., & Gardea-Torresdey, J. L. (2012). Transport and retention behavior of ZnO nanoparticles in two natural soils: Effect of surface coating and soil composition. *Journal of Nano Research, 17*, 229–242. https://doi.org/10.4028/www.scientific.net/JNanoR.17.229

Zhao, X., Zhang, W., He, Y., Wang, L., Li, W., Yang, L., & Xing, G. (2021). Phytotoxicity of Y2O3 nanoparticles and Y3+ ions on rice seedlings under hydroponic culture. *Chemosphere, 263*, 127943. https://doi.org/10.1016/j.chemosphere.2020.127943

Zhou, D., Jin, S., Li, L., Wang, Y., & Weng, N. (2011). Quantifying the adsorption and uptake of CuO nanoparticles by wheat root based on chemical extractions. *Journal of Environmental Sciences, 23*(11), 1852–1857. https://doi.org/10.1016/S1001-0742(10)60646-8

Zwieniecki, M. A., & Newton, M. (1995). Roots growing in rock fissures: Their morphological adaptation. *Plant and Soil, 172*(2), 181–187. https://doi.org/10.1007/BF00011320

CHAPTER

8

Biointeractions of plants—microbes—engineered nanomaterials

Ileana Vera-Reyes[1], *Edgar Vázquez-Núñez*[2], *Laura E. Castellano*[2], *Diana Ivette Aparicio Bautista*[3], *José Humberto Valenzuela Soto*[1], *and Jessica Denisse Valle-García*[4]

[1] Departamento de Biociencias y Agrotecnología, CONACYT-Centro de Investigación en Química Aplicada, Saltillo, Coah, Mexico; [2] Universidad de Guanajuato, Departamento de Ingenierías Química, Electrónica y Biomédica, División de Ciencias e Ingenierías, Campus León, León, Guanajuato, Mexico; [3] Laboratorio de Estructura de Proteínas, Instituto Nacional de Medicina Genómica, INMEGEN, Ciudad de México, México; [4] Programa de Doctorado Transdisciplinario en Desarrollo Científico y Tecnológico para la Sociedad, Cinvestav Unidad Zacatenco, Ciudad de México, Mexico

8.1 Introduction

Engineered nanomaterials (ENMs) interact with soil microbial communities in expected and unexpected ways. Thus, it is important to elucidate the biological interactions of plants, microorganisms, and ENMs when they enter the soil system (Vishwakarma et al., 2017). The rhizosphere is a highly biologically active region containing elevated concentrations of bacteria interacting with surrounding soil. It is in this root—soil—microorganisms zone where crucial processes occur, such as the nitrogen and carbon cycles, which support life in soil (Zhang et al., 2017). The disruption of these vital interactions by means of nanomaterials (NMs) might permanently affect the functionality of soils and, ultimately, the entire ecosystem.

Physicochemical Interactions of Engineered Nanoparticles and Plants
https://doi.org/10.1016/B978-0-323-90558-9.00001-2

8.2 Biological factors: microbes in the soil

8.2.1 Soil microbial communities

NMs interact with soil microorganisms using different pathways. However, some preferred interactions have been described. For example, silica nanoparticles are preferentially associated to fungal spores over microalgae and bacteria (Singh et al., 2019). The mechanisms through which the nanostructures can affect microorganisms remain unclear. In this context, some scientific approaches have studied morphological alterations under in vitro conditions.

Several studies have reported the antimicrobial activity of a wide range of ENMs. The antimicrobial mechanisms of ENMs may vary depending on the microorganisms and the nature of the NMs; however, it is recognized that ENMs have multidimensional effects on microbes including damage to the cell membrane and deactivation of membrane-bound enzymes, as well as indirect damages caused by reactive oxygen species (ROS) inside the cells (Li et al., 2017). There has been substantial research analyzing the influence of ENMs on microbial communities of diverse environments (Mohanty et al., 2014) including natural or engineered ecosystems. What has been highlighted is the role of microbial communities on the biogeochemical cycles, agriculture production, and plant diversity (Wagg et al., 2021). The effect of ENMs on the soil microbial community structure has also been studied. For this reason, the impact of the physicochemical properties of soils, i.e., pH, humidity, water-holding capacity, and the content of organic matter and texture, have been widely analyzed.

It has been reported that the transport rate of NMs in soils varies, and their impact on microbial communities and the soil type affects both phenomena (Shah et al., 2017). For example, microbial communities in sandy or loamy soils can respond differently to the same type of NMs. Similarly, different ENMs can display a different impact on microorganisms inhabiting the same type of soil. The scientific community notes that the study of ENMs on terrestrial ecosystems, especially microbial communities, should review the composition and structure, in addition to their function and implications in terms of their biointeractions with plants under natural conditions.

Finally, it is essential to mention that the nutrient cycling by microbes should be considered when studying the effect of ENMs on soil functions. Several authors have reported the inhibition of the nitrogen microbial cycling process when ENMs such as Ag NPs enter in contact with soil microorganisms (Masrahi et al., 2014). The majority of studies report the organism-specific toxicity of ENMs. For example, multiwalled carbon nanotubes (MWCNTs) can reduce the population of a few phyla (e.g.,

Proteobacteria and *Verrucomicrobia*) and increase the cell count of several other phyla (e.g., *Bacteroidetes* and *Firmicutes*) (Khodakovskaya et al., 2013). It was also reported that MWCNTs might increase the population of a few bacterial genera such as *Rhodococcus*, *Cellulomonas*, *Nocardioides*, and *Pseudomonas*, which are degraders of aromatic compounds in soil (Vázquez Núñez et al., 2009).

8.3 ENMs and their interactions with microorganisms and plants

8.3.1 Microbial communities as ENMs receptors: routes of exposure and contact

The phyllosphere is the soil region where symbiotic and pathogenic interactions occur. Different extracellular polymeric substances (EPSs) are produced by some microorganisms, and they work as quorum-sensing signal molecules (Arya & Harel, 2019, pp. 39–65; de la Rosa et al., 2021). It has been proven that EPSs promote the dissolution of NPs, as well as the uptake of metal ions by plant leaves. A contrary effect can be observed in some cases, for example, when extracellular substances form a protective barrier to avoid the contact and entrance of ENMs into plant tissues.

The interaction between ENMs and the biological interface of several species is defined by the bioavailability of these NMs where speciation plays an extremely important role since it affects the chemical transformation and aggregation of ENMs (Cornelis et al., 2014). Aggregation is associated with the formation of ENMs clusters, specifically motivated by the adhesion of ENMs to each other and/or to other particles (Wigger et al., 2020).

After reaching living organisms in the soil, ENMs may be transformed due to the local physical and chemical conditions and also by the enzymes and chelating agents produced by microorganisms and plants. There are few studies on the interaction between microorganisms and plants in porous media such as soil. The whole response of this interaction in ecosystems continues to remain unclear.

Regarding microorganisms, it has been reported that bacteria are an excellent sorption site for ENMs due to the bacteria-negative surface charge (McKee & Filser, 2016). The results of a field experiment reported a 10-fold increase of Ag NPs in forest soil bacteria within 60 days of exposure (Carbone et al., 2014). Other experiments under laboratory conditions reported that *Pseudomonas aeruginosa* can take up intact CdSe quantum dots (QDs) from the growth medium with 75 mg CdSe QD (Priester et al., 2009).

A similar study demonstrated the bioaccumulation and biomagnification of ENMs in the trophic chain (Werlin, 2011). The results showed that bare CdSe QD accumulated in *Pseudomonas aeruginosa* can be transferred to and biomagnified in its predator *Tetrahymena thermophila*. Cd concentration in the protozoal predator was approximately five times higher than that in their bacterial prey. The study concluded that QD-treated bacteria were differentially toxic to the protozoa, as they inhibited their own digestion in the protozoal food vacuoles.

The uptake and bioaccumulation of ENMs in other ecosystems has been also investigated. For instance, the bioaccumulation, biomagnification, and ecotoxicity of cadmium telluride QD were confirmed in microbial food chains under laboratory conditions (Gupta et al., 2017). In these experiments, the simulation of an aquatic ecosystem using bacteria (*Escherichia coli*) as prey and a ciliated protozoan (*Paramecium caudatum*) as predator was carried out. The authors observed that 29% of the bacterivore potential of *Paramecium* was lost, including a ~12-h of delay in doubling time on exposure to 25 mg/L of CdTe QD (~4 nm), compared to the control. The fluorescence-based stoichiometric analysis revealed that 65% of the QD bioaccumulated when paramecia were exposed to 25 mg/L of QD at 24 h.

In addition to the previously described studies, there is still a lack of understanding concerning the interaction of ENMs with microorganisms in natural environments. Thus, it is necessary to perform more experiments to entirely elucidate the effects of NMs on microorganisms in soils and in other ecosystems.

In addition to microorganisms, other soil organisms can be affected by ENMs, for instance, invertebrates (Ribeiro et al., 2019). Although soil mesofauna (acari, diplurans, enchytraeids, springtails, and symphylans) and soil macrofauna (earthworms, insects, mollusks, and spiders) are important to soils acting as degraders, there is still little information concerning the effects of ENMs on these communities (Hermes et al., 2020). Recent studies have focused their interest on the effects of NMs on mesofauna. For instance, the toxicity of Ag NMs (particles, wires, plates, NPs, NW, and NPL) on *Caenorhabditis elegans* in soil media (Moon et al., 2019) has been reported. The study showed that Ag NPs and Ag PL inhibited the growth and reproduction of *C. elegans*, whereas Ag NW had a negligible effect. The study concluded that the shape of Ag NPs plays a significant role in their toxicity.

In an interesting experiment, the response of maize (*Zea mays* L.) seedlings and terrestrial isopods (*Armadillidium vulgare*, Latreille) to TiO_2 nanoparticles under greenhouse conditions was assessed (Pérez-Hernández et al., 2021). Results demonstrated an inhibitory effect of TiO_2 NPs on plant length and root size after 21 days in alkaline soils, but not in acid soils. In addition, it was reported that an increase in Ti concentrations in the soil organic matter harmed the survival and weight of the terrestrial isopods when they consumed the organic matter.

Regarding carbon nanotubes (CNT) toxicity and bioaccumulation, the potential toxic effects of NiZn CNT on *Caenorhabditis elegans* have been evaluated (Kong et al., 2021). Herein, three physiological parameters were studied, including the effect of NiZn CNT on *C. elegans* feed behavior, reproductive ability, and the overall lifespan of the worms. No significant difference was noted between the feeding rate of worms treated with NiZn CNT and the control population. Also, there was no reduction in the production of progenies in NiZn CNT–treated worms, which indicated that the tested nanoparticles did not negatively affect the reproductive system of the animal.

Herein, we recommend to carry out more experiments to increase our understanding upon the effects of ENMs on soil microorganisms, and the relationship of soil biota and plants with these materials. Fig. 8.1 represents the interaction among ENMs–soil microorganisms–plants and their effects.

8.3.2 ENMs and their effect on the physiological and morphological response of plants biointeracting with microbes

ENMs can undergo chemical transformation, aggregation, and disaggregation. The interplay between these processes and the NM transportation in porous media such as soil determines the fate and ecotoxicological potential of these materials (Peijnenburg et al., 2016) and, finally, their role as a modifier of the soil microbiome–plant interactions.

The rhizosphere extends a few millimeters from the root surface (Balestrini et al., 2015) and contains up to 10^{11} microbial cells per gram of

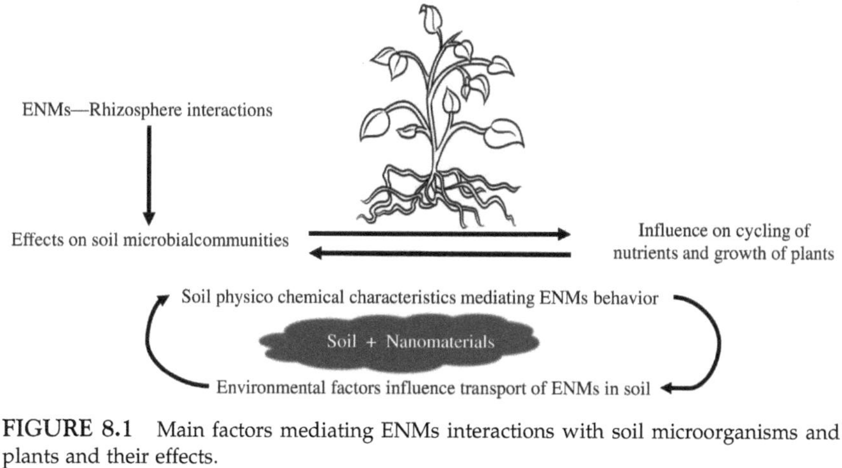

FIGURE 8.1 Main factors mediating ENMs interactions with soil microorganisms and plants and their effects.

root (Berendsen et al., 2012). The health of the plant deeply depends on the root microbiome, and this can be observed in disease-suppressive soils in which crop plants suffer less from specific soil-borne pathogens than expected. Despite the great interest in this phenomenon, the microbes and mechanisms involved are yet largely unknown.

The previously mentioned interactions occur naturally in soils in the absence of exogenous materials, such as ENMs. However, during the last years, the incorporation of these potentially toxic materials into the trophic chain affecting human health has raised concerns. In this regard, different studies have analyzed the toxic effect of ZnO, SiO_2, TiO_2, and CeO_2 NPs at 1 mg/g on agricultural soils (Chai et al., 2015). In this experiment, the toxicity was evaluated by measuring thermal metabolism, the abundance of functional bacteria, and enzymatic activity. The study reported that ZnO and CeO_2 NPs hindered thermogenic metabolism, reduced the number of soil *Azotobacter*, P-solubilizing and K-solubilizing bacteria, and inhibited enzymatic activities. On the other hand, TiO_2 NPs reduced the abundance of functional bacteria and enzymatic activities. SiO_2 nanoparticles slightly boosted the soil's microbial activity. These findings demonstrated that the combined approach of monitoring thermal metabolism, functional bacteria, and enzymatic activity is feasible for testing the ecotoxicity of NPs in agricultural soils.

The relevance of long-term studies to determine the environmental effects of different concentrations of Ag NPs on microbial biomass and their relationship with the N cycle in soil was evaluated by Grün et al. (2018). This study was performed under laboratory conditions for one year, and tested Ag NPs concentrations ranging from 0.01 to 1 ng Ag NPs/kg soil. The results demonstrated that the short-term effects of Ag NPs were limited. However, after one year of exposure to 0.01 mg Ag NPs/kg, significant negative effects on soil microbial biomass (quantified as DNA) and bacterial ammonia oxidizers (quantified by *AmoA* gene copy numbers) were observed. These results proved that even very low concentrations of Ag NPs affected the microbial soil community especially in nitrogen cycling processes, which comprises an element essential for plant growth, health, and productivity.

8.4 Biotic and abiotic factors driving plant–microbes interactions in the presence of ENMs

Abiotic factors can be defined as those inherent to the environment, which include extreme temperatures, water stress (drought or flooding), high salinity, pressure, changes in pH, radiation, cold, and sunlight, among others. Biotic factors are characterized as those caused by living organisms, especially viruses, bacteria, fungi, nematodes, insects, arachnids, and weeds (Echevarria-Molinar, 2017; Lowry et al., 2019).

These factors can generate stress on environmentally important plants and microorganisms. Stress from biotic and abiotic factors is a constraint to crop production and yield as well as to food security (Kumar & Verma, 2018) as they affect agricultural soils and decrease global production causing millions of dollars in losses (Ho et al., 2017). Physiological, biochemical, and molecular parameters of plants are affected by drought, flooding, extreme temperatures, mineral toxicity, and pathogenic diseases. Consequently, it is estimated that more than 50% of crop plants worldwide are affected by stress (Lata & Gond, 2019). It has been projected that drought conditions caused by climate change will increase drastically in future decades causing severe damage to plants (Duo et al., 2018).

On the other hand, the association between plants and microorganisms that play an important role in plant health is well established. Several studies emphasize the importance of considering the influence of biotic and abiotic factors on microbial diversity, as they affect soil and plants (Santoyo et al., 2017; Wille et al., 2019). Because of this, there is an urgent need to halt and to adapt these systems to stress factors. A promising alternative to solve these problems is the use NMs (Elsakhawy et al., 2018).

ENMs display a surface area that facilitates contact with the environment, where they can be taken up by organisms and undergo many transformations including dissolution, aggregation, and reaction with biomolecules. These processes affect uptake rates and biological responses. For example, in plants, various factors including mycorrhizae, root hairs, membrane porosity, and nodes affect the internalization of ENMs (Vitali et al., 2019; Yang Liu et al., 2020). Thus, it is important to understand the mechanisms by which ENMs interact with plants, microorganisms, and the soil media, since this is the region where most of these processes take place.

8.4.1 Soil–plant–microbes–ENMs interactions

The structure and function of soil are quite complex, and a clear indicator of the soil quality are the microorganisms that also influence the mineralization of organic matter and biogeochemical cycles (Pan et al., 2020). Therefore, it is essential to understand how soil microorganisms respond to ENMs and to consider that they can have physical, chemical, and/or biological transformations depending on interactions with matrix components (Pan et al., 2020; Usman et al., 2020). The pH and organic matter are two factors that have substantial effects on the behavior, adsorption, and stabilization of ENMs. These factors may decrease ENMs mobility and bioavailability (Wang et al., 2020). Water is a vital component for soil-dwelling plants as it is required for nutrient transport. Thus, water deficiency causes drought stress that negatively impacts the ecosystem. The physicochemical characteristics of water are very similar to those of soil and include temperature, pH, electrolytic conductivity, dissolved oxygen, alkalinity, and turbidity, among others (Kumari et al., 2012). Appropriate water management practices are needed to ensure crop productivity but also the quality of interactions between environmental systems and, particularly, the efficiency between soils, microorganisms, and plants.

8.4.2 Type of plant

Symbiotic microorganisms in plants increase tolerance to biotic and abiotic stress. The majority of these root-associated organisms exhibit beneficial effects by improving soil and plant health through the modification of root properties when stress conditions are present (Ho et al., 2017; Mokarram-Kashtiban et al., 2019). Numerous studies have addressed the role of ENMs in the control and care of plants undergoing abiotic stress or growth inhibition. The controlled release of these ENMs to crops can modify the response of the plant to abiotic stresses such as drought, salinity, and the presence of heavy metals in the soil, rendering them more resilient. By releasing them in the right place, time, and dose, many stressors could be bypassed or even combated (Lowry et al., 2019) as NMs may induce the synthesis of molecules and regulate defense genes that affect the resistance to biotic and abiotic stress (Wang et al., 2020). ENMs also enhance tolerance to water stress by improving the hydraulic capacity of roots, allowing the entrance of the nutrients necessary for proper metabolic function (Das et al., 2018). On the other hand, the application of nanobiofertilizers promotes plant growth, nutritional quality, and reduces disease vulnerability (Kumar & Verma, 2018).

The effects and impacts of ENMs in plants are related to their type, size, and concentrations, as well as the interaction with the microorganisms present in the soil, root, and plant type. For example, the symbiotic association among rhizobia, arbuscular mycorrhizal fungi (AMF), and magnetite ENMs in chickpea plants presents positive effects on chlorophyll when plants are affected by soil salinity (Abd-Alla et al., 2019). SiO_2 NPs promote seed germination, increase biomass and chlorophyll content in tomato and squash under salinity stress (Siddiqui & Al-Whaibi, 2014). They also reduce the effects of drought stress deterioration in wheat (Pei et al., 2010) and alleviate arsenic-induced oxidative stress (OS) in rice (Cui et al., 2020). nZVI has been used in conjunction with rhizosphere microorganisms and AMF in white willow, and at low concentrations promoted plant growth (Mokarram-Kashtiban et al., 2019).

It has also been reported that 30 mg/kg Ag NPs aided in combatting canker disease stress in mandarin. In addition, biotic stress was reduced compared to other Ag treatments that did not combat the disease and that significantly decreased fruit yield (Hussain et al., 2019). The use of copper nanoparticles (Cu NPs) to reduce biotic stress generated by a fungus has been explored. According to research results, an increase in watermelon productivity was obtained (Borgatta et al., 2018). In this experiment, Cu was expected to modulate microbial communities involved in nutrient assimilation in plants counteracting biotic factors affecting crops. The accumulation of NPs in plant tissues directly affects soil and rhizosphere microbial diversity; thus, it is a reciprocal relationship (Vitali et al., 2019).

Regardless of the plant type, NPs interact with soil components before entering and affecting the plants. For this reason, their bioavailability may be affected. Moreover, there are many microorganisms accumulated in the rhizosphere that directly affect the relationship between plants and NMs, but that end up being a key factor in combating the various problems faced by plants under biotic and abiotic stress.

8.4.3 Microbial groups

Interactions between microorganisms (bacteria, fungi, or viruses) and plants can be classified into three types: commensal, beneficial, or pathogenic. In the case of bacteria, the phyla that tend to dominate the rhizosphere are *Actinobacteria, Proteobacteria, Chloroflexi,* and *Firmicutes.* In the case of viruses, bacteriophages impact the plant microbiome through gene transfer and they can present negative effects on bacteria. At the same time, they are important for the evolution of plants in disease resistance (Zhang et al., 2021). Archaea also interact with plants and fungi affecting nutrient delivery (Taffner et al., 2018). With respect to fungi, they have been studied to a lesser extent as compared to bacteria. Nonetheless, they play special roles, vary from host to host, and are strongly dependent on soil type (Zhang et al., 2021).

These microorganisms are known as plant growth–promoting microbes (PGPMs) since they stimulate plant growth through hormone regulation (Kumar & Verma, 2018). PGPMs include those that are free-living, those that form specific relationships with plants, endophytes that can colonize tissues, and cyanobacteria. The association of PGPM with plants reduces certain abiotic stresses. They can exist in the rhizosphere or rhizoplane (extracellular) or can colonize nodular structures (intracellular) (Ho et al., 2017; Sashidhar et al., 2019).

On the other hand, AMF participate in symbiotic processes between the plant and microorganisms, promote plant growth, regulate hormones, and improve the antioxidant system and water supply under abiotic stress conditions, increasing plant tolerance to different stressors (Kumar & Verma, 2018). It has been demonstrated that inoculation with *Glomus intraradices* AMF increased the growth and photosynthetic efficiency of rice plants under drought conditions (Ruiz-Sánchez et al., 2010).

Thus, PGPM and AMF hold promise as successful tools to promote plant growth under different environmental stressors. As mentioned earlier, microorganisms play an important role in the interaction with ENMs, since they may be helpful in inducing their biotransformation, which can be utilized as a strategy for controlled release (Lowry et al., 2019). However, depending on the nature of the NP, NP size, dose, and concentration, as well as on the soil type and other characteristics, NMs may be toxic to soil microorganisms (Usman et al., 2020). Therefore, a deeper understanding of the interactions between plants, ENMs, PGPM, and AMF is required.

PGPM and AMF may promote plant health through the change in root properties under stress conditions. For instance, the application of microorganisms in the rhizosphere (rhizobacteria and arbuscular mycorrhizae) mitigates the impacts generated by certain Zn ENMs and enhances plant growth parameters (Mokarram-Kashtiban et al., 2019).

A variety of studies report the relationship between ENMs and plant-associated microorganisms. Some of these indicate the influence of *Rhizobium* spp. activated with magnetite ENPs on chickpea nodulation under saline conditions (Abd-Alla et al., 2019). In this specific experiment, magnetite was employed at a concentration of 150 μg/mL, and *Rhizobium*, AMF, and *Stenotrophomonas maltophilia* mitigated the effect of salinity by enhancing nodule formation and in consequence, nitrogen fixation. On the other hand, a decrease in plant growth–promoting rhizobacteria (PGPR) was observed when exposed to 30 mg/kg CeO_2 NPs (Pan et al., 2020). The application of NPs on leaves and roots to evaluate the effect of NMS on plant microbiota showed interesting results. The leaves and roots of poplar trees were exposed to Ag nanoparticles. When Ag nanoparticles were applied to the leaves, there was a reduction in fungal and bacterial groups, while in the roots, the biodiversity was reduced (Vitali et al., 2019). Further studies will help to identify the mechanisms of interaction between ENMs, microorganisms, plants, and the environment.

8.5 Mechanisms involved in the biointeraction of plant–microbe mediated by ENMs

ENMs are considered a powerful tool with multiple benefits for crops. Therefore, it is necessary to evaluate all possible effects on the environment, specifically toxicity in plants, bacterial, and fungal communities (phyllosphere and rhizosphere). The persistence of ENMs depends on the soil and its chemical characteristics; others, such as the antimicrobial properties of ENMs, may act detrimentally on microbial communities (i.e. nanoparticles based on Ag, Cu, and Zn). Soil microbiome alteration by

ENMs also impaired the microbial process of nutrients in the soil, reducing beneficial rhizobacteria activities for nitrogen fixation, biocontrol, and plant–growth promotion. In this respect, it is of the utmost necessity to investigate all the possible interactions between ENMs and plant–microorganisms in the environment. The data obtained will support the mechanisms related to ENMs regulation and could contribute valuable information for the proper use of ENMs in sustainable agriculture.

It has been reported that Ag NPs (size range, 5–50 nm) applied to *Vicia faba* (800 μg/kg soil) caused detrimental effects in associated faba bean–*Rhizobium-Glomus* spp. The seed germination and rhizobial growth was reduced, while nitrogen fixation and mycorrhizal colonization was impaired by Ag NPs treatments (Abd-Alla et al., 2016). Additionally, Ag NPs supplemented to the soil alterated the microbial rhizosphere community in maize (*Z. mays*). A dose of 100 mg/kg Ag NPs in the soil increased plant biomass and alterated the bacterial communities that showed less enzymatic activity and carbon use alteration. Conversely, fungal communities were less affected by Ag NPs and their composition was slightly modified during exposition (Sillen et al., 2015). Similar results were observed in wheat plants treated with TiO_2 NPs. The effects on the microbial communities were evaluated after three months with the application of 1, 100, and 1000 mg/kg to the soil. Results indicated that TiO_2 NPs changed the bacterial community; however, the fungi community remained unaltered (Moll et al., 2017). No negative effects were observed in wheat growth and the arbuscular mycorrhizal colonization, which indicated that prokaryotes are more susceptible to TiO_2 NPs as compared with fungi (Moll et al., 2017). However, the authors argued that prokaryotes are sensitive to NPs, possibly by interaction with the bacterial surface.

8.5.1 Rhizosphere as the interface mediating the plants–microbe interactions

Associations between plants and beneficial microorganisms occur in the rhizosphere (Brencic & Winans, 2005; van Dam & Bouwmeester, 2016; Venturi & Keel, 2016). Among soil microorganisms in the rhizosphere, arbuscular mycorrhizal fungi (AMF) are the most common organisms that can establish symbiosis with 80% of plant families (Jamiołkowska et al., 2017). The effects of AMF on the growth and development of fenugreek (*Trigonella foenum-gracum*) under different doses (0, 125, 250, 375, and 500 μg/g) of ZnO NPs were evaluated for 60 days (Ghasemi Siani et al., 2017). AMF colonization was significantly reduced at 375 μg/g of ZnO NPs, compared to controls. In this case, AMF inoculation ameliorated the inhibitory effects of ZnO NPs by producing the glycoprotein glomalin, a potent metal chelator. These results suggested that AMF inoculation may protect plants from nanotoxicity through glomalin secretion (Ghasemi Siani et al., 2017). On the other hand, PGPR are associated with the rhizoplane. PGPR are known for their ability to promote growth directly and indirectly in crop plants through the activation of different phytostimulatory mechanisms (Nayana et al., 2020). Different publications reported the effects of PGPR in host plants treated with ENMs. For example, silica nanoparticles in combination with *Bacillus megaterium, Bacillus brevis, Pseudomonas fluorescens,* and *Azotobacter vinelandi* promoted seed germination in maize (*Z. mays),* compared to conventional silica (Karunakaran et al., 2013). AgNPs increased root area, root length, and root-shoot in maize plants with the increase in PGPR (*P. fluorescens* and *Bacillus cereus*) (Khan & Bano, 2016). TiO_2 NPs enhanced colonization and performance of *Bacillus thuringiensis* AZP2 and *Paenibacillus polymyxa* (Timmusk et al., 2018). Also, ZnO NPs increased the number of nodules, plant height, and grain yield in soybean (*Glycine max* L.) inoculated with *Bradyrhizobium japonicum, Pseudomonas putida,* and *Azospirillum lipoferum* (Sharifi & Khoramdel, 2015).

The association plant–microorganisms modifies the behavior and fate of ENMs in the environment and the chemical reactions with the surface of the biological targets (Maurer-Jones et al., 2013).

8.6 Analyzing the effect of ENM–plant-associated microorganisms

8.6.1 Alteration of biogeochemical cycles

ENMs application or dissemination into the soil may suppress the fundamental activity of soil microorganisms and soil enzymes that participate in C and N cycles. This will affect the capability of soils to perform such critical environmental functions as participation in biogeochemical cycles. Numerous enzymes are involved in nutrient cycling in soils: b-glucosidases are related to an indicator of C cycling in soils, while urease, phosphatase, and sulfatase are enzymes responsible for generating available N, P, and S. Most of the studies on the influence of ENMs on soil health refer to the soil enzymatic activities associated with C, N, and P cycling.

Regarding soil enzymes, some studies have been carried out to elucidate the effect of NPs on the cycling of soil nutrients. A study evaluated the activity of phenol oxidase and arylamidase and their interaction with Ag NPs of three coatings and sizes. Both enzymes were inhibited in a concentration-dependent manner by Ag NPs treatments. The arylamidase activity was severely affected probably because of the interaction of Ag NPs with the active site of the enzyme and the reduction of free thiols available for enzyme function. On the other hand, phenoloxidase activity was recovered over time, probably due to the low bioavailability of NPs and tolerance of the microbial species to the NP toxicity. Results evidenced that the toxic effects of ENMs were due to their presence and not to the dissolution products (Mishra et al., 2021). Similar effects were reported by Galaktionova et al. (2019). The application of 50 and 100 mg/kg soil of Zn and Cu NPs increased invertase activity, which was possibly related to the stimulation of microbial activity involved in the cycling of C. However, at concentrations higher than 200 mg/kg soil, the EMNs aggregated and formed microparticles. For this reason, their ecotoxicological properties decreased. Various studies have indicated that EMNs applied at higher concentrations repressed the activities of soil enzymes such as dehydrogenase, urease, phosphatase, cellobiohydrolase, β-1,4-xylosidase, β-1,4-N-acetylglucosaminidase, arylsulfatase, and β-glucosidase (Asadishad et al., 2018; EIVAZI et al., 2018; Li et al., 2017). Another study reported the effects of Cu NPs on the soil nitrification process. The results evidenced that Cu NPs affected nitrification kinetics due to the sorption of Cu on soil particles and the oxidization into Cu(II)-oxide and Cu(II)-hydroxides (Vandevoort & Arai, 2018). This results in the release of Cu^{2+} and the increase in ROS production, which inhibits the kinetic nitrification rate. More studies are needed to determine the long-term impacts of ENM at concentrations that benefit plant growth.

Several studies have demonstrated that NP interactions with the rhizosphere can benefit crop plants. The application of TiO_2 nanoparticles to oilseed rape (*Brassica napus*) increased the adhesion of a plant growth–promoting bacterium (*Bacillus amyloliquefaciens*) to the roots. Consequently, the plants were more resistant to infection with *Alternaria brassicicola* (Palmqvist et al., 2015). Nevertheless, the effects of NPs depend on the particle size/aggregation. Three types of CeO_2 NPs (two different sizes, 31 ± 18 nm and 3.5 ± 0.5 nm and citrate-coated, 3.9 ± 1.8 nm) were evaluated to determine their effect on the development and growth of canola plants. The authors observed that CeO_2 NPs supplied to the soil at 1 mg/kg did not affect plant growth. However, they impacted microbial activities and C, N, and P cycles. It has been determined that changes in bacterial communities in the rhizosphere were dependent on particle aggregation and coating. Small NPs did not exhibit any effect on microbiota composition. On the other hand, it has been reported that large size particles of ENMs form aggregates that modify the bacterial community structure. NPs mobility in the soil is also affected by small organic molecules on the NPs surface that modify their interaction with root exudates and enzymes (Hamidat et al., 2016). It has been also demonstrated that TiO_2 NPs, relatively low concentrations of MWCNT, and CeO_2 NPs did not affect nitrogen fixation and P content in red clover. However, 3000 mg/kg MWCNT increased biological nitrogen fixation by 8% (Moll et al., 2016). In addition, the vesicular root colonization of red clover was not affected by NM treatments. Other studies confirmed that MWCNT did not affect the nitrogen fixation activity of *Azotobacter chroococcum*, indicating that CNT are safe to nitrogen-fixing bacteria (Ouyang et al., 2021).

In mung bean plants, the application of ZnO NPs increased phosphatase (acid and alkaline) and phytase enzymatic activity in soil. Consequently, P content increased by 10.85%. Also, ZnO NPs increased root volume, soil microbial population (fungi, bacteria, and actinomycetes), and dehydrogenase activity, essentials factors for various nutrient cycles (Raliya et al., 2016). Yue et al. (2020) revealed that La_2O_3, Nd_2O_3, and Gd_2O_3 NPs applied to soil were able to regulate the biogeochemical cycling of nitrogen through the reduction of the activities of ammonia-oxidizing microorganisms. This process was dependent on incubation time and NP types. Thus, it could improve the efficiency of nitrogen cycling by inhibiting ammonia oxidation and limiting the nitrification process in nitrate.

The preseed treatment combined with microbial inoculants and Mo NPs modified root exudates composition and the taxonomic diversity of microorganisms in chickpea plant's rhizosphere. Mo NPs contributed to the increase of the genetic diversity of microorganisms. In combination with nodule bacteria, Mo increased the number of symbiotic and associative nitrogen-fixing bacteria and activated the natural pool of the *Rhizobiales* bacteria in the soil's microbiome. It is likely that this change was induced by variations in the proportion of organic acids (citric, succinic, and malic) and sugars in root exudates (Shcherbakova et al., 2017). Similarly, exposure of *Cajanus cajan* plants to Cu NPs combined with the growth-promoting fungus *Piriformospora indica* displayed a synergistic activity. Cu NPs improved growth parameters and demonstrated the role of the fungal mycelium for transporting NPs into the roots. Plants treated with Cu NPs + *P. indica* showed maximal chl-a fluorescence, vitality, number of activated reaction centers, fresh and dry weight of shoots and roots, and the highest root length, followed by treatment with Cu NPs alone and *P. indica* (Rajak et al., 2017).

New evidence indicates that the combination of calcium phosphate nanoparticles with endosymbiotic and AMF improved vitality and enhanced growth and root proliferation in *Z. mays* (Rane et al., 2015). On the contrary, another study reported that ZnO NPs inhibited the root colonization rate of AMF (*Funneliformis mosseae*) in maize plants, although these treatments increased the shoot/root P concentration and the Zn content in soil and plants (Wang et al., 2018). In a recent study, it was reported that three ENPs (MWCNTs, metal oxide, and hydroxyapatite nanoparticles) improved cell growth and viability, indole-3-acetic acid (IAA) content, siderophore production, and increased P solubilization (Kaur et al., 2021). In addition, the physiology and biochemistry of the endosymbiotic bacteria also showed improved parameters. It has also been demonstrated that single-walled carbon nanotubes enhanced the microorganism's metabolism and modified microbial community structure, but the diversity was not affected. Consequently, they improved soil fertility by enhancing N and P availability (Qian et al., 2018).

CuO and ZnO NPs act as micronutrients for soil microbial communities, increasing the abundance and community diversity of soil bacteria or fungi and the α-diversity indices. At high concentrations (>500 mg/kg soil), these NPs may negatively affect soil structure without disturbing the microbial community (Liu et al., 2021). Likewise, the effects of CuO NPs on the prokaryotic rhizospheric soil community in the pecan tree (*Carya illinoinensis*) were analyzed, and it was found that the community structure was not disturbed. This community included *Caulobacterales, Burkholderiales, Xanthomonadales,* and *Clostridiales* (Salas-Leiva et al., 2021). It was also observed that the metabolic profile correlated to nitrogen, sulfur, and halogen cycles was not affected by 1000 mg CuO NPs/kg soil. The differences between the results of these experiments are correlated with the type of soil used for the assay, which impacts NPs interaction with the rhizosphere.

On the other hand, nanozeolite and nanochitosan may improve the rhizospheric bacterial population and soil health. Some studies have evaluated the effects of these nanocompounds (alone or in combination with microorganisms) on plants; for example, the effect of *Bacillus* spp. on the *Z. mays* rhizosphere bacterial community has been determined by (Chaudhary et al., 2021). The results indicated a higher bacterial diversity dominated by the genera *Actinobacteria, Bacteroidetes, Acidobacteria,* and *Chloroflexi* in soils treated with nanocompounds. Nevertheless, combined treatment with *Bacillus* spp. revealed a superior result in the composition of rhizospheric microbiota. Previous reports demonstrated that nanoclay, nanochitosan, and nanozeolite are suitable alternatives for agricultural applications due to increased microbial activity and improved soil health (Priyanka Khati et al., 2017). Nanozeolites increased bacterial species and the diversity associated with nutrient cycling, residue decomposition, and xenobiosis (Khati et al., 2019). Similarly, nanoformulations of Zn (zincated nanoclay polymer composites) increased Zn use efficiency in rice rhizosphere and the soil microbial population as a consequence of augmented soil's enzymatic activities (Dubey et al., 2021). According to these results, nanotechnology is a promising field for agricultural applications. However, additional investigation is needed in the permanency, environmental fate, and toxicity of such nanoformulations prior to their use as agro-inputs on sustainable agriculture.

8.6.2 Adverse effects of ENMs on microbes—plants interactions: implications and possible consequences

A variety of nano-based products are currently offered on the market for plant disease control or as an alternative to conventional agrochemicals. Although nanotechnology offers benefits, there are concerns about the accumulation and fate of NMs. The unregulated exposure of nanoparticles to the soil may negatively affect the soil microbiota, and the prolonged exposure may impair soil fertility (Javed et al., 2019). We cannot overlook potential effects of ENMs on soil enzymatic activities or soil communities because several factors including the type, size, coatings, concentration, exposure time, and the physicochemical properties of the soil can modulate their impact. For example, the use of copper-based nanopesticides in agroecosystems may reduce nontarget soil microbial communities due to their broad antimicrobial spectrum (Simonin et al., 2018). The impact of commercial $Cu(OH)_2$-nanopesticides in edaphic organisms (the isopod *Porcellionides pruinosus*) and in microbial communities (Peixoto et al., 2021) was evaluated in a long-term exposure experiment. The study concluded that the nanopesticide application reduced the bacterial- and fungal-community structure but increased the microbial ability of carbon utilization; however, the presence of *P. pruinosus* relieved the effects of nanopesticides on soil microbial communities.

As a critical remark, the addition of pesticides negatively affects the abundance of soil microorganisms. Instead, the gradual release of pesticide formulations has a subtle effect on these microorganisms (Prudnikova et al., 2021). The development of nanomatrix to gradually deliver pesticides could be an alternative to conventional products as this will redcue xenobiotics and the negative environmental impacts associated with their application.

8.6.3 Disruption of ecological niches

A soil microecosystem was designed to study the effects of metallic NPs on soil ecosystems. This ecosystem included *Arabidopsis thaliana* plants, soil microorganisms, and *Eisenia fetida* earthworms. The soil was irrigated with water containing TiO_2 NPs and/or Ag NPs, simulating wastewater with metallic NPs (Liu et al., 2019). The supplementation of TiO_2 NP in the irrigation water was not toxic to the soil microecosystem. Even when TiO_2 NP were added in combination with AgNPs, the treatments did not exhibit toxicity to *A. thaliana* and earthworms. The relative abundances of bacteria were not affected by the treatments, revealing an average of around 43.7%–49.1% of proteobacteria, 10.8%–16.4% of bacteroidetes, and 11.0%–13.6% of verrucomicrobia. In contrast, the Ag NPs (NPs type or residual ion at 20 μg/L) negatively affected the soil microbial biomass, plant, and earthworm growth. The TiO_2 supplied to the Ag mixture relieved the toxicity of Ag on the microecosystem.

It has been reported that Ag NPs cause weight loss in earthworms, which is associated to OS. In addition, Ag NPs reduced P and S availability by means of Ag_2S or Ag_3PO_4 formation and microbial growth in soil (Das et al., 2018). Similar results were reported when Ag NPs were assayed in a complete system, where decreased soil microbial metabolic activity, nitrification potential, and an abundance of bacteria and ammonia-oxidizing bacteria were observed (He et al., 2016).

Regarding other soil species, some studies have analyzed the sensitivity of earthworms to nanoparticles, for instance, to Cu and Zn ENMs. In an interesting investigation, it was found that earthworms were sensitive to Cu and Zn NPs. In this case, Cu NPs increased the earthworm's weight, and this weight decreased with Zn NPs exposure. It was also found that soil enzymes' activities were inhibited. For instance, urease was inhibited with Cu NP, catalase with Zn NPs at a dosage above 50 mg/kg, and invertase with both NPs at 200 mg/kg (Galaktionova et al., 2019). Additionally, it has been demonstrated that the aggregation and agglomeration of ENMs affect their availability and toxicity, which is also associated with the chemical transformation in the soils and the absorption of the soil's organic matter (Shen et al., 2020). An important study demonstrated that interactions of earthworms (*Eisenia fetida*) with Au and Ag NPs suppressed the expression of genes involved in the relief of OS without causing genotoxicity (Bourdineaud et al., 2021). The study concluded that the low bioavailability of Ag NPs resulted in a reduced negative effect. However, inside the cells, they cause equal or more damage than their ionic form through different molecular and cell signaling pathways.

8.6.4 Alteration of flora patterns

Studies on ENMs revealed that they are able to interfere with agricultural systems in the long term, impacting the archaeal community and affecting the nitrification process. Moreover, enhanced growth of phytopathogenic fungi may occur because of the harmful effects of Ag NPs on antagonistic microorganisms, which play a role as biocontrol agents. Moreover, Ag NPs caused a hormetic dose-response that resulted in a growth increase of maize plant-root (Sillen et al., 2020). It was also reported that Ag NPs reduced microbial biomass, soil respiration, and microbe-released enzymes in sandy loam soil (Das et al., 2018). In another study, Ag and CuO NPs exhibited a negative impact on soil bacterial diversity, community structure, and the population of PGPR. Although CuO NPs exhibited more toxicity than the Ag NPs at phylum and genus taxonomic levels, the phylum was nonetheless recovered over time due to the transformation of NPs or to the defense mechanisms of bacteria (Abdulsada et al., 2021). It has been reported that the AMF symbiosis relieved the adverse effects of Ag NPs on plants by varying the expression of Ag transporters and collaborating with the soil bacterial community (Cao et al., 2020). On the other hand, a recent research indicated that ZnO NPs decreased the total bacterial population in saline–alkali soil and affected bacterial community composition including the *Bacilli, Alphaproteobacteria*, and *Gammaproteobacteria* classes (You et al., 2018).

8.7 Molecular-based methodologies to determine the impact of engineered nanomaterials on plant–microbe biointeractions

The use of ENMs as a strategy to control plant diseases requires, in the first instance, the evaluation of their impact on plant–microbe interaction. This evaluation may include the application of molecular methods with high sensitivity and specificity. Nowadays, the most recent technologies for the study of plant–microbe interaction include: (a) polymerase chain reaction (PCR); (b) real-time PCR or quantitative PCR (qPCR); (c) metagenomic analysis; (d) metabarcoding; (e) high-throughput sequencing technologies; (f) nanopore sequencing; (f) CRISPR/Cas; (g) omics. A complete explanation of these techniques can be found in the literature.

8.8 Functional elucidation of the impact of ENMs on plant–microbe interactions: the omics sciences approach

Complete knowledge of the biological systems at multiple levels is critical for exploiting the potential of ENMs (Yin et al., 2018). The impact of ENMs on plant–microbe interaction remains mostly unknown. However, the development of sequencing technologies and a variety of "omics" tools has amazingly increased research in this area. Herein, we discuss the application of omic tools at the transcriptomic, proteomic, and metabolomic level for the study of ENMs interactions with plants and microbes.

8.8.1 Transcriptomic level

Gene expression analysis represents a robust approach to identify the molecular mechanisms of toxicity and the physiological consequences on plant–microbe interaction when they are exposed to ENMs. It has been shown that NMs are able to enter living organisms, spread through organs and tissues, and exert toxic effects at the cellular level, including plasma membrane damage, protein inactivation, DNA damage, disruption of energy transfer, the formation of ROS, inhibition of cell growth, damage to the cell wall, and the release of toxic substances (Van Aken, 2015). Recent in vivo studies have employed transcriptomic analysis to evaluate gene expression in plants in response to ENMs. These investigations were carried out using different plant species including barley (Petrova et al., 2021), *Arabidopsis thaliana* (Landa et al., 2017), tomato (Sun, Wang, Ju, & Xu, 2020), zucchini (Pagano et al., 2017), and tobacco (Frazier et al., 2014; Van Aken, 2015), among others.

8.8.2 Proteomic level

The genome of the plant codes for 36,795 proteins, and the average size of a plant protein is 436 amino acid residues, 27% smaller than the average animal protein (Ramírez-Sánchez et al., 2016). Proteins play a very important role in plant signaling and stress response, and they participate in cellular homeostasis. Characterization of the plant proteome reveals the functional analysis of the translated regions of the genome; consequently, proteomics complements transcriptomic and metabolomics for the understanding of the cellular mechanism in plants. Proteomic analysis provides us with an important tool for crop improvement, protection, and phytotoxicity, since it can identify crucial regulatory proteins, post-translational modifications, and protein–protein interactions in plants exposed to ENMs (Majumdar & Keller, 2021). Analyzing plant–ENMs interaction through proteomic studies allows understanding the positive and negative effects of nanoparticles at the plant's functional level. In addition, the affected proteins can be identified, and in consequence, cellular processes involved in plant response to ENMs can be elucidated. A few studies have used proteomics to identify specific plant proteins involved in ENMs-related cellular signaling or stress response. Studied NMs include Ag, CeO_2, nAl_2O_3, Fe, and Cu NPs (Hossain et al., 2020; Majumdar & Keller, 2021; Salehi et al., 2018).

8.8.3 Metabolomic level

Metabolites are intermediate or end products of cellular regulatory processes that reveal the last response of an organism to external stimuli. Plant metabolic pathway databases report 4544 compounds implicated in 1123 pathways among 350 plant species (Schläpfer et al., 2017). The metabolites produced by plants are classified into two main categories: primary and secondary metabolites (Dixon, 2001). Primary metabolites are essential for plant growth and development and include carbohydrates, amino acids, vitamins, organic acids, and fatty acids. Secondary metabolites are synthesized from the primary metabolites that participate in the adaptation and defense response in plants (Fiehn, 2002). The most important classes of secondary metabolites include polyketides, terpenoids, steroids, phenylpropanoids, alkaloids, and glucosinolates, which display specific biogenetic pathways and thousands of intermediates metabolites (Hounsome et al., 2008; Majumdar & Keller, 2021).

The exposure of plants to ENMs can stimulate the production of signaling molecules that trigger the activation or inhibition of metabolic pathways in the plant. In order to identify the essential role of the affected metabolites in plant response to ENMs, it is important to study this interaction using metabolomic analysis. Some of these studies have been performed in tomato, maize, *Arabidopsis thaliana*, barley, rice, wheat, and beans and have included the following ENMs: ZnO, Fe_3O_4, CuO, CdO, SiO_2, TiO_2, and CeO_2 NPs (Chavez Soria et al., 2019; Majumdar & Keller, 2021; Rico et al., 2020; Sun, Wang, Wang, et al., 2020; Večeřová et al., 2016; Wu et al., 2017; Yan et al., 2020; Zhao et al., 2019).

8.9 Molecular modifications in soil fungi and bacteria induced by ENMs

The impact of ENMs on soil organisms depends on soil properties and complexity (Zhang et al., 2020). These nanoparticles can affect soil microbiota, soil fertility, and crops (Ameen et al., 2021). Beneficial soil microbes such as bacteria and fungi participate in maintaining both soil and plant health (Jacoby et al., 2017; Lambers et al., 2009). To evaluate the impact of NPs on beneficial soil bacteria and fungi, it is important to analyze changes in the gene expression, proteome, and metabolome of bacteria and fungi. The effect of Ag NPs and Ag^+ on two fungal ecotypes (metal-polluted At72 and nonpolluted At61) of *Articulospora tetracladia* was assessed based on antioxidant enzymatic and proteomic responses. The treatment of Ag NPs and Ag^+ affected only 172 proteins of the At72 and 260 proteins of At61 hierarchical protein clusters. In At72, gene expression analysis showed that Ag^+ increased the production of proteins involved in proteostasis and decreased those related with vesicle-mediated transport. In addition, Ag NPs stimulated the production of proteins involved in DNA repair and energy production. In At61, Ag NPs resulted in increased number of proteins involved in energy production and protein biosynthesis, while both Ag forms promoted proteins related to cell redox and protein homeostasis, ascospore formation, fatty acid biosynthesis, and nucleic acids (Ameen et al., 2021; Barros et al., 2019; Santos et al., 2009).

It has also been shown that ENMs affect gene expression and transcriptome in soil bacteria. In vitro analysis was performed to assess the impact of Al_2O_3 and Ag NPs on *B. cereus* and *Pseudomonas stutzeri*, two common soil bacteria (Fajardo et al., 2014). Results also indicated that after 3 h of exposure of *B. cereus* to each NP, the transcriptional response of katB, pykA, nirS, and narG, which are genes related to cellular activity, exhibited an insignificant change in the expression ratio. However, a remarkable increase (8.2-fold) in katB (catalase enzyme-related gene) gene expression was observed in *P. stutzeri* after exposure to Al_2O_3 NPs. This result may be related to an increase in the OS defense mechanism in this bacterium. This study demonstrated a slower response of *B. cereus* versus *P. stutzeri* after NPs exposure (Ameen et al., 2021; Fajardo et al., 2014).

In another research, the gene expression in *Desulfovibrio vulgaris* after exposure to CuO NPs was determined. Transcriptomic analysis indicated that genes encoding for flagellar assembly and certain genes involved in electron transfer and respiration were down-regulated, while genes for the ferric uptake regulator (Fur) were up-regulated. In addition, CuO NPs exposure significantly increased the expression of genes that code for proteins involved in ATP synthesis, probably to overcome the stress caused by the overproduction of ROS after CuO NPs treatment. These results suggest that CuO NPs inhibited energy conversion, cell mobility, and iron starvation in *D. vulgaris* (Ameen et al., 2021; Chen et al., 2019).

8.10 Final remarks

Many questions regarding the effects of ENMs on microorganisms–plants interactions are still unanswered. However, novel methodological and conceptual advances in the field will increase our understanding of the soil microbiome system and its interaction with plants under stressing environments, including the presences of ENMs. It is of paramount importance to elucidate these effects and to identify specific links between soil microbial processes and the microbial taxa that are altered by NMs. The consequences of these interactions at the rhizosphere and plant physiology level need to be clarified. Vast literature has indicated that plants have an important direct or indirect effect on soil microbial communities. However, it is often difficult to predict their effect on microbial taxa because of the high complexity in the parameters and interactions involved, including the spatial and temporal variability of the components such as soil type, microbial communities, plants, and in some cases, the presence of nanomaterials.

References

Abd-Alla, M. H., Nafady, N. A., Bashandy, S. R., & Hassan, A. A. (2019). Mitigation of effect of salt stress on the nodulation, nitrogen fixation and growth of chickpea (*Cicer arietinum* L.) by triple microbial inoculation. *Rhizosphere, 10*. https://doi.org/10.1016/j.rhisph.2019.100148

Abd-Alla, M. H., Nafady, N. A., & Khalaf, D. M. (2016). Assessment of silver nanoparticles contamination on faba bean-Rhizobium leguminosarum bv. viciae-Glomus aggregatum symbiosis: Implications for induction of autophagy process in root nodule. *Agriculture, Ecosystems and Environment, 218*, 163–177. https://doi.org/10.1016/j.agee.2015.11.022

Abdulsada, Z., Kibbee, R., Örmeci, B., DeRosa, M., & Princz, J. (2021). Impact of anaerobically digested silver and copper oxide nanoparticles in biosolids on soil characteristics and bacterial community. *Chemosphere, 263*, 128173. https://doi.org/10.1016/j.chemosphere.2020.128173

Ameen, F., Alsamhary, K., Alabdullatif, J. A., & ALNadhari, S. (2021). A review on metal-based nanoparticles and their toxicity to beneficial soil bacteria and fungi. *Ecotoxicology and Environmental Safety, 213*, 112027. https://doi.org/10.1016/j.ecoenv.2021.112027

Arya, G. C., & Harel, A. (2019). *Phyllosphere and its potential role in sustainable agriculture* (pp. 39–65). Springer Science and Business Media LLC. https://doi.org/10.1007/978-981-13-8739-5_3

Asadishad, B., Chahal, S., Akbari, A., Cianciarelli, V., Azodi, M., Ghoshal, S., & Tufenkji, N. (2018). Amendment of agricultural soil with metal nanoparticles: Effects on soil enzyme activity and microbial community composition. *Environmental Science and Technology, 52*(4), 1908–1918. https://doi.org/10.1021/acs.est.7b05389

Balestrini, R., Lumini, E., Borriello, R., & Bianciotto, V. (2015). *Plant-soil biota interactions* (pp. 311–338). Elsevier BV. https://doi.org/10.1016/b978-0-12-415955-6.00011-6

Barros, D., Pradhan, A., Mendes, V. M., Manadas, B., Santos, P. M., Pascoal, C., & Cássio, F. (2019). Proteomics and antioxidant enzymes reveal different mechanisms of toxicity induced by ionic and nanoparticulate silver in bacteria. *Environmental Science: Nano, 6*(4), 1207–1218. https://doi.org/10.1039/c8en01067f

Berendsen, R. L., Pieterse, C. M. J., & Bakker, P. A. H. M. (2012). The rhizosphere microbiome and plant health. *Trends in Plant Science, 17*(8), 478–486. https://doi.org/10.1016/j.tplants.2012.04.001

Borgatta, J., Ma, C., Hudson-Smith, N., Elmer, W., Plaza Pérez, C. D., De La Torre-Roche, R., Zuverza-Mena, N., Haynes, C. L., White, J. C., & Hamers, R. J. (2018). Copper based nanomaterials suppress root fungal disease in watermelon (*Citrullus lanatus*): Role of particle morphology, composition and dissolution behavior. *ACS Sustainable Chemistry and Engineering, 6*(11), 14847–14856. https://doi.org/10.1021/acssuschemeng.8b03379

Bourdineaud, J. P., Štambuk, A., Šrut, M., Radić Brkanac, S., Ivanković, D., Lisjak, D., Sauerborn Klobučar, R., Dragun, Z., Bačić, N., & Klobučar, G. I. V. (2021). Gold and silver nanoparticles effects to the earthworm Eisenia fetida—the importance of tissue over soil concentrations. *Drug and Chemical Toxicology, 44*(1), 12–29. https://doi.org/10.1080/01480545.2019.1567757

Brencic, A., & Winans, S. C. (2005). Detection of and response to signals involved in host-microbe interactions by plant-associated bacteria. *Microbiology and Molecular Biology Reviews, 69*(1), 155–194. https://doi.org/10.1128/MMBR.69.1.155-194.2005

Cao, J., Feng, Y., Lin, X., & Wang, J. (2020). A beneficial role of arbuscular mycorrhizal fungi in influencing the effects of silver nanoparticles on plant-microbe systems in a soil matrix. *Environmental Science and Pollution Research, 27*(11), 11782–11796. https://doi.org/10.1007/s11356-020-07781-w

Carbone, S., Vittori Antisari, L., Gaggia, F., Baffoni, L., Di Gioia, D., Vianello, G., & Nannipieri, P. (2014). Bioavailability and biological effect of engineered silver nanoparticles in a forest soil. *Journal of Hazardous Materials, 280*, 89–96. https://doi.org/10.1016/j.jhazmat.2014.07.055

Chai, H., Yao, J., Sun, J., Zhang, C., Liu, W., Zhu, M., & Ceccanti, B. (2015). The effect of metal oxide nanoparticles on functional bacteria and metabolic profiles in agricultural soil. *Bulletin of Environmental Contamination and Toxicology, 94*(4), 490–495. https://doi.org/10.1007/s00128-015-1485-9

Chaudhary, P., Sharma, A., Chaudhary, A., Khati, P., Gangola, S., & Maithani, D. (2021). Illumina based high throughput analysis of microbial diversity of maize rhizosphere treated with nanocompounds and Bacillus sp. *Applied Soil Ecology, 159*, 103836. https://doi.org/10.1016/j.apsoil.2020.103836

Chavez Soria, N. G., Bisson, M. A., Atilla-Gokcumen, G. E., & Aga, D. S. (2019). High-resolution mass spectrometry-based metabolomics reveal the disruption of jasmonic pathway in *Arabidopsis thaliana* upon copper oxide nanoparticle exposure. *Science of the Total Environment, 693*, 133443. https://doi.org/10.1016/j.scitotenv.2019.07.249

Chen, Z., Gao, S. h., Jin, M., Sun, S., Lu, J., Yang, P., Bond, P. L., Yuan, Z., & Guo, J. (2019). Physiological and transcriptomic analyses reveal CuO nanoparticle inhibition of anabolic and catabolic activities of sulfate-reducing bacterium. *Environment International, 125*, 65–74. https://doi.org/10.1016/j.envint.2019.01.058

Cornelis, G., Hund-Rinke, K., Kuhlbusch, T., van den Brink, N., & Nickel, C. (2014). Fate and bioavailability of engineered nanoparticles in soils: A review. *Critical Reviews in Environmental Science and Technology, 44*(24), 2720–2764. https://doi.org/10.1080/10643389.2013.829767

Cui, J., Li, Y., Jin, Q., & Li, F. (2020). Silica nanoparticles inhibit arsenic uptake into rice suspension cells via improving pectin synthesis and the mechanical force of the cell wall. *Environmental Science: Nano, 7*(1), 162–171. https://doi.org/10.1039/c9en01035a

van Dam, N. M., & Bouwmeester, H. J. (2016). Metabolomics in the rhizosphere: Tapping into belowground chemical communication. *Trends in Plant Science, 21*(3), 256–265. https://doi.org/10.1016/j.tplants.2016.01.008

Das, P., Barua, S., Sarkar, S., Chatterjee, S. K., Mukherjee, S., Goswami, L., Das, S., Bhattacharya, S., Karak, N., & Bhattacharya, S. S. (2018). Mechanism of toxicity and transformation of silver nanoparticles: Inclusive assessment in earthworm-microbe-soil-plant system. *Geoderma, 314*, 73–84. https://doi.org/10.1016/j.geoderma.2017.11.008

Dixon, R. A. (2001). Natural products and plant disease resistance. *Nature, 411*(6839), 843–847. https://doi.org/10.1038/35081178

Dubey, A. N., Chattopadhyaya, N., & Mandal, N. (2021). Variation in soil microbial population and soil enzymatic activities under zincated nanoclay polymer composites (ZNCPCs), nano-ZnO and Zn solubilizers in rice rhizosphere. *Agricultural Research, 10*(1), 21–31. https://doi.org/10.1007/s40003-020-00488-x

Duo, L. A., Liu, C. X., & Zhao, S. L. (2018). Alleviation of drought stress in turfgrass by the combined application of nano-compost and microbes from compost. *Russian Journal of Plant Physiology, 65*(3), 419–426. https://doi.org/10.1134/S102144371803010X

Echevarria-Molinar, A. (2017). *Efecto de factores abióticos y bióticos sobre la estructura de la comunidad microbiana del suelo en un ambiente oligotrófico 117*.

Eivazi, F., Afrasiabi, Z., & Jose, E. (2018). Effects of silver nanoparticles on the activities of soil enzymes involved in carbon and nutrient cycling. *Pedosphere, 28*(2), 209–214. https://doi.org/10.1016/S1002-0160(18)60019-0

Elsakhawy, T., Omara, A. E.-D., Alshaal, T., El-Ramady, H., Ghazi, A., El-Nahrawy, S., & Elhawat, N. (2018). Nanomaterials and plant abiotic stress in agroecosystems. *Environment, Biodiversity and Soil Security, 50–55*. https://doi.org/10.21608/jenvbs.2018.3897.1030

Faizan, M., Faraz, A., Yusuf, M., Khan, S. T., & Hayat, S. (2018). Zinc oxide nanoparticle-mediated changes in photosynthetic efficiency and antioxidant system of tomato plants. *Photosynthetica, 56*(2), 678–686. https://doi.org/10.1007/s11099-017-0717-0

Fajardo, C., Saccà, M. L., Costa, G., Nande, M., & Martin, M. (2014). Impact of Ag and Al2O3 nanoparticles on soil organisms: In vitro and soil experiments. *Science of the Total Environment, 473–474*, 254–261. https://doi.org/10.1016/j.scitotenv.2013.12.043

Fiehn, O. (2002). Metabolomics - the link between genotypes and phenotypes. *Plant Molecular Biology, 48*(1–2), 155–171. https://doi.org/10.1023/A:1013713905833

Frazier, T. P., Burklew, C. E., & Zhang, B. (2014). Titanium dioxide nanoparticles affect the growth and microRNA expression of tobacco (*Nicotiana tabacum*). *Functional and Integrative Genomics, 14*, 75–83. https://doi.org/10.1007/s10142-013-0341-4

Galaktionova, L., Gavrish, I., & Lebedev, S. (2019). Bioeffects of Zn and Cu nanoparticles in soil systems. *Toxicology and Environmental Health Sciences, 11*(4), 259–270. https://doi.org/10.1007/s13530-019-0413-5

Ghasemi Siani, N., Fallah, S., Pokhrel, L. R., & Rostamnejadi, A. (2017). Natural amelioration of Zinc oxide nanoparticle toxicity in fenugreek (*Trigonella foenum-gracum*) by arbuscular mycorrhizal (*Glomus intraradices*) secretion of glomalin. *Plant Physiology and Biochemistry, 112*, 227–238. https://doi.org/10.1016/j.plaphy.2017.01.001

Grün, A. L., Straskraba, S., Schulz, S., Schloter, M., & Emmerling, C. (2018). Long-term effects of environmentally relevant concentrations of silver nanoparticles on microbial biomass, enzyme activity, and functional genes involved in the nitrogen cycle of loamy soil. *Journal of Environmental Sciences (China), 69*, 12–22. https://doi.org/10.1016/j.jes.2018.04.013

Gupta, G. S., Kumar, A., Senapati, V. A., Pandey, A. K., Shanker, R., & Dhawan, A. (2017). Laboratory scale microbial food chain to study bioaccumulation, biomagnification, and ecotoxicity of cadmium telluride quantum dots. *Environmental Science and Technology, 51*(3), 1695–1706. https://doi.org/10.1021/acs.est.6b03950

Hamidat, M., Barakat, M., Ortet, P., Chanéac, C., Rose, J., Bottero, J. Y., Heulin, T., Achouak, W., & Santaella, C. (2016). Design defines the effects of nanoceria at a low dose on soil microbiota and the potentiation of impacts by the canola plant. *Environmental Science and Technology, 50*(13), 6892–6901. https://doi.org/10.1021/acs.est.6b01056

He, S., Feng, Y., Ni, J., Sun, Y., Xue, L., Feng, Y., Yu, Y., Lin, X., & Yang, L. (2016). Different responses of soil microbial metabolic activity to silver and iron oxide nanoparticles. *Chemosphere, 147*, 195–202.

Hermes, P. H., Fabián, F. L., Esperanza, H. L., Jorge, M. V., José David, A. S., Edilberto, H. G., Javier Francisco, V. M., & Marcos, P. S. (2020). The first evidence of accumulation and avoidance behavior of macroinvertebrates in a forest soil spiked with human-made iron nanoparticles: A field experiment. *Heliyon, 6*(9). https://doi.org/10.1016/j.heliyon.2020.e04860

Ho, Y.-N., Mathew, D. C., & Huang, C.-C. (2017). *Plant-microbe ecology: Interactions of plants and symbiotic microbial communities*. IntechOpen. https://doi.org/10.5772/intechopen.69088

Hossain, Z., Yasmeen, F., & Komatsu, S. (2020). Nanoparticles: Synthesis, morphophysiological effects, and proteomic responses of crop plants. *International Journal of Molecular Science, 21*(9), E3056. https://doi.org/10.3390/ijms21093056

Hounsome, N., Hounsome, B., Tomos, D., & Edwards-Jones, G. (2008). Plant metabolites and nutritional quality of vegetables. *Journal of Food Science, 73*(4), R48–R65. https://doi.org/10.1111/j.1750-3841.2008.00716.x

Husen, A., & Siddiqi, K. S. (2014). Carbon and fullerene nanomaterials in plant system. *Journal of Nanobiotechnology, 12*(1). https://doi.org/10.1186/1477-3155-12-16

Hussain, M., Raja, N. I., Mashwani, Z. U. R., Iqbal, M., Ejaz, M., & Aslam, S. (2019). Green synthesis and evaluation of silver nanoparticles for antimicrobial and biochemical profiling in Kinnow (*Citrus reticulata* L.) to enhance fruit quality and productivity under biotic stress. *IET Nanobiotechnology, 13*(3), 250–256. https://doi.org/10.1049/iet-nbt.2018.5049

Jacoby, R., Peukert, M., Succurro, A., Koprivova, A., & Kopriva, S. (2017). The role of soil microorganisms in plant mineral nutrition—current knowledge and future directions. *Frontiers in Plant Science, 8*. https://doi.org/10.3389/fpls.2017.01617

Jamiołkowska, A., Księżniak, A., Hetman, B., Kopacki, M., Skwaryło-Bednarz, B., Gałązka, A., & Thanoon, A. H. (2017). Interactions of arbuscular mycorrhizal fungi with plants and soil microflora. *Acta Scientiarum Polonorum Hortorum Cultus, 16*(5), 89–95. https://doi.org/10.24326/asphc.2017.5.9

Javed, Z., Dashora, K., Mishra, M., D. Fasake, V., & Srivastva, A. (2019). Effect of accumulation of nanoparticles in soil health- a concern on future. *Frontiers in Nanoscience and Nanotechnology*. https://doi.org/10.15761/fnn.1000181

Karunakaran, G., Suriyaprabha, R., Manivasakan, P., Yuvakkumar, R., Rajendran, V., Prabu, P., & Kannan, N. (2013). Effect of nanosilica and silicon sources on plant growth promoting rhizobacteria, soil nutrients and maize seed germination. *IET Nanobiotechnology, 7*(3), 70–77. https://doi.org/10.1049/iet-nbt.2012.0048

Kaur, H., Kalia, A., & Sharma, S. P. (2021). Multi-Wall Carbon nanotubes, metal oxide and hydroxy-apatite nanoparticles enhanced plant growth promoting capabilities of root endosymbionts of cowpea (*Vigna unguiculata* L.) walp. *Journal of Nanoscience and Nanotechnology, 21*, 3634–3649. https://doi.org/10.1166/jnn.2021.18995

Khan, N., & Bano, A. (2016). Role of plant growth promoting rhizobacteria and Ag-nano particle in the bioremediation of heavy metals and maize growth under municipal wastewater irrigation. *International Journal of Phytoremediation, 18*(3), 211–221. https://doi.org/10.1080/15226514.2015.1064352

Khati, P., Sharma, A., Chaudhary, P., Singh, A. K., Gangola, S., & Kumar, R. (2019). High-throughput sequencing approach to access the impact of nanozeolite treatment on species richness and evenness of soil metagenome. *Biocatalysis and Agricultural Biotechnology, 20*. https://doi.org/10.1016/j.bcab.2019.101249

Khati, Priyanka, Sharma, A., Gangola, S., Kumar, R., Bhatt, P., & Kumar, G. (2017). Impact of agri-useable nanocompounds on soil microbial activity: An indicator of soil health. *CLEAN - Soil, Air, Water, 45*(5), 1600458. https://doi.org/10.1002/clen.201600458

Khodakovskaya, M. V., Kim, B. S., Kim, J. N., Alimohammadi, M., Dervishi, E., Mustafa, T., & Cernigla, C. E. (2013). Carbon nanotubes as plant growth regulators: Effects on tomato growth, reproductive system, and soil microbial community. *Small, 9*(1), 115–123. https://doi.org/10.1002/smll.201201225

Kong, C., Azizan, S. S. S., Kakarla, A. B., Kong, I., & Kong, W. (2021). In vivo toxicity assessment of nickel zinc carbon nanotubes in a living host model. *Solid State Phenomena, 317*, 227–232. https://doi.org/10.4028/www.scientific.net/ssp.317.227

Kumari, M., Ernest, V., Mukherjee, A., & Chandrasekaran, N. (2012). In vivo nanotoxicity assays in plant models. *Methods in Molecular Biology, 926*, 399–410. https://doi.org/10.1007/978-1-62703-2-1_26

Kumar, A., & Verma, J. P. (2018). Does plant—microbe interaction confer stress tolerance in plants: A review? *Microbiological Research, 207*, 41–52. https://doi.org/10.1016/j.micres.2017.11.004

Lambers, H., Mougel, C., Jaillard, B., & Hinsinger, P. (2009). Plant-microbe-soil interactions in the rhizosphere: An evolutionary perspective. *Plant and Soil, 321*(1–2), 83–115. https://doi.org/10.1007/s11104-009-0042-x

Landa, P., Dytrych, P., Prerostova, S., Petrova, S., Vankova, R., & Vanek, T. (2017). Transcriptomic response of *Arabidopsis thaliana* exposed to CuO nanoparticles, bulk material, and ionic copper. *Environmental Science Technology, 51*, 10814–10824. https://doi.org/10.1021/acs.est.7b02265

Lata, R., & Gond, S. K. (2019). *Plant growth-promoting microbes for abiotic stress tolerance in plants* (pp. 89–105). Elsevier BV. https://doi.org/10.1016/b978-0-12-817004-5.00006-3

Li, W. R., Sun, T. L., Zhou, S. L., Ma, Y. K., Shi, Q. S., Xie, X. B., & Huang, X. M. (2017). A comparative analysis of antibacterial activity, dynamics, and effects of silver ions and silver nanoparticles against four bacterial strains. *International Biodeterioration and Biodegradation, 123*, 304–310. https://doi.org/10.1016/j.ibiod.2017.07.015

Liu, Y., Li, Y., Pan, B., Zhang, X., Zhang, H., Steinberg, C. E. W., Qiu, H., Vijver, M. G., & Peijnenburg, W. J. G. M. (2021). Application of low dosage of copper oxide and zinc oxide nanoparticles boosts bacterial and fungal communities in soil. *Science of the Total Environment, 757*. https://doi.org/10.1016/j.scitotenv.2020.143807

Liu, Yang, Pan, B., Li, H., Lang, D., Zhao, Q., Zhang, D., Wu, M., Steinberg, C. E. W., & Xing, B. (2020). Can the properties of engineered nanoparticles be indicative of their functions and effects in plants? *Ecotoxicology and Environmental Safety, 205*, 111128. https://doi.org/10.1016/j.ecoenv.2020.111128

Liu, J., Williams, P. C., Goodson, B. M., Geisler-Lee, J., Fakharifar, M., & Gemeinhardt, M. E. (2019). TiO_2 nanoparticles in irrigation water mitigate impacts of aged Ag nanoparticles on soil microorganisms, *Arabidopsis thaliana* plants, and *Eisenia fetida* earthworms. *Environmental Research, 172*, 202–215. https://doi.org/10.1016/j.envres.2019.02.010

Lowry, G. V., Avellan, A., & Gilbertson, L. M. (2019). Opportunities and challenges for nanotechnology in the agri-tech revolution. *Nature Nanotechnology, 14*(6), 517–522. https://doi.org/10.1038/s41565-019-0461-7

Majumdar, S., & Keller, A. A. (2021). Omics to address the opportunities and challenges of nanotechnology in agriculture. *Critical Reviews in Environmental Science and Technology*, 1–42. https://doi.org/10.1080/10643389.2020.1785264

Masrahi, A., VandeVoort, A. R., & Arai, Y. (2014). Effects of silver nanoparticle on soil-nitrification processes. *Archives of Environmental Contamination and Toxicology, 66*(4), 504–513. https://doi.org/10.1007/s00244-013-9994-1

Maurer-Jones, M. A., Gunsolus, I. L., Murphy, C. J., & Haynes, C. L. (2013). Toxicity of engineered nanoparticles in the environment. *Analytical Chemistry, 85*(6), 3036–3049. https://doi.org/10.1021/ac303636s

McKee, M. S., & Filser, J. (2016). Impacts of metal-based engineered nanomaterials on soil communities. *Environmental Science: Nano, 3*(3), 506–533. https://doi.org/10.1039/c6en00007j

Mishra, P., Xue, Y., Eivazi, F., & Afrasiabi, Z. (2021). Size, concentration, coating, and exposure time effects of silver nanoparticles on the activities of selected soil enzymes. *Geoderma, 381*, 114682. https://doi.org/10.1016/j.geoderma.2020.114682

Mohanty, A., Wu, Y., & Cao, B. (2014). Impacts of engineered nanomaterials on microbial community structure and function in natural and engineered ecosystems. *Applied Microbiology and Biotechnology, 98*(20), 8457–8468. https://doi.org/10.1007/s00253-014-6000-4

Mokarram-Kashtiban, S., Hosseini, S. M., Tabari Kouchaksaraei, M., & Younesi, H. (2019). The impact of nanoparticles zero-valent iron (nZVI) and rhizosphere microorganisms on the phytoremediation ability of white willow and its response. *Environmental Science and Pollution Research, 26*(11), 10776–10789. https://doi.org/10.1007/s11356-019-04411-y

Moll, J., Gogos, A., Bucheli, T. D., Widmer, F., & van der Heijden, M. G. A. (2016). Effect of nanoparticles on red clover and its symbiotic microorganisms. *Journal of Nanobiotechnology, 14*(1). https://doi.org/10.1186/s12951-016-0188-7

Moll, J., Klingenfuss, F., Widmer, F., Gogos, A., Bucheli, T. D., Hartmann, M., & van der Heijden, M. G. A. (2017). Effects of titanium dioxide nanoparticles on soil microbial communities and wheat biomass. *Soil Biology and Biochemistry, 111*, 85–93. https://doi.org/10.1016/j.soilbio.2017.03.019

Moon, J., Kwak, J. I., & An, Y. J. (2019). The effects of silver nanomaterial shape and size on toxicity to *Caenorhabditis elegans* in soil media. *Chemosphere, 215*, 50–56. https://doi.org/10.1016/j.chemosphere.2018.09.177

Nayana, A. R., Joseph, B. J., Jose, A., & Radhakrishnan, E. K. (2020). *Nanotechnological advances with PGPR applications* (pp. 163–180). Springer Science and Business Media LLC. https://doi.org/10.1007/978-3-030-33996-8_9

Ouyang, B., Yilihamu, A., Liu, D., Ouyang, P., Zhang, D., Wu, X., & Yang, S.-T. (2021). Toxicity and environmental impact of multi-walled carbon nanotubes to nitrogen-fixing bacterium Azotobacter chroococcum. *Journal of Environmental Chemical Engineering, 9*(4), 105291. https://doi.org/10.1016/j.jece.2021.105291

Pagano, L., Pasquali, F., Majumdar, S., Torre-Roche, R. D. la, Zuverza-Mena, N., Villani, M., Zappettini, A., Marra, R. E., Isch, S. M., Marmiroli, M., Maestri, E., Dhankher, O. P., White, J. C., & Marmiroli, N. (2017). Exposure of *Cucurbita pepo* to binary combinations of engineered nanomaterials: Physiological and molecular response. *Environmental Science: Nano, 4*, 1579–1590. https://doi.org/10.1039/C7EN00219J

Palmqvist, N. G. M., Bejai, S., Meijer, J., Seisenbaeva, G. A., & Kessler, V. G. (2015). Nano titania aided clustering and adhesion of beneficial bacteria to plant roots to enhance crop growth and stress management. *Scientific Reports, 5*. https://doi.org/10.1038/srep10146

Pan, C., Bao, Y., Guo, A., & Ma, J. (2020). Environmentally relevant-level CeO2NP with ferrous amendment alters soil bacterial community compositions and metabolite profiles in rice-planted soils. *Journal of Agricultural and Food Chemistry, 68*(31), 8172–8184. https://doi.org/10.1021/acs.jafc.0c03507

Peijnenburg, W., Praetorius, A., Scott-Fordsmand, J., & Cornelis, G. (2016). Fate assessment of engineered nanoparticles in solids dominated media – current insights and the way forward. *Environmental Pollution, 218*, 1365–1369. https://doi.org/10.1016/j.envpol.2015.11.043

Pei, Z. F., Ming, D. F., Liu, D., Wan, G. L., Geng, X. X., Gong, H. J., & Zhou, W. J. (2010). Silicon improves the tolerance to water-deficit stress induced by polyethylene glycol in wheat (*Triticum aestivum* L.) seedlings. *Journal of Plant Growth Regulation, 29*(1), 106–115. https://doi.org/10.1007/s00344-009-9120-9

Peixoto, S., Henriques, I., & Loureiro, S. (2021). Long-term effects of $Cu(OH)_2$ nanopesticide exposure on soil microbial communities. *Environmental Pollution, 269*, 116113. https://doi.org/10.1016/j.envpol.2020.116113

Pérez-Hernández, H., Huerta-Lwanga, E., Mendoza-Vega, J., Álvarez-Solís, J. D., Pampillón-González, L., & Fernández-Luqueño, F. (2021). Assessment of TiO_2 nanoparticles on maize seedlings and terrestrial isopods under greenhouse conditions. *Journal of Soil Science and Plant Nutrition*. https://doi.org/10.1007/s42729-021-00515-y

Petrova, A., Plaksenkova, I., Kokina, I., & Jermaļonoka, M. (2021). Effect of Fe_3O_4 and CuO nanoparticles on morphology, genotoxicity, and mirna expression on different barley (*Hordeum vulgare* L.) Genotypes. *The Scientific World Journal, 2021*, e6644689. https://doi.org/10.1155/2021/6644689

Priester, J. H., Stoimenov, P. K., Mielke, R. E., Webb, S. M., Ehrhardt, C., Zhang, J. P., Stucky, G. D., & Holden, P. A. (2009). Effects of soluble cadmium salts versus cdse quantum dots on the growth of planktonic pseudomonas aeruginosa. *Environmental Science and Technology, 43*(7), 2589–2594. https://doi.org/10.1021/es802806n

Prudnikova, S., Streltsova, N., & Volova, T. (2021). The effect of the pesticide delivery method on the microbial community of field soil. *Environmental Science and Pollution Research, 28*(7), 8681–8697. https://doi.org/10.1007/s11356-020-11228-7

Qian, H., Ke, M., Qu, Q., Li, X., Du, B., Lu, T., Sun, L., & Pan, X. (2018). Ecological effects of single-walled carbon nanotubes on soil microbial communities and soil fertility. *Bulletin of*

Environmental Contamination and Toxicology, 101(4), 536–542. https://doi.org/10.1007/s00128-018-2437-y

Rajak, J., Bawaskar, M., Rathod, D., Agarkar, G., Nagaonkar, D., Gade, A., & Rai, M. (2017). Interaction of copper nanoparticles and an endophytic growth promoter Piriformospora indica with *Cajanus cajan*. *Journal of the Science of Food and Agriculture, 97*(13), 4562–4570. https://doi.org/10.1002/jsfa.8324

Raliya, R., Tarafdar, J. C., & Biswas, P. (2016). Enhancing the mobilization of native phosphorus in the mung bean rhizosphere using ZnO nanoparticles synthesized by soil fungi. *Journal of Agricultural and Food Chemistry, 64*(16), 3111–3118. https://doi.org/10.1021/acs.jafc.5b05224

Ramírez-Sánchez, O., Pérez-Rodríguez, P., Delaye, L., & Tiessen, A. (2016). Plant proteins are smaller because they are encoded by fewer exons than animal proteins. *Genomics, Proteomics and Bioinformatics, 14*(6), 357–370. https://doi.org/10.1016/j.gpb.2016.06.003

Rane, M., Bawskar, M., Rathod, D., Nagaonkar, D., & Rai, M. (2015). Influence of calcium phosphate nanoparticles, Piriformospora indica and Glomus mosseae on growth of *Zea mays*. *Advances in Natural Sciences: Nanoscience and Nanotechnology, 6*(4), 045014. https://doi.org/10.1088/2043-6262/6/4/045014

Ribeiro, M. J., Amorim, M. J. B., & Scott-Fordsmand, J. J. (2019). Cell in vitro testing with soil invertebrates—challenges and opportunities toward modeling the effect of nanomaterials: A surface-modified cuo case study. *Nanomaterials, 9*(8). https://doi.org/10.3390/nano9081087

Rico, C. M., Wagner, D., Abolade, O., Lottes, B., & Coates, K. (2020). Metabolomics of wheat grains generationally-exposed to cerium oxide nanoparticles. *Science of the Total Environment, 712*, 136487. https://doi.org/10.1016/j.scitotenv.2019.136487

de la Rosa, G., Vázquez-Núñez, E., Molina-Guerrero, C., Serafín-Muñoz, A. H., & Vera-Reyes, I. (2021). Interactions of nanomaterials and plants at the cellular level: Current knowledge and relevant gaps. *Nanotechnology for Environmental Engineering, 6*(1). https://doi.org/10.1007/s41204-020-00100-1

Ruiz-Sánchez, M., Aroca, R., Muñoz, Y., Polón, R., & Ruiz-Lozano, J. M. (2010). The arbuscular mycorrhizal symbiosis enhances the photosynthetic efficiency and the antioxidative response of rice plants subjected to drought stress. *Journal of Plant Physiology, 167*(11), 862–869. https://doi.org/10.1016/j.jplph.2010.01.018

Salas-Leiva, J., Salas-Leiva, D. E., Tovar-Ramírez, D., Herrera-Pérez, G., Tarango-Rivero, S., Luna-Velasco, A., & Orrantia-Borunda, E. (2021). Copper oxide nanoparticles slightly affect diversity and metabolic profiles of the prokaryotic community in pecan tree (*Carya illinoinensis*) rhizospheric soil. *Applied Soil Ecology, 157*. https://doi.org/10.1016/j.apsoil.2020.103772

Salehi, H., Chehregani, A., Lucini, L., Majd, A., & Gholami, M. (2018). Morphological, proteomic and metabolomic insight into the effect of cerium dioxide nanoparticles to *Phaseolus vulgaris* L. under soil or foliar application. *Science of the Total Environment*. https://doi.org/10.1016/j.scitotenv.2017.10.159

Santos, P. M., Simões, T., & Sá-Correia, I. (2009). Insights into yeast adaptive response to the agricultural fungicide mancozeb: A toxicoproteomics approach. *Proteomics, 9*(3), 657–670. https://doi.org/10.1002/pmic.200800452

Santoyo, G., Hernández-Pacheco, C., Hernández-Salmerón, J., & Hernández-León, R. (2017). The role of abiotic factors modulating the plant-microbe-soil interactions: Toward sustainable agriculture. A review. *Spanish Journal of Agricultural Research, 15*(1). https://doi.org/10.5424/sjar/2017151-9990

Sashidhar, P., Dubey, M. K., & Kochar, M. (2019). Sensing soil microbes and interactions: How can nanomaterials help?. In *Nanotechnology in the life sciences* (pp. 213–236). Springer Science and Business Media B.V. https://doi.org/10.1007/978-3-030-16534-5_11

Schläpfer, P., Zhang, P., Wang, C., Kim, T., Banf, M., Chae, L., Dreher, K., Chavali, A. K., Nilo-Poyanco, R., Bernard, T., Kahn, D., & Rhee, S. Y. (2017). Genome-wide prediction of metabolic enzymes, pathways, and gene clusters in plants. *Plant Physiology, 173*(4), 2041–2059. https://doi.org/10.1104/pp.16.01942

Shah, S. N. A., Shah, Z., Hussain, M., & Khan, M. (2017). Hazardous effects of titanium dioxide nanoparticles in ecosystem. *Bioinorganic Chemistry and Applications, 2017*. https://doi.org/10.1155/2017/4101735

Sharifi, & Khoramdel, S. (2015). Effects of Nano-zinc oxide and seed inoculation by plant growth promoting rhizobacteria (PGPR) on yield, yield components and grain filling period of soybean (*Glycine max* L.). *Iranian Journal of Field Crop Research, 13*, 738–753. https://doi.org/10.22067/gsc.v13i4.32491

Shcherbakova, E. N., Shcherbakov, A. V., Andronov, E. E., Gonchar, L. N., Kalenskaya, S. M., & Chebotar, V. K. (2017). Combined pre-seed treatment with microbial inoculants and Mo nanoparticles changes composition of root exudates and rhizosphere microbiome structure of chickpea (*Cicer arietinum* L.) plants. *Symbiosis, 73*(1), 57–69. https://doi.org/10.1007/s13199-016-0472-1

Shen, Z., Zhang, Z., Li, T., Yao, Q., Zhang, T., & Chen, W. (2020). Facet-dependent adsorption and fractionation of natural organic matter on crystalline metal oxide nanoparticles. *Environmental Science and Technology, 54*(14), 8622–8631. https://doi.org/10.1021/acs.est.9b06111

Siddiqui, M. H., & Al-Whaibi, M. H. (2014). Role of nano-SiO_2 in germination of tomato (*Lycopersicum esculentum* seeds Mill.). *Saudi Journal of Biological Sciences, 21*(1), 13–17. https://doi.org/10.1016/j.sjbs.2013.04.005

Sillen, W. M. A., Thijs, S., Abbamondi, G. R., De La Torre Roche, R., Weyens, N., White, J. C., & Vangronsveld, J. (2020). Nanoparticle treatment of maize analyzed through the metatranscriptome: Compromised nitrogen cycling, possible phytopathogen selection, and plant hormesis. *Microbiome, 8*(1). https://doi.org/10.1186/s40168-020-00904-y

Sillen, W. M. A., Thijs, S., Abbamondi, G. R., Janssen, J., Weyens, N., White, J. C., & Vangronsveld, J. (2015). Effects of silver nanoparticles on soil microorganisms and maize biomass are linked in the rhizosphere. *Soil Biology and Biochemistry, 91*, 14–22. https://doi.org/10.1016/j.soilbio.2015.08.019

Simonin, M., Colman, B. P., Tang, W., Judy, J. D., Anderson, S. M., Bergemann, C. M., Rocca, J. D., Unrine, J. M., Cassar, N., & Bernhardt, E. S. (2018). Plant and microbial responses to repeated Cu $(OH)_2$ nanopesticide exposures under different fertilization levels in an agro-ecosystem. *Frontiers in Microbiology, 9*, 1769.

Singh, J., Vishwakarma, K., Ramawat, N., Rai, P., Singh, V. K., Mishra, R. K., Kumar, V., Tripathi, D. K., & Sharma, S. (2019). Nanomaterials and microbes' interactions: A contemporary overview. *3 Biotech, 9*(3). https://doi.org/10.1007/s13205-019-1576-0

Sun, L., Wang, R., Ju, Q., & Xu, J. (2020). Physiological, metabolic, and transcriptomic analyses reveal the responses of *Arabidopsis seedlings* to carbon nanohorns. *Environmental Science Technology, 54*, 4409–4420. https://doi.org/10.1021/acs.est.9b07133

Sun, L., Wang, Y., Wang, Ruling, Wang, Ruting, Zhang, P., Ju, Q., & Xu, J. (2020). Physiological, transcriptomic, and metabolomic analyses reveal zinc oxide nanoparticles modulate plant growth in tomato. *Environmental Science: Nano, 7*, 3587–3604. https://doi.org/10.1039/D0EN00723D

Taffner, J., Erlacher, A., Bragina, A., Berg, C., Moissl-Eichinger, C., & Berg, G. (2018). What is the role of archaea in plants? New insights from the vegetation of alpine bogs. *mSphere, 3*(3). https://doi.org/10.1128/mSphere.00122-18

Timmusk, S., Seisenbaeva, G., & Behers, L. (2018). Titania (TiO_2) nanoparticles enhance the performance of growth-promoting rhizobacteria. *Scientific Reports, 8*(1). https://doi.org/10.1038/s41598-017-18939-x

Usman, M., Farooq, M., Wakeel, A., Nawaz, A., Cheema, S. A., Rehman, H.u., Ashraf, I., & Sanaullah, M. (2020). *Nanotechnology in agriculture: Current status, challenges and future opportunities* (Vol. 721). Science of the Total Environment. https://doi.org/10.1016/j.scitotenv.2020.137778

Van Aken, B. (2015). Gene expression changes in plants and microorganisms exposed to nanomaterials. *Current Opinion in Biotechnology, 33*, 206–219. https://doi.org/10.1016/j.copbio.2015.03.005

Vandevoort, A. R., & Arai, Y. (2018). Macroscopic observation of soil nitrification kinetics impacted by copper nanoparticles: Implications for micronutrient nanofertilizer. *Nanomaterials, 8*(11). https://doi.org/10.3390/nano8110927

Vázquez Núñez, E., García Gaytán, A., Luna-Guido, M., Marsch, R., & Dendooven, L. (2009). Impact of moisture dynamic and sun light on anthracene removal from soil. *Biodegradation, 20*(2), 191–198. https://doi.org/10.1007/s10532-008-9212-4

Večeřová, K., Večeřa, Z., Dočekal, B., Oravec, M., Pompeiano, A., Tříska, J., & Urban, O. (2016). Changes of primary and secondary metabolites in barley plants exposed to CdO nanoparticles. *Environmental Pollution, 218*, 207–218. https://doi.org/10.1016/j.envpol.2016.05.013

Venturi, V., & Keel, C. (2016). Signaling in the rhizosphere. *Trends in Plant Science, 21*(3), 187–198. https://doi.org/10.1016/j.tplants.2016.01.005

Vishwakarma, K., Shweta, Upadhyay, N., Singh, J., Liu, S., Singh, V. P., Prasad, S. M., Chauhan, D. K., Tripathi, D. K., & Sharma, S. (2017). Differential phytotoxic impact of plant mediated silver nanoparticles (AgNPs) and silver nitrate ($AgNO_3$) on Brassica sp. *Frontiers in Plant Science, 8*. https://doi.org/10.3389/fpls.2017.01501

Vitali, F., Raio, A., Sebastiani, F., Cherubini, P., Cavalieri, D., & Cocozza, C. (2019). Environmental pollution effects on plant microbiota: The case study of poplar bacterial-fungal response to silver nanoparticles. *Applied Microbiology and Biotechnology, 103*(19), 8215–8227. https://doi.org/10.1007/s00253-019-10071-2

Wagg, C., Hautier, Y., Pellkofer, S., Banerjee, S., Schmid, B., & van der Heijden, M. G. A. (2021). Diversity and asynchrony in soil microbial communities stabilizes ecosystem functioning. *eLife, 10*. https://doi.org/10.7554/eLife.62813

Wang, F., Jing, X., Adams, C. A., Shi, Z., & Sun, Y. (2018). Decreased ZnO nanoparticle phytotoxicity to maize by arbuscular mycorrhizal fungus and organic phosphorus. *Environmental Science and Pollution Research, 25*(24), 23736–23747. https://doi.org/10.1007/s11356-018-2452-x

Wang, Z., Yue, L., Dhankher, O. P., & Xing, B. (2020). Nano-enabled improvements of growth and nutritional quality in food plants driven by rhizosphere processes. *Environment International, 142*, 105831. https://doi.org/10.1016/j.envint.2020.105831

Werlin, R. A. (2011). *Investigations on Pollutant and nanoparticle toxicity, and quantum dot biomagnification with the ciliated protozoan tetrahymena thermophila.*

Wigger, H., Kägi, R., Wiesner, M., & Nowack, B. (2020). Exposure and possible risks of engineered nanomaterials in the environment—current knowledge and directions for the future. *Reviews of Geophysics, 58*(4). https://doi.org/10.1029/2020rg000710

Wille, L., Messmer, M. M., Studer, B., & Hohmann, P. (2019). Insights to plant—microbe interactions provide opportunities to improve resistance breeding against root diseases in grain legumes. *Plant Cell and Environment, 42*(1), 20–40. https://doi.org/10.1111/pce.13214

Wu, B., Zhu, L., & Le, X. C. (2017). Metabolomics analysis of TiO_2 nanoparticles induced toxicological effects on rice (*Oryza sativa* L.). *Environmental Pollution, 230*, 302–310. https://doi.org/10.1016/j.envpol.2017.06.062

Yan, L., Li, P., Zhao, X., Ji, R., & Zhao, L. (2020). Physiological and metabolic responses of maize (*Zea mays*) plants to Fe_3O_4 nanoparticles. *Science of the Total Environment, 718*, 137400. https://doi.org/10.1016/j.scitotenv.2020.137400

Yin, J., Wang, Y., & Gilbertson, L. M. (2018). Opportunities to advance sustainable design of nano-enabled agriculture identified through a literature review. *Environmental Science: Nano, 5*(1), 11–26. https://doi.org/10.1039/c7en00766c

You, T., Liu, D., Chen, J., Yang, Z., Dou, R., Gao, X., & Wang, L. (2018). Effects of metal oxide nanoparticles on soil enzyme activities and bacterial communities in two different soil types. *Journal of Soils and Sediments, 18*(1), 211–221. https://doi.org/10.1007/s11368-017-1716-2

Yue, Y., Qi, L., Li, Y., Wang, J., Shen, C., & Ge, Y. (2020). Negative effects of rare earth oxide nanoparticles of La_2O_3, Nd_2O_3, and Gd_2O_3 on the ammonia-oxidizing microorganisms. *Journal of Soils and Sediments, 20*(8), 3114–3123. https://doi.org/10.1007/s11368-020-02627-x

Zhang, J., Cook, J., Nearing, J. T., Zhang, J., Raudonis, R., Glick, B. R., Langille, M. G. I., & Cheng, Z. (2021). Harnessing the plant microbiome to promote the growth of agricultural crops. *Microbiological Research, 245*, 126690. https://doi.org/10.1016/j.micres.2020.126690

Zhang, P., Guo, Z., Zhang, Z., Fu, H., White, J. C., & Lynch, I. (2020). Nanomaterial transformation in the soil–plant system: Implications for food safety and application in agriculture. *Small, 16*(21), 2000705. https://doi.org/10.1002/smll.202000705

Zhang, R., Vivanco, J. M., & Shen, Q. (2017). The unseen rhizosphere root–soil–microbe interactions for crop production. *Current Opinion in Microbiology, 37*, 8–14.

Zhao, L., Zhang, H., White, J. C., Chen, X., Li, H., Qu, X., & Ji, R. (2019). Metabolomics reveals that engineered nanomaterial exposure in soil alters both soil rhizosphere metabolite profiles and maize metabolic pathways. *Environmental Science: Nano, 6*, 1716–1727. https://doi.org/10.1039/C9EN00137A

CHAPTER

9

Chemical transformation and mechanisms of ENMs transport in plants

Guadalupe de la Rosa-Alvarez[1], *and Jose R. Peralta-Videa*[2]

[1] Department of Chemical, Electrical, and Biomedical Engineering, Universidad de Guanajuato, León, Guanajuato, Mexico; [2] Department of Chemsitry and Biochemistry, The University of Texas at El Paso, El Paso, TX, United States

9.1 Uptake and biotransformation of ENMs at the root level: interactions with biomolecules and cellular structures

Although there are hundreds of papers discussing the effects of ENMs in plants, their uptake and biotransformation processes within the plant body are not yet well understood. Of special significance is the interaction of nanomaterials with biomolecules and cellular structures at root level which, in some way, drives the compartmentalization of biotransformation products in the entire plant. So far, only a few studies have explained the uptake mechanism and ENMs biotransformation in roots. The most abundant details belong to CeO_2 and CuO ENPs, two of the most produced and used ENMs.

Cerium oxide nanoparticles—These NPs have special properties due to the high oxygen storage capability and the unique surface redox chemistry of the Ce(IV)/Ce(III) redox couple (Cornelis et al., 2011; Singh et al., 2020). An increase in Ce(IV) provides CeO_2 ENMs with catalase mimetic activity, and a decrease in Ce(III)/Ce(IV) ratio reduces the superoxide dismutase mimetic activity (Dahle, 2013). Both enzymes are important in the physiology of terrestrial plants. CeO_2 ENPs are among the most stable ENMs. Cornelis et al. (2011) performed a series of experiments to

Physicochemical Interactions of Engineered Nanoparticles and Plants
https://doi.org/10.1016/B978-0-323-90558-9.00009-7

determine the dissolution of this ENM in artificial soil solutions at pH values of 4, 7, and 9. At pH 7 and 9, the dissolution was negligible, while at pH 4, it was 3.1% after 28 days of equilibrium. These, and other results, were the basis to hypothesize that these ENPs were absorbed with no chemical modifications by plant roots.

The first evidence that roots accumulate CeO_2 in nanoparticulate form was provided by López-Moreno et al. (2010). These researchers exposed soybean (*Glycine max*) seeds in Petri dishes to 5 mL of CeO_2 ENPs at a concentration of 4000 mg/L. The seedlings were harvested when 65% of control roots were 5 mm long. Then, the roots were analyzed using X-ray absorption spectroscopy (XAS) at Stanford Synchrotron Radiation Laboratory. In the soybean samples, speciation of cerium in root tissues was determined by comparing XANES spectra from model compounds and root samples. In the mentioned study, cerium nitrate (CeN_3O_9) and CeO_2 ENPs were used as model compounds. By comparing the spectra of the root sample and the model compounds, it was observed that the root spectrum was very similar to the spectrum of the CeO_2 ENPs (Fig. 9.1). This suggested that the soybean root absorbed the nanoparticles. For more information about XAS analyses, the reader is invited to look at the

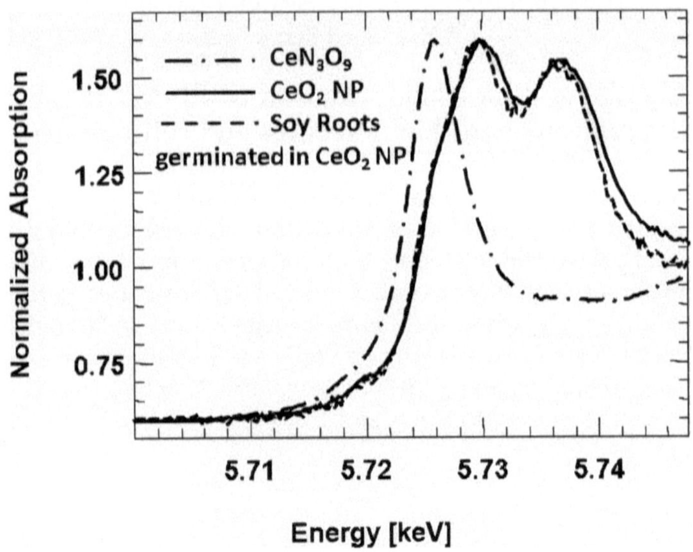

FIGURE 9.1 XANES L_{III} edge spectra (5723 eV) of CeO_2 NPs and CeN_3O_9 as model compounds, and spectra from soybean roots germinated in 4000 mg/L CeO_2 NPs. *Reprinted with permission from López-Moreno, M. L., De La Rosa, G., Hernández-Viezcas, J. A., Castillo-Michel, H., Botez, C. E., Peralta-Videa, J. R. & Gardea-Torresdey, J. L. (2010). Evidence of the differential biotransformation and genotoxicity of ZnO and CeO₂ nanoparticles on soybean* (Glycine max) *plants.* Environmental Science and Technology, *44(19), 7315–7320. Copyright (2010) American Chemical Society.*

summary about XANES theory published by LibreTexts (XANES – Theory. (2020, August 15). (Retrieved March 12, 2021, from https://chem.libretexts.org/@go/page/1869).

The experiments of López-Moreno et al. (2010) left no doubt that the soybean plants absorbed the CeO_2 ENPs; though, the uptake pathway was still elusive. In a subsequent study, corn (*Zea mays*) roots were exposed for 24 h to CeO_2 ENPs stained with fluorescein isothiocyanate. Micro-X-ray fluorescence (μ-XRF) images of a cross-section of the maize root (Fig. 9.2) showed CeO_2 ENPs aggregated in the cell walls of the cortex tissue (image B in Fig. 9.2) and a few of them in the xylem (image D in Fig. 9.2) (Zhao et al., 2012). This suggested the radial movement of the CeO_2 ENPs/aggregates from the surface to the center of the root through the apoplast (space through cell walls). However, the entrance of the CeO_2 aggregates into xylem was not proved with this assay. The answer came from an experiment with kidney bean (*Phaseolus vulgaris*) plants. Seven-

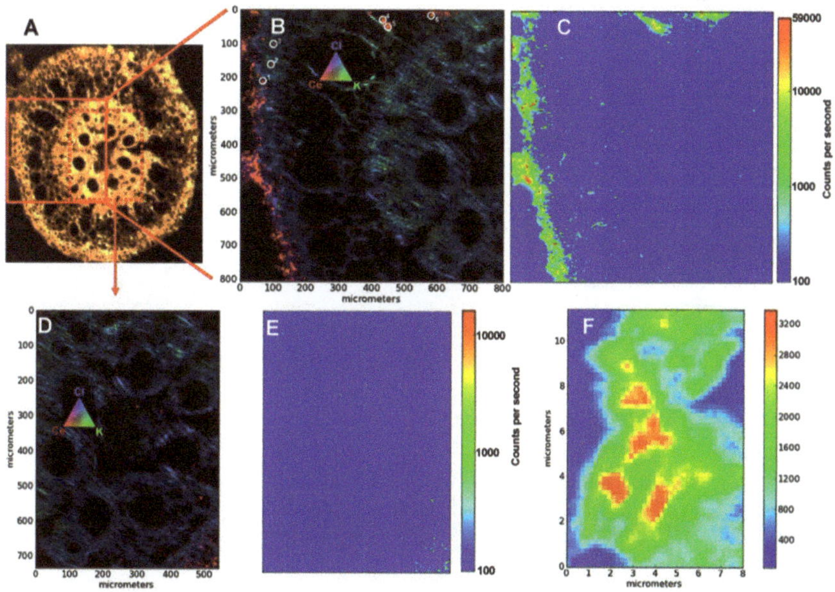

FIGURE 9.2 μXRF maps of a corn root cross-section exposed to CeO_2 NPs. (A) optical image, (B) μXRF tricolor map epidermis + vascular region, (C) temperature map epidermis + vascular region of the Ce-L_{III} fluorescence line, (D) μXRF tricolor map vascular region, (E) temperature map vascular region of the Ce-L_{III} fluorescence line, (F) temperature map of Ce particles with 0.2 μm pixel size (maximum resolution). Maps acquired at 5.8 keV with a 0.2 × 0.8 μm^2 focused beam and 3 μm^2 pixel size (except map F). *Reprinted from Zhao, L., Peralta-Videa, J. R., Varela-Ramirez, A., Castillo-Michel, H., Li, C., Zhang, J., Aguilera, R. J., Keller, A. A. & Gardea-Torresdey, J. L. (2012). Effect of surface coating and organic matter on the uptake of CeO_2 NPs by corn plants grown in soil: Insight into the uptake mechanism.* Journal of Hazardous Materials, *131–138. Copyright (2012) with permission from Elsevier.*

day-old seedlings were hydroponically cultivated for 15 days with 500 mg/L CeO_2 ENPs, and tissues were analyzed using synchrotron μ-XRF and micro-XANES techniques. Fig. 9.3 depicts the speciation of Ce in the bean root. The μ-XRF map (B) of the selected area (A) in Fig. 9.3 shows the emission point of a lateral root, where some CeO_2 ENPs (red dots) were observed. To form secondary roots, the cambium (a meristematic tissue responsible for increasing the diameter) undergoes periclinal cell divisions that disrupt the epidermis, cortex, and endodermis (Strock & Lynch, 2020). The cambium grows from the pericycle, which surrounds the xylem. Thus, at the time of formation of secondary roots, the Caspary band presents some gaps used by the CeO_2 ENPs to enter into the xylem without passing through the symplast (protoplast of the cells) (Majumdar

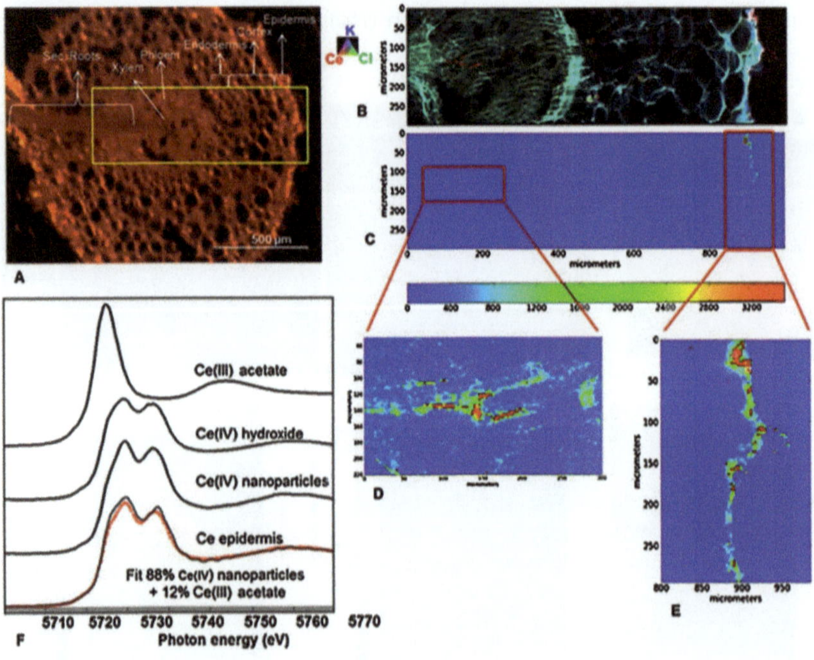

FIGURE 9.3 Images of a transverse section of kidney bean roots exposed to 500 mg/L $nCeO_2$. (A) Optical image of the root cross-section, (B) Tricolor μ-XRF map of the selected area from the cross-section (red = Ce, green = Cl, blue = K, (C) μ-XRF temperature map of the selected region, showing normalized Ce intensity; red represents higher Ce intensity, and dark blue represents the absence of a Ce signal. (D) Temperature map of Area 1 (vascular region), (E) Temperature map of Area 2 (epidermis). (F) Ce $L_{(III)}$ μ-XANES from root epidermis (*solid*, black) and the spectra from the linear combination fitting (*dotted*, red). *Reprinted from Majumdar, S., Peralta-Videa, J. R., Bandyopadhyay, S., Castillo-Michel, H., Hernandez-Viezcas, J. A., Sahi, S., & Gardea-Torresdey, J. L. Exposure of cerium oxide nanoparticles to kidney bean shows disturbance in the plant defense mechanisms.* Journal of Hazardous Materials, 278, 279–287. *Copyright (2014). With permission from Elsevier.*

et al., 2014). It has also been proposed that ENMs can enter the root through root-apex meristem, where the vascular cylinder is surrounded by a single-layered pericycle. In this root section the cells are not fully elongated and the Casparian strip is not well established (Barberon, 2017). Thus, the emission points of secondary roots and root tips are considered the entry point of ENMS into the xylem tissue (Barberon & Geldner, 2014).

Data on the biotransformation of CeO_2 ENMs at root level is not abundant. Zhang, Ma, Zhang, He, Guo, et al. (2012) and Zhang, Ma, Zhang, He, Zhang, et al. (2012) exposed 10-day-old cucumber (*Cucumis sativus*) seedlings to 2000 mg CeO_2 ENPs/L in ¼ strength Hoagland solution. Using XANES studies, these researchers demonstrated that, after 3 weeks of treatment, a portion of the CeO_2 ENPs accumulated as $CePO_4$ in the epidermis and intercellular spaces of cucumber roots. The ENPs attached to the root surface through the action of root exudates and transformed to Ce(III) (Ma, Zhang, Zhang, He, Li, et al., 2015; Ma, Zhang, Zhang, He, Zhang, et al., 2015). However, the transformation varied with the type of particle. More and faster transformations were observed when rod-shaped ENPs were involved in the experiments, as compared to spherical ENPs (Zhang, Ma, et al., 2017; Zhang, Xie, et al., 2017). Similar results were observed in soybean plants cultivated to full maturity in farm soil amended with 1000 mg CeO_2 ENPs/kg. In this study, μ-XANES analysis performed in the epidermis of root nodules showed that most of the CeO_2 ENPs (79%) remained as such, while a portion was biotransformed to Ce(III) (Hernandez-Viezcas et al., 2013). More details were obtained from radish (*Raphanus sativus*) seedlings cultivated for 2 weeks in ¼ strength Hoagland solution and exposed for 5 days to 10 mg CeO_2 ENPs/L (Zhang, Dan, et al., 2017). These researchers analyzed, separately fine roots and storage roots. Low molecular weight organic acids in root exudates, like succinic acid, enhanced the dissolution of CeO_2 ENPs; thus, both the Ce ions and the CeO_2 ENPs were taken up by the root. While Ce was found as free ions or in the form of $CePO_4$, the CeO_2 ENPs remained as such within the root tissues (Zhang, Ma, et al., 2017; Zhang, Xie, et al., 2017). In an interesting experiment, Spielman-Sun et al. (2017) demonstrated that surface chemistry plays important role in the allocation and biotransformation of CeO_2 NPs in roots. Four-day-old wheat seedlings were cultivated for 7 days in 25% strength Hoagland solution amended with CeO_2 NPs at 20 mg Ce/L. Seedlings were exposed to dextran-coated CeO_2 NPs either uncharged (0), functionalized with diethylaminoethyl groups (+), or with carboxymethyl groups (−). The three NPs penetrated the root, but at different rates. The CeO_2 (+) NPs accumulated the more on root surface retained by negative charges of the cell plasma membranes. Consequently, the highest NPs accumulation occurred in roots. XANES spectra showed that in roots exposed to CeO_2 (0), about 15% of the Ce was present as Ce(III) and the rest remained in the same chemical form as present in the CeO_2 NP treatment. No Ce(III) was observed in roots exposed to CeO_2 (+) NPs. While in roots exposed to CeO_2 (−), most of the Ce remained in nanoparticulate form within the vein, but a small portion was found outside veins in reduced form, or Ce(III). The authors stated that the lack of Ce(III) signal in CeO_2 (+) treated plants could be due to the huge amount of CeO_2 NPs that concealed the Ce(III) signal.

Copper-based ENMs—the biotransformation and uptake of copper-based ENPs have been studied in a few plant species. In a hydroponics experiment, maize seedlings (Yudan No. 8 variety) were exposed to 100 mg/L of CuO ENPs for 15 days. TEM images of the root tip showed CuO ENPs in intercellular spaces and cytoplasm of cortical cells. The presence of endosomes (membrane-bound vesicles) in cortical cells suggested that the CuO ENPs were absorbed via endocytosis. Although Cu ions were observed inside the roots, there was no clear explanation if they were released from the absorbed ENPs or if entered from the root surface (Wang et al., 2012). μ-XRF and μ-XANES studies of hydroponically cultivated rice (cv. Xiuyou No. 5) exposed for 2 weeks to 100 mg CuO ENPs/L showed CuO ENPs and Cu ions mainly in the root epidermis and exodermis (Peng et al., 2015). μ-XANES analysis of the cortex tissue showed CuO ENPs (28.11%), Cu-cysteine (22.5%), Cu-citrate (25%), and $Cu_3(PO_4)_2$ (24.4%). Speciation of Cu in the endodermis showed CuO ENPs (27.6%), Cu-cysteine (54%), and Cu_2O (18.4%). However, in the central cylinder (transport system) there were no ENPs but Cu-cysteine (26.4%), Cu-citrate (31.3%), and Cu_2O (42.3%) (Peng et al., 2015). On the other hand, experiments performed in soil have shown some differences. Rice seeds from the Wuwangnong group were sown in a soil that had a pH of 5.3, organic matter of 4.15%, and 6.71 cmol/kg cation exchange capacity. The soil was amended with CuO ENPs (spherical, 43 ± 9 nm diameter) at 50–1000 mg/kg. Plants were harvested at several stages including maturation stage (88 days). After 21 days under flooded conditions, 34.7% of Cu was in the form of Cu_2S in the soil solution. In subsequent days, the soil was under flooding-drying regime and at the maturation stage, there were no CuO ENPs in the soil, but Cu_2S and Cu were bound to goethite (Peng et al., 2017). However, a small portion of CuO and Cu_2O was found in the roots. At that time, there was no justification for the presence of the CeO_2 ENPs in the root since they were not present in the soil solution. The explanation came from an experiment performed in Standard Lufa 2.2 soil (loamy sand, 1.6 wt% organic matter), where ~7 day-old wheat (*Triticum aestivum*) seedlings were exposed for 2 weeks to freshly prepared CuO ENP suspensions or ENPs aged for 28 days in the Lufa soil (500 mg Cu/kg). By using enhanced dark-field hyperspectral imaging (DF-HSI), Gao et al. (2018) found CuO ENPs in the root tip mucilage of plants exposed to either fresh suspension or soil aged treatments. The result suggested that CuO ENPs were absorbed through root tips at the early growth stage of the plants.

Besides CeO_2 and CuO, the uptake and biotransformation of other ENMs have been reported–Cowpea (*Vigna unguiculata* L.) was grown in soil spiked with 125–1000 mg/kg Cu NPs (Ogunkunle et al., 2019). XANES spectra of root samples and model compounds showed the presence of Cu NPs, Cu_2O, and Cu(II) as Cu-nicotianamine and Cu-glutathione. However, it is likely that other Cu complexes were formed in the roots. To prove this hypothesis, more model compounds should be used.

A few studies have shown the uptake and biotransformation of TiO_2 ENMs by terrestrial plants. Larue et al. (2011) exposed for 7 days wheat seedlings to 12 ± 3 nm spherical TiO_2 NPs composed of 95% anatase and 5% rutile suspended in ultrapure water. XAS spectra of root samples showed TiO_2 in parenchyma and vascular cylinder. The size and crystalline form of the NPs did not change within the root tissue, suggesting no biotransformation. Servin et al. (2012) exposed for 15 days 1-week-old cucumber seedlings to 500 mg/L of TiO_2 ENPs. The ENPs (Evonik) were semi-spherical (27 ± 4 nm) and composed of 82% anatase and 18% rutile crystalline phase. They were suspended in Hoagland solution and adjusted to pH 5.8 because at that pH, the solubility of TiO_2 ENPs is very low. XANES data showed TiOs ENPs in the root dermis, cortex, and xylem at the same composition as the supplied particles, indicating no transformation. A recent study performed in soil amended with TiO_2 ENPs (anatase or rutile), where pea (*Pisum sativum*) plants were cultivated for 28 days after germination, corroborated the lack of biotransformation of TiO_2 ENPs in the roots of terrestrial plants (Muccifora et al., 2021). This study showed that anatase particles predominated over rutile and accumulated, mainly, in the root cortex near the epidermis.

Data on the biotransformation of Ag NPs has been obtained from hydroponic culture of common grass (*Lolium multiflorum*) and lettuce (*Lactuca sativa*). Common grass seedlings were exposed for 7 days to 1–40 mg/L of gum Arabic coated AgNPs (6 and 25 nm) (Yin et al., 2011). On the other hand, lettuce seedlings were exposed for 9 days in Hoagland solution to sodium citrate (2 mM) stabilized silver nanoparticles (citrate-AgNPs) of 60 and 100 nm; polyvinylpyrrolidone-coated Ag NPs (PVP-AgNPs) of 75 and 100 nm also stabilized with sodium citrate (2 mM), or polyethylene glycol-coated AgNPs (PEG-AgNPs) of 100 nm dissolved in water (Torrent et al., 2020). In common grass roots, μXANES spectra showed Ag complexed with sulfur and oxygen/nitrogen ligands. Similarly, in lettuce roots, Ag was reported as complexed with sulfur or oxygen. Although the results were similar, more plant species and more model compounds are needed to have a more general picture of biotransformation of AgNPs at root level.

In addition to CeO_2, the biotransformation of other rare earth nanomaterials has been reported. In-house fabricated Yb_2O_3 NPs were exposed to cucumber seeds and after 5 days, the seedlings were analyzed by synchrotron-based scanning transmission X-ray microscopy (STXM) coupled to near edge X-ray absorption fine structure (NEXAFS), also called XANES. Spectra showed $YbPO_4$ in the intercellular spaces of roots (Zhang, Ma, Zhang, He, Guo, et al., 2012; Zhang, Ma, Zhang, He, Zhang, et al., 2012). Similar results were reported by Ma et al. (2011) who exposed La_2O_3 NPs to cucumber seedlings. The intercellular spaces of the roots showed the presence of $LaPO_4$. Neither report presented evidences on the absorption mechanisms of Yb_2O_3 or La_2O_3. It is possible that both ENPs were absorbed in a similar pathway as that of CeO_2 ENPs. Thus, it is likely that they were adsorbed to the root surface and dissociated by roots exudates, absorbed either through the apoplast or symplast (or through both), and complexed within the roots.

In summary, current literature suggests that ENPs are absorbed through the root tips and the emission points of secondary roots. Some stable nanomaterials, such as TiO_2 and CeO_2, are accumulated mainly in nanoparticulate form. No TiO_2 ENPs biotransformation has been observed in roots. Other ENPs are partially reduced by low molecular weight organic acids found in root exudates, and the respective ions are taken up by the roots and complexed by polyatomic ions. However, in the case of CeO_2 NPs, it is not clear if positively charged particles are reduced outside or within the roots.

9.2 Biotransformation of ENMs at stem level: interaction with biomolecules and cellular structures

Similar to the biotransformation of ENMs at root level, the biotransformation at stem level has been studied only in a few NMs. Most of these studies have been performed at synchrotron facilities, which provide limited access. Additionally, the number of model compounds, which help during the identification of biotransformation products, has been also limited.

Most studies on ENMs' biotransformation at stem level have been performed in plants exposed to CeO_2, followed by CuO, ZnO, and $Ni(OH)_2$. The literature shows that the biotransformation in stems is driven by various factors including NP crystallinity, surface chemistry, presence of specific molecules, and plant species.

The effects of plant species on NPs biotransformation have been observed in mesquite, soybean, rice, and Shiny Elsholtzia. For instance, two portions of the same ZnO ENPs were applied to mesquite and soybean. While mesquite stems showed mostly the presence of $Zn(NO_3)_2$, in soybeans stems only 10% of the Zn was biotransformed to $Zn(NO_3)_2$, and the remaining 90% was found as Zn citrate ($C_6H_5O_7^{-3}$) (Hernandez-Viezcas et al., 2016). Another example of the effect of plant species on the biotransformation in stems was observed in rice and Shiny Elsholtzia. Both plant species were treated with spherical-ellipsoidal CuO NPs of 43 ± 9 nm and stem samples were analyzed using XANES. In rice, 6.2% of Cu remained as CuO NPs, 19% was transformed to Cu-acetate, 20.3% to Cu-cysteine, and 54.6% to Cu-citrate (Peng et al., 2017). On the other hand, results indicated that in Shiny Elsholtzia stems, 35% of Cu remained as CuO NPs, 16.9% was transformed to Cu-alginate, and both Cu-cysteine and Cu-oxalate were found at 23.9% (Shi et al., 2014).

The biotransformation in stems of CeO_2 ENPs seems to be affected by the experimental conditions including the NP shape, plant species, and growth medium. For example, Zhang, Ma, et al. (2017) and Zhang, Xie et al. (2017) treated hydroponically grown cucumber with octahedral, cubic, and rod-shaped CeO_2 NPs. Results of these experiments indicated that, in the case of treatments with octahedral and cubic particles, 80% of Ce remained unchanged, whereas the residual 20% was found as $CePO_4$ and cerium carboxylates. On the other hand, when rod NPs were applied, 62% of the Ce remained as CeO_2 NPs and 38% as $CePO_4$ and cerium carboxylates. Another factor that seems to affect the biotransformation of CeO_2 NPs is the presence of polyatomic ions, such as PO_4^{3-}. Rui et al. (2015) exposed cucumber plants to octahedral CeO_2 NPs in hydroponic solution amended/unamended with PO_4^{3-}. They found that stems of plants exposed to NPs in the presence of PO_4^{3-} had 81.3% of the Ce in the form of CeO_2 NPs and 21% as $Ce(CH_3COO)_3$. On the other hand, in the absence of PO_4^{3-}, only 32.5% Ce was present in the form of CeO_2 NPs and 67.5% as $Ce(CH_3COO)_3$.

In summary, the biotransformation of ENMs at stem level is affected by several factors and only a few experiments have been performed to clarify the role of each factor in the biotransformation process.

9.3 Biotransformation of ENMs at leaf level: interaction with biomolecules and cellular structures

Data related to the biotransformation of ENMs in leaves is very reduced. Current literature mostly refers to three common ENMs CeO_2, CuO, and TiO_2 and five plant species: cucumber, lettuce, wheat, rice, and Shiny Elsholtzia (Peng et al., 2015, 2017; Servin et al., 2012; Shi et al., 2014; Spielman-Sun et al., 2017; Zhang, Ma, et al., 2017; Zhang, Xie, et al., 2017). One of the first reports was published by Servin et al. (2012) who used micro X-ray fluorescence (μ-XRF) and micro X-ray absorption spectroscopy (μ-XAS, μ-XANES) to track the presence and chemical speciation of Ti in cucumber leaves. The μ-XANES spectra showed that Ti in leaves was in the rutile crystalline form. It is hypothesized that the anatase crystalline phase is biotransformed to the rutile crystalline phase (Fig. 9.4). However, more studies are needed to test this hypothesis.

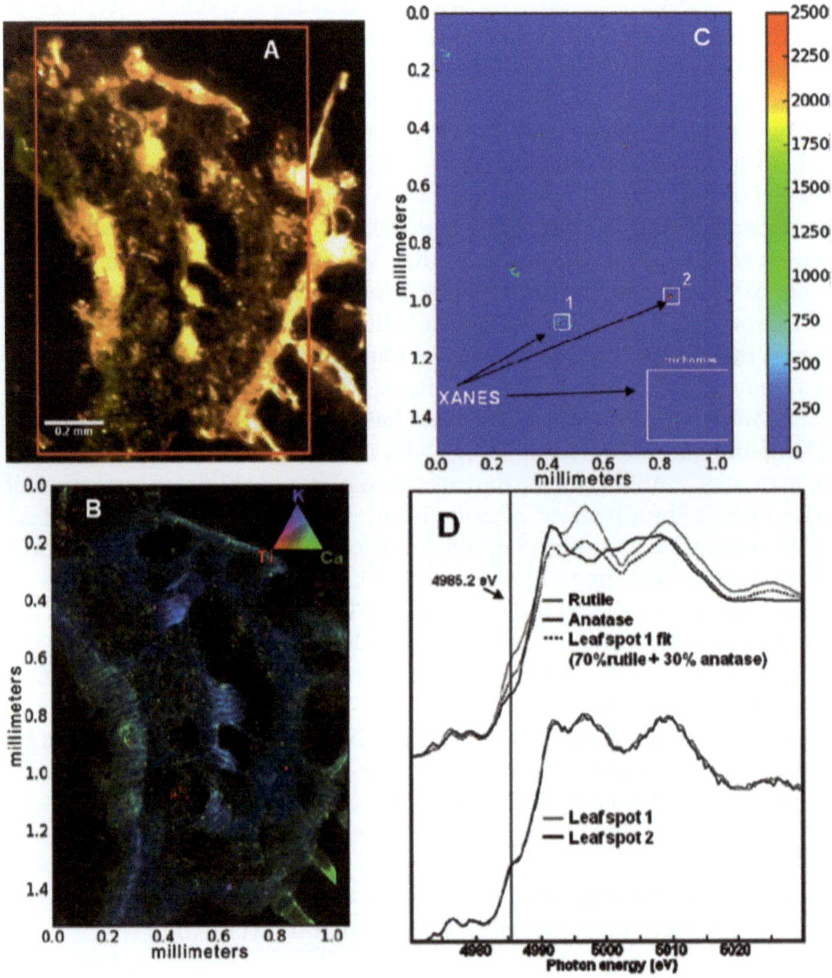

FIGURE 9.4 Images of transverse sections of cucumber leaf treated for 15 days with 500 mg/L TiO_2 NPs. (A) Video microscope image of the cucumber leaf transverse section. (B) Tricolor micro-XRF map. Red color stands for titanium, green for calcium, and blue for potassium. The XRF map was acquired at 5.1 KeV, 200 ms dwell time, 3 μm^2 pixel size, and $0.33 \times 0.65\ \mu m^2$ beam size. (C) Ti temperature map, color scale units are raw intensity. The scale was adjusted to 2500 in order to enhance the medium range intensity regions; the max in some of the red spots was ~8000. White box marked areas indicate where micro-XANES was acquired with the use of a 50 μm pinhole. (D) Micro-XANES spectra of reference materials and spot 1 and 2 from Fig. 9.3C. *Reprinted with permission from Servin, A. D., Castillo-Michel, H., Hernandez-Viezcas, J. A., Diaz, B. C., Peralta-Videa, J. R., & Gardea-Torresdey, J. L. Synchrotron micro-XRF and micro-XANES confirmation of the uptake and translocation of* TiO_2 *nanoparticles in cucumber* (Cucumis sativus) *plants.* Environmental Science and Technology, *46(14), 7637–7643. Copyright (2012) American Chemical Society.*

The biotransformation in leaves of CuO NPs was studied in rice and Shiny Elsholtzia exposed for 14 days to 100–1000 mg/L CuO NPs under controlled conditions. XANES spectra showed that leaves of Shiny Elsholtzia contained Cu as CuO NPs (43.5%), Cu-cysteine (26.1%), Cu-oxalate (27.2%), and Cu-alginate (3.3%) (Shi et al., 2014). On the other hand, rice plants depicted some changes associated with leaf age. In young leaves, about ¼ of the Cu (24.2%) was present as CuO NPs, 27.7% as Cu-cysteine, 21.7 as Cu-citrate, 18.4% as Cu_2O, and 0.8% as $Cu_2(PO_4)_2$. In mature leaves, more than half (53.5%) of the Cu was CuO NPs, 23.2% Cu-citrate, 11.8% Cu-cysteine, and 11.4% Cu_2O (Peng et al., 2015). In addition, Peng et al. (2017) exposed rice plants to the same CuO NPs, but this time the support was soil, and the plants were cultivated for the full life cycle. After 88 days of growth, they found out that leaves contained 18.4% of Cu as CuO NPs, 34.6% as citrate, 25.4% as Cu-cysteine, and 21.5% as Cu-acetate. Concerning the higher accumulation of CuO NPs in mature leaves, compared to young leaves in the hydroponically grown rice, the authors suggested that the high concentration of Cu damaged the antioxidant defense and the detoxification capacity; thus, there was a limited dissolution of the CuO NPs in older leaves. Since the youngest leaves had less CuO NPs, this could result in less Cu(I) released as effect of NPs reduction, in consequence, diminishing the amount of Cu_2O (Peng et al., 2015). As per the soil cultivated plants, the exposure time allowed a higher dissolution of the NPs within tissues. In this case, Cu(II) was transformed to Cu(I), which was complexed with cysteine and acetate.

A study with CeO_2 NPs showed that particle surface chemistry affected the biotransformation of these NPs in leaves of wheat plants. Spielman-Sun et al. (2017) exposed for 34 h wheat seedlings in a hydroponic solution amended with uncharged, positively, or negatively charged CeO_2 NPs. XANES spectra of the leaves showed that the biotransformation in leaves was driven by the surface charge. Plants exposed to uncharged NPs had Ce–Ca association in leaf tissue (outside veins); those exposed to negatively charged CeO_2 (−), remained in the xylem as CeO_2 NPs and ~20% was reduced to Ce(III), while positively charged CeO_2 (+) were not detected in the leaves. Studies have shown that negatively charged plasma membrane in root cells attracts positive particles and repels negatively charged particles (Wang et al., 2014). The presence of Ce–Ca agglomerate found in leaves of plants exposed to neutral particles was related to the defense mechanisms against excess of reactive oxygen species created by the CeO_2 NPs (Spielman-Sun et al., 2017).

In summary, our knowledge about the biotransformation of ENMs in living plants is at its infancy. Both the number of ENMs and plant species studied are limited. In addition, few data are available on ENMs whose main application would be in agricultural activities. The number of plant species studied is low, and the experimental conditions (mainly from hydroponics) do not allow to predict responses on natural environments. Additionally, very few investigations have been planned to answer specific questions related to experimental variables such as the growth medium, plant species, and the type of particle.

9.4 Potential implications of ENMs biotransformation within plant body

As previously stated, some ENPs remain untransformed within the plant systems, at least partially, while others suffer different chemical modifications. Both situations may impact plant health and in consequence, performance and survival. In the case of untransformed ENPs, it is relatively easy to identify potential interactions with plant biomolecules and structures. However, when transformed, these mechanisms become complex. Biotransformation may occur when ENPs are converted to other type of compounds, chemical species, or when crystalline structure changes.

In Section 9.3 we mentioned the experiments performed by Servin et al. (2012), where they exposed cucumber plants to TiO_2 NPs formed by 82 anatase and 18% rutile. At the end of their experiments, these researchers found out that cucumber roots still contained TiO_2 NPs. However, in stems and leaves, the anatase/rutile ratios were 25/75, and 15/85, respectively. As shown, the percentage of anatase decreased as NPs were transported from root to stems and leaves (Fig. 9.4). The mechanism for this phenomenon is unclear, and at least two possible pathways were proposed by this research group: (a) rutile is preferentially translocated to stems and leaves; and (b) anatase phase is dissolved, translocated to stems and leaves, and precipitated in the rutile phase. The implications of these transformations are also unclear, as it is necessary to determine which biomolecules are involved and if they are affected or not. These authors also indicated that in their experiments, TiO_2 NPs reached the fruit organ, which is the part consumed by humans. Thus, cucumber plants containing this type of NPs may become a source of entry into the food chain.

Different effects have been observed on ENMs biotransformation. For example, Li et al. (2015) proved the internalization of nZVI in peanut seeds and seedlings and suggested that these NPs promoted water absorption by plant seeds. This may have occurred because nZVI generated ROS, which oxidized the seed coat allowing the entrance of water. In this case, the implication of nZVI transformation was positive, since seed germination was facilitated. It was also reported that nZVI, at relatively low concentration (up to 80 μmol/L), increased root growth. However, at rather high concentrations (320 μmol/L), the plant growth was negatively affected. Fe is a plant micronutrient necessary for the proper function of plant system. However, it may also be toxic to plants because of its high reactivity. Fe in nZVI may be oxidized to Fe^{2+} and Fe^{3+} (Chen et al., 2017), and subsequently, these ions are released into media (Eq. 9.1). However, Fe^{2+} is the one that is carried inside cells by different transporters (Morrissey & Guerinot, 2009). Moreover, through Fenton reactions, Fe produce ROS that may cause cell membrane disruption and other damages, which was the case in seed coat rupture of peanut seeds exposed to nZVI by Li et al. (2015).

$$nZVI(Fe^{o}) \rightarrow Fe^{2+} + Fe^{3+} + e^{-} \tag{9.1}$$

The research group led by Ghosh et al. (2017) performed an exhaustive study on the uptake and toxicity of nZVI in *Allium cepa* roots. These researchers identified NPs adsorbed onto root surface and observed significant damage to root cell structures including separation of epidermal cell layers, root tip bending, rupture and perforation, increase in membrane permeability, decrease in cell viability, and increased cell death. Moreover, they determined that nZVI promoted oxidative stress, apoptosis, and DNA damage. These researchers proposed that both, Fe ions and nZVI particles, participated in the mechanisms leading to cell damage.

In other studies, Ma, Zhang, Zhang, He, Li, et al. (2015) and Ma, Zhang, Zhang, He, Zhang, et al. (2015) found out that CeO_2 and La_2O_3 NPs showed differential toxicity in cucumber plants. While CeO_2 NPs did not affect plant growth parameters, La_2O_3 NPs significantly diminished plant elongation and biomass production. These NPs also promoted the generation of ROS and death of root cells. These authors stated that differential toxicity was probably the effect of dissimilarities in NPs solubility. According to their X-ray absorption spectroscopy studies, these researchers concluded that a portion of CeO_2 NPs was biotransformed to produce Ce(III)–carboxyl complexes, while a higher amount of La_2O_3 was solubilized, compared to CeO_2. Subsequently, lanthanum (La) formed La-phosphate (La^{3+}-PO_4^{3-}) and La-carboxylic (La^{3+}-$(COOH)^{-}{}_{3}$) complexes. Whether these metallic species are later released or not into the media will depend on stability constants of the formed species. Reported solubility for La_2O_3 NPs in water is 0.07 mg/L at 20°C (https://www.sigmaaldrich.com/MX/en/sds/aldrich/634271). On the other hand, CeO_2 NPs are reported as insoluble in water (https://www.sigmaaldrich.com/MX/en/sds/aldrich/544841). However, different researchers have reported differences in solubility as pH varies. Schwabe et al. (2014) indicated that at pH values below 4.6, these NPs were able to release Ce ions to the media. This information is relevant since different plant organs may present differences in pH. For example, Mitreva (1989) reported differences in pH in roots, stems, and leaves of sunflower and maize plants. Thus, it is likely that CeO_2 NPs solubility changes as they move throughout the plant.

Schwabe et al. (2014) investigated the chemical transformation of CeO_2 NPs in different media. Even when they performed the experiments using the Hoagland growth medium, these findings may help to understand what happens inside the plant structure. These researchers proposed that chemical transformation of these NPs occurred in the following order: (a) dissolution; (b) surface reduction; (c) complexation; and (d) precipitation. It is likely that same mechanisms applies to other types of ENMs.

(a) Dissolution.

As previously stated, Ce(III) is more soluble than Ce(IV). Thus, this chemical species is more likely to be absorbed by roots. Zhang, Ma, et al. (2017) and Zhang, Xie, et al. (2017) identified Ce(III) in stems and leaves of cucumber plants hydroponically exposed to CeO_2 NPs. However, it was unclear if Ce(III) was transported in this chemical form, or if it was the product of Ce(IV) reduction in different plant parts. In any case, in the dissolution process, Ce(III) ions leave the crystal lattice as they provide relative instability to the CeO_2 NP (Schwabe et al., 2014). In addition, these researchers found that, the smaller the NP, the highest the dissolution rate. Moreover, the presence of gum Arabic (GA), Fe(III)-EDTA, and EDTA influenced CeO_2 NP dissolution. These processes are represented in Eqs. (9.2)–(9.4)

$$Ce(III)\ in\ CeO_2\ NP + GA \rightarrow Ce(III) - GA \tag{9.2}$$

$$Ce(III)\ in\ CeO_2\ NP + Fe(III) - EDTA \rightarrow Ce(III) - EDTA \tag{9.3}$$

$$Ce(III)\ in\ CeO_2\ NP + EDTA \rightarrow Ce(III) - EDTA \tag{9.4}$$

(b) Surface reduction.

The catalytic activity of CeO_2 NP allows their interaction with different physiological molecules including H_2O_2. One of the first explanations for this interaction was provided by Saito and Bielski (1961), who proposed that Ce(IV) reacts with hydrogen peroxide as shown in Eq. (9.5).

$$Ce(IV) + H_2O_2 \rightarrow Ce(III) + H^+ + O_2 \tag{9.5}$$

Thus, CeO_2 NPs decompose ROS, including super oxide radicals (Singh et al., 2020), which represent a threat to living systems.

(c) Complexation.

A complex is formed when two or more compounds interact and are maintained together through physicochemical and ionic forces (Vaccari et al., 2006). Formation of the Ce-EDTA complex is shown in Eq. (9.4). Complex formation changes the toxicity of the ions influencing the nanoparticle dissolution. In this context, Schwabe et al. (2014) proved that EDTA increased CeO_2 NPs dissolution when present in Hoagland solution. In addition, complexes are relatively stable and can be maintained in solution.

(d) *Precipitation.*

When dissolved Ce(III) reacts with certain ions, the resulting product may be an insoluble species. This is the case for Ce(III)PO_4 (Eq. 9.6). When this precipitate is formed, it may remain attached to different plant parts. In addition, Ce(III)PO_4 precipitate may also form new NPs, which may differ in toxicity, compared to CeO_2 NPs. Liu et al. (2021) exposed lettuce plants to these two types of nanomaterials, and determined that CeO_2 NPs affected root elongation and caused cell death, while Ce(III)PO_4 NPs showed no toxicity to this plant. Using XANES and LCF (linear combination fitting), these researchers also found out that CeO_2 NPs partially released Ce(III), while Ce(III)PO_4 NPs did not. This occurred because of differences in solubility. They concluded that the toxicity to plants was mainly due to the presence of Ce(III) ions.

$$Ce(III)\ in\ CeO_2\ NP + KH_2PO_4 \rightarrow Ce(III)PO_4 \quad (9.6)$$

Franziska Schwabe et al. (2015) reported that the interaction of Zr/CeOx NPs with sunflower plants resulted in NP dissolution in the rhizosphere and probably reprecipitation at leaf level. However, more studies are needed to confirm this hypothesis.

Similar to CeO_2 NPs, MnO_x NPs possess enzyme like activities. For this reason, they can catalyze a series of biological reactions. Jiang et al. (2020) studied the catalytic behavior of three different MnO_x NPs: (a) MnO_2; (b) Mn_2O_3; and (c) Mn_3O_4. These researchers reported that the three NPs presented oxidase-, catalase-, and SOD-like activities, while they lacked peroxidase-like behavior as well as ability to scavenge hydroxyl radicals. Relatively low concentrations of these NPs proved beneficial for human cells, since they reduced the presence of $H_2O_2^-$, in consequence reducing cell toxicity. However, relatively high concentrations reduced cell viability. Even when these results come from experiments with human cells, they help to understand potential interactions with plant cells. More research is needed to determine potential differences in specific mechanisms of interaction with different living systems.

9.4.1 Implications of ENMs biotransformation at the genetic level

9.4.1.1 *Genotoxicity of ENMs in plants*

Genotoxicity refers to the process where a chemical causes direct or indirect DNA damage. Given that these mutations may result in abnormalities, it is important to assess to what extent ENMs may or may not be genotoxic and use this information for decision-making purposes. One of the first reports on NPs genotoxicity was published by López-Moreno et al. (2010). These researchers identified genotoxic effects in soybean seedlings obtained from seeds exposed to ZnO and CeO_2 NPs. However, a much higher ZnO NPs concentration was needed to cause DNA damage, compared to CeO NPs. The researchers concluded that Zn^{2+} might have caused DNA damage and mutations because of its ability to produce ROS, as it has been the case for other transition metals. The same may apply to the mechanism of DNA damage by CeO_2 NPs. Further studies have provided additional insights into potential mechanisms of ENMs genotoxicity in plants.

Ghosh et al. (2017) proved that two forms of nZVI, which differed in surface chemistry, presented differential genotoxic effects on *Allium cepa* roots. Specifically, nZVI-NH_2 contained amino groups at the NP surface, while carboxylic groups were present in the surface of nZVI-COOH. According to their results, nZVI-NH_2 provoked more DNA damage, micronuclei formation, as well as aberration of chromosomes and nuclei, compared to nZVI-COOH. Low zeta potential and conductivity, as well as high polydispersity index in nZVI-NH_2, caused the increased injury. Zeta potential is related to NPs stability and provides information on the effective electric charge on the surface of the particle. Thus, low zeta potential indicates low nanoparticle stability, which explains the higher reactivity of nZVI-NH_2, compared to nZVI-COOH. Later, Ghosh et al. (2019) reported that MnO NPs induced genetic damage in *Physcomitrella patens*.

According to Casillas-Figueroa et al. (2020), Ag NPs genotoxicity depends on whether or not coating is present, its nature, method of synthesis, and NP size. It seems that Ag NPs obtained through biosynthesis are more toxic than those obtained using other methodologies. In the case of coating, another important factor in genotoxicity is the ratio Ag:coating agent. In this context, Casillas-Figueroa et al. (2020) exposed *Allium cepa* plants to Ag-PVP NPs with Ag:PVP ratios of 6:94, and compared the results with those of others researchers that used ratios of 34:66 (Scherer et al., 2019), and 40:60 (Vinković Vrček et al., 2015). According to their data, a ratio of 6:94 did not cause genotoxic effects; however, the other ratios did it. Thus, a relatively high amount of PVP molecules present in the 6:94 ratio significantly stabilized the Ag NPs, avoiding their solubilization and further ROS generation.

Mehrian and De Lima (2016) published an exhaustive review on NPs toxicity in plants. They analyzed data reported for different plant species and different ENMs and proposed a model that details the entrance of NPs into plant systems and the pathways for genotoxicity. In summary, ENMs enter the cell via: (a) diffusion through cell membrane; (b) endocytosis; and (c) transport assisted by membrane proteins. Once inside the cell, NPs interact with different biomolecules and organelles. For example, they are able to trespass mitochondrial and nuclear membranes. In mitochondria, NPs cause the overproduction of H_2O_2, which eventually react with other ENMs to generate ROS. These species are directly responsible for damage of DNA and DNA-related proteins. In nuclei, ENMs interact with chromosomes at different levels as well as with histone proteins, centrioles, and centromeres. These pathways are shown in Fig. 9.5.

Detailed mechanisms of DNA damage by free radicals in humans were reviewed by Dizdaroglu and Jaruga (2012). However, similar principles may apply to plants as the chemical structure of DNA is fundamentally the same in terms of basic components. DNA contains the information for the synthesis of biomolecules and is formed by (a) purine and pyrimidine bases; (b) pentose; and (c) phosphate groups.

The formation of free radicals after the reaction of ENMs with H_2O_2 has been proven (He et al., 2012, Eq. 9.7). In addition, it is well known that transition metal ions produced after ENMs dissolution are also able to generate ROS (Halliwell et al., 2000, Eq. 9.8).

$$Ag\ \ NPs + H_2O_2 \rightarrow Ag^+ + OH^\bullet + O_2 \tag{9.7}$$

$$Fe^{2+} + H_2O_2 \rightarrow Fe^{3+} + OH^\bullet + OH^- \tag{9.8}$$

According to Dizdaroglu and Jaruga (2012), these ROS oxidize purine and pyrimidine bases, as well as the pentose moiety. Eventually, oxidation may result in either bound rupture within the molecule or DNA strand rupture. These alterations cause changes in the genetic information carried by DNA, and implications may vary from reduced production of biomolecules to cell death.

Similar to ROS, reactive nitrogen species (RNS) are the result of cell stress, and also participate in different physiological processes including cell defense against pathogens, germination, and reproduction, among others (Corpas, 2016). At proper concentrations, these radicals are beneficial. However, at high concentrations, RNS also cause cell damage. Some of the RNS that have been detected in living cells include peroxynitrite ($O=NOO-$), nitrogen dioxide (NO_2), nitrogen monoxide (NO), and the free radical •NO (Corpas, 2016; Ronzio, 2020, pp. 731–751). However, to

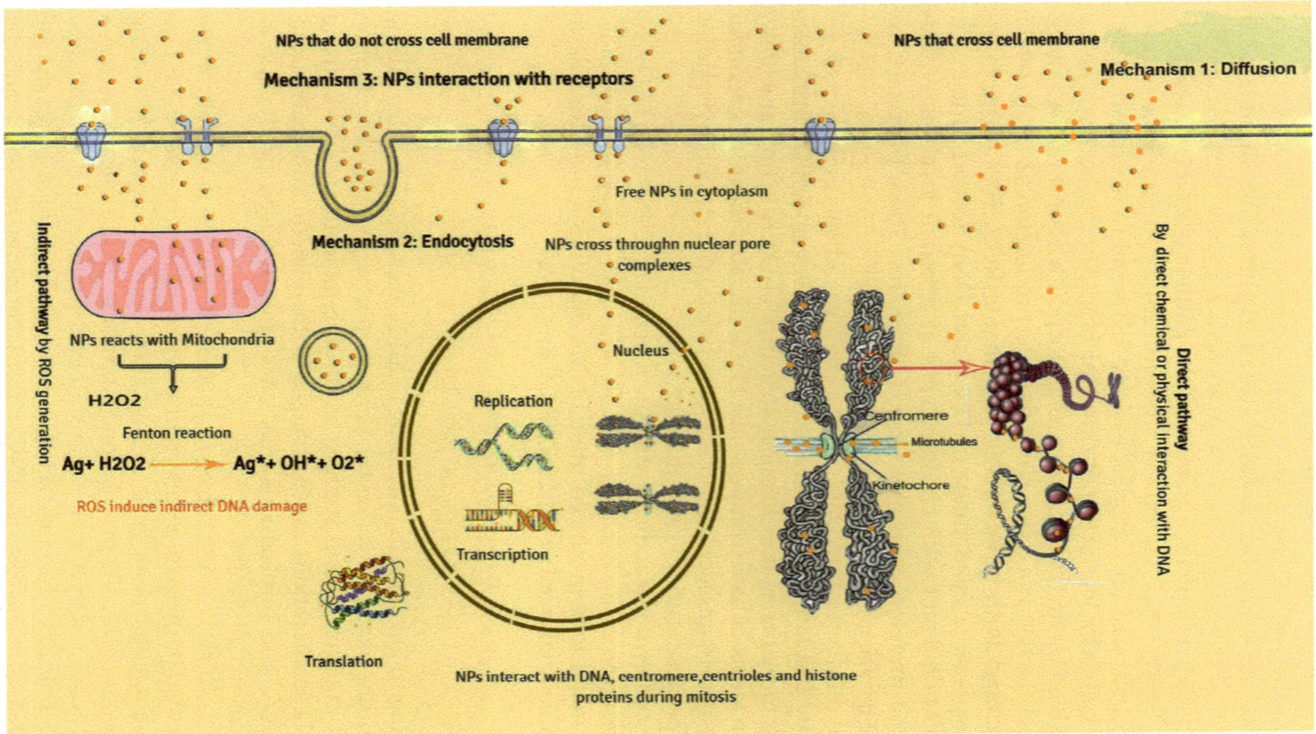

FIGURE 9.5 Direct and indirect genotoxicity caused by ENMs. *Reprinted from Mehrian, S. K., & De Lima, R. (2016). Nanoparticles cyto and genotoxicity in plants: Mechanisms and abnormalities.* Environmental Nanotechnology, Monitoring and Management, *6, 184–193. Copyright (2016). With permission from Elsevier.*

the best of authors knowledge, most investigation on nanotoxicity in plants has focused on ROS. It has been reported that the tyrosine amino acid in proteins can be nitrated by the radical peroxynitrite to produce 3-nitrotyrosine and dityrosine (van der Vliet et al., 1995). The implications of these modifications have been mainly studied in humans where it has been reported that high levels of 3-nitrotyrosine have been related to inflammation, aging, as well as cardiovascular and neurodegenerative diseases (Campolo et al., 2020). In plants, Sakihama et al. (2003) proved that peroxynitrite was also able to nitrate the phenolic compound p-coumaric acid, a plant metabolite. According to these researchers, peroxynitrite is the RNS with the highest damaging power, and is able to oxidize amino acids, lipids, and DNA. Thus, it is likely that RNS also produce genotoxic effects in plants. However, this is to be studied and proved.

9.4.1.2 *Effect of ENMs at gene expression level*

The effect of ENMs at the gene expression level has been studied. Some of the first reports in this field correspond to Landa et al. (2012) and Kaveh et al. (2013). In this respect, Landa et al. (2012) exposed thale cress (*Arabidopsis thaliana*) plants to 100 mg/L of either ZnO or TiO_2 NPs. Results indicated that these ENMs upregulated 660 and 80 genes, respectively. In addition, 826 and 74 genes, correspondingly, were downregulated. These researchers also found that gene expression was affected at a larger extent by ZnO NPs, compared to TiO_2 NPs (Landa et al., 2012). Furthermore, ZnO NPs affected expression of genes related to stress response, including those associated with water scarcity, wounding, pathogen defense, biogenesis, and nucleosome assembly, among others. In the case of TiO_2 NPs, they influenced genetic response to biotic and abiotic stimulus. Later, Kaveh et al. (2013) performed experiments with thale cress plants but this time they used Ag NPs. These ENMs upregulated 286 genes (response to oxidative stress and metals) and downregulated 81 (response to pathogens and hormonal stimuli).

TiO_2 and CeO_2 NPs differentially modified gene expression at the tissue level in thale cress plants (Tumburu et al., 2017). In total, TiO_2 upregulated more genes in leaves and roots, compared to CeO_2 NPs. However, CeO_2 NPs preferentially affected gene expression in root tissues, and TiO_2 in leaves. These researchers also reported that the metabolism was upregulated by nano-titania, but not by nano-ceria. Gene expressions affected by nano-titania and nano-ceria were associated to photosynthesis and activating transcription factor, respectively. Lately, Adhikari et al. (2020) identified 3 upregulated and 77 downregulated miRNAs in leaves of maize treated with 800 mg/L ZnO NPs. In addition, they determined that potential target genes are involved in different processes including heavy metal transport and detoxification, synthesis of chlorophyll, photosynthethic pathways, ROS balance, and plant growth, among others.

Relevant results in the agro field were obtained by Ramezani et al. (2020). These researchers found out that Stevia (*Stevia rebaudiana*) plants treated with 0–40 mM Ag NPs modified the transcription of different genes involved in glycoside production. According to these researchers, 10 and 20 mM Ag NPs resulted in gene underexpression, while the 40 mM treatment provoked an overexpression. In consequence, at 40 mM Ag NPs levels, production of steviol glycosides increased. This result is relevant since these compounds are the ones that provide the sweet flavor to the leaves, which are used by humans to replace sugar cane.

According to our bibliographical research, most investigations of the effects of ENMs on plants at the gene expression level have been performed with thale cress, which is a model plant. However, further studies involving crops are extremely important as they form the basis of the food chain.

9.4.2 Effect of ENMs at the energy production level

Different biomolecules involved in the photosynthetic process and energy production contain or involve the participation of transition metal species. For example, some metalloproteins participate in the electron transport chain (Yruela, 2013). Throughout the years, it has been proven that ENMs affect photosynthesis in plants. Some of the NPs that have been studied include SiO_2, TiO_2, ZnO, and Ag, among others. According to Tighe-Neira et al. (2018), TiO_2 may positively influence photosynthetic efficiency in plants, since these NPs may increase the flux of electrons between the photosynthetic systems I and II. In addition, they may increase Rubisco activity in the Benson and Calvin cycle. However, this is still to be proven. Another ENM that has been studied for potential effects on photosynthetic activity in plants is CeO_2. Wu et al. (2017) proved differential effects of CeO_2 NPs with surface charge. These researchers determined that negatively charged nanoceria were uptaken into chloroplasts at a higher rate as compared to positively charged nanoparticles. In addition, anionic, spherical (<11 nm) ceria NPs with low Ce^{3+}/Ce^{4+} ratio were able to scavenge significant ratios of ROS in plants. As a consequence, leaves containing these NPs presented increased quantum yield of PSII, compared to controls. Further explanation of complete mechanisms for other NPs is still needed.

Faizan et al. (2018) reported the enhanced photosynthetic efficiency in tomato plants exposed to ZnO NPs (0–16 mg/L). These researchers measured chlorophyll and proline concentration, net photosynthetic rate (P_N), stomatal conductance (gs), intercellular concentration of carbon dioxide (Ci), and transpiration rate (E), among other parameters. In all cases, values were significantly higher as compared to controls. These authors concluded that ZnO NPs treatments incremented proline content in tomato plants, which positively influenced photosynthetic activity. Similarly, Ruiz-Torres et al. (2021) reported that 100–400 mg/kg ZnO NPs increased chlorophyll content in plants grown in soil. However, plants also increased activity of CAT, APX, and POD enzymes, indicating an increment in ROS in plant cell. On the other hand, higher ZnO NPs concentrations produced dissimilar effects in maize, as 800 mg/L ZnO NPs resulted in diminished chlorophyll content and augmented MDA production (Adhikari et al., 2020). Thus, before field application, it is important to determine the range of tolerance, as an inadequate dose may result in plant damage.

Zn is one of the essential micronutrients. This element is part of different metalloproteins and metalloenzymes and provides diverse structural, catalytic, protein-interface and cocatalytic properties (Castillo-González et al., 2018). Some examples include zinc-finger proteins involved in DNA organization (Ferl & Paul, 2000), alcohol dehydrogenase, RNA polymerase, and carbonic anhydrase, among others (Castillo-González et al., 2018). Thus, Zn deficiency may result in reduced enzymatic activity and structural changes in proteins. During the photosynthetic process, β-carbonic anhydrase catalyzes the reaction between CO_2 and H_2O to produce carbonic acid (H_2CO_3), which is later dissociated to the ion bicarbonate (HCO_3^-) and a proton (H^+). This process facilitates the diffusion of CO_2 present in the atmosphere to the interior of the chloroplast and provides (HCO_3^-) as substrate for the enzyme phosphoenolpyruvate carboxylase. However, as previous examples showed, high Zn^{2+} concentrations result in overproduction of ROS, which destroy chlorophyll and other pigments (Adhikari et al., 2020). It has also been determined that Zn^{2+} can replace Mg in the porphyrin moiety, which may lead to biomolecule denaturation and reduced capacity for electron transport. The final effect is a diminished photosynthetic activity (Küpper & Andresen, 2016). According to these authors, the mechanism also applies to other heavy metals that substitute Mg in the chlorophyll biomolecule.

9.4.3 Implications of ENMs transformations within plant body as related to human health

In the past, TiO_2 materials were considered quasi-inert. However, recent research has suggested that they may display carcinogenic properties when inhaled. In fact, in October 2019, the Risk Assessment Committee (RAC) of the European Chemical Agency (ECHA) classified TiO_2 powder with at least a 1% of particles with an aerodynamic diameter below 10 μm (European Commission, 2020) as carcinogen by inhalation. This information is important when considering other routes of exposure such as oral and absorption through skin. According to this, a potential implication of NP accumulation in plant systems is the risk they represent for humans and other living systems. Brand et al. (2020) compared human clinical and histopathological data with results of oral animal studies and postmortem Ti concentrations in order to determine to which extent TiO_2 NPs may result in damage to humans after oral exposure. These authors concluded that adverse effects on human liver and intestine include inflammation, ROS production, oxidative stress, epithelial injury, DNA impairment in intestinal cells, and generation of intestinal tumors, among others. For complementary information on TiO_2 NPs effect in humans, readers are encouraged to read the review published by Bischoff et al. (2021).

ENMs have been detected in fruits and seeds of plants exposed to different NPs. For example, Servin et al. (2012) identified TiO_2 NPs in cucumber fruits. Also, Wang et al. (2016) for the first time observed the presence of CuO NPs in harvested seeds of thale cress plants. With this observation they confirmed the transfer or CuO NPs from parent to progeny. Considering the known and unknown toxic effects, the presence of this type of NPs in edible species should not be underestimated.

9.5 Conclusions

A few publications have described in detail the uptake mechanisms, transport, and biotransformation within plants of selected ENMs. However, with slight exceptions, studies have focused on specific plant parts, which make it difficult to have a complete picture of ENM–plant interactions. Even so, the information available is extremely important in identifying effects on plants and possible risks to human health. The literature has shown that the chemistry (type of metal/metal oxide), physicochemical surface properties (particle charge, coating, ratio of oxidation states), shape, crystallinity, and particle size determine the ENM–plant interactions and the extent of their biotransformation. The first step of biotransformation is determined by the solubility of the ENM. Then, solubilized ions may be adsorbed to the root surface or complexed by different metabolites and cell structures. Another determining factor for the biotransformation is the particle charge. This determines the uptake rate since positively charged materials will be preferentially retained in root surface, while negatively charged particles will be repelled. Depending on the chemistry and concentration, ENMs/solubilized ions generate excess of ROS molecules that alter different biomolecules including enzymes, chlorophyll, and DNA. Damages in DNA may affect gene expression, ending up in abundance or reduction of important proteins. Energy production and plant protection are also important metabolic traits that are affected by the presence of ENMs and their solubilized products. The biotransformation of ENMs within plants has been mainly studied using XANES techniques. However, identification of biotransformation products has been limited due to the reduced number of model compounds used in these studies. Herein we have shown specific chemical reactions that take place during the interaction of ENMs with plants. There are few studies showing the presence of ENMs/biotransformation products in plants' edible portion; however, their implications in human health have not been yet determined.

References

Adhikari, S., Adhikari, A., Ghosh, S., Roy, D., Azahar, I., Basuli, D., & Hossain, Z. (2020). Assessment of ZnO-NPs toxicity in maize: An integrative microRNAomic approach. *Chemosphere, 249*, 126197. https://doi.org/10.1016/j.chemosphere.2020.126197

Barberon, M. (2017). The endodermis as a checkpoint for nutrients. *New Phytologist, 213*(4), 1604–1610. https://doi.org/10.1111/nph.14140

Barberon, M., & Geldner, N. (2014). Radial transport of nutrients: The plant root as a polarized epithelium1. *Plant Physiology, 166*(2), 528–537. https://doi.org/10.1104/pp.114.246124

Bischoff, N. S., de Kok, T. M., Sijm, D. T. H. M., van Breda, S. G., Briedé, J. J., Castenmiller, J. J. M., Opperhuizen, A., Chirino, Y. I., Dirven, H., Gott, D., Houdeau, E., Oomen, A. G., Poulsen, M., Rogler, G., & van Loveren, H. (2021). Review possible adverse effects of food additive e171 (Titanium dioxide) related to particle specific human toxicity, including the immune system. *International Journal of Molecular Sciences, 22*(1), 1–35. https://doi.org/10.3390/ijms22010207

Brand, W., Peters, R. J. B., Braakhuis, H. M., Maślankiewicz, L., & Oomen, A. G. (2020). Possible effects of titanium dioxide particles on human liver, intestinal tissue, spleen and kidney after oral exposure. *Nanotoxicology, 14*(7), 985–1007. https://doi.org/10.1080/17435390.2020.1778809

Campolo, N., Issoglio, F. M., Estrin, D. A., Bartesaghi, S., & Radi, R. (2020). 3-Nitrotyrosine and related derivatives in proteins: Precursors, radical intermediates and impact in function. *Essays in Biochemistry, 64*(1), 111–133. https://doi.org/10.1042/EBC20190052

Casillas-Figueroa, F., Arellano-García, M. E., Leyva-Aguilera, C., Ruíz-Ruíz, B., Luna Vázquez-Gómez, R., Radilla-Chávez, P., Chávez-Santoscoy, R. A., Pestryakov, A., Toledano-Magaña, Y., García-Ramos, J. C., & Bogdanchikova, N. (2020). Argovit^{TM} silver nanoparticles effects on Allium cepa: Plant growth promotion without cyto genotoxic damage. *Nanomaterials*, 1386. https://doi.org/10.3390/nano10071386

Castillo-González, J., Ojeda-Barrios, D., Hernández-Rodríguez, A., González-Franco, A. C., Robles-Hernández, L., & López-Ochoa, G. R. (2018). Zinc metalloenzymes in plants. *Interciencia, 43*(4), 242–248. https://www.interciencia.net/wp-content/uploads/2018/05/242-6131-OJEDA-43-04.pdf.

Chen, X., Ji, D., Wang, X., & Zang, L. (2017). Review on nano zerovalent iron (nZVI): From modification to environmental applications. *IOP Conference Series: Earth and Environmental Science, 51*(1), 012004. https://doi.org/10.1088/1742-6596/51/1/012004

Cornelis, G., Ryan, B., McLaughlin, M. J., Kirby, J. K., Beak, D., & Chittleborough, D. (2011). Solubility and batch retention of CeO2 nanoparticles in soils. *Environmental Science and Technology, 45*(7), 2777–2782. https://doi.org/10.1021/es103769k

Corpas, F. J. (2016). Reactive nitrogen species (RNS) in plants under physiological and adverse environmental conditions: Current view. In F. M. Cánovas, U. Lüttge, &

R. Matyssek (Eds.) (1st ed.,, *78. Progress in botany* (p. 437). Cham: Springer. https://doi.org/10.1007/124_2016_3

Dahle, J. (2013). *The effects of physicochemical properties of CeO_2 nanoparticles on toxicity to soil denitrification processes.*

Dizdaroglu, M., & Jaruga, P. (2012). Mechanisms of free radical-induced damage to DNA. *Free Radical Research, 46*(4), 382–419. https://doi.org/10.3109/10715762.2011.653969

European Commission. (2020). Commission delegated regulation (EU) 2020/217 of 4 October 2019 amending, for the purposes of its adaptation to technical and scientific progress, Regulation (EC) No 1272/2008 of the European Parliament and of the Council on classification, labelling and packaging of substances and mixtures and correcting that regulation. *Official Journal of the European Union, 44*. https://eur-lex.europa.eu/legal-content/EN/TXT/PDF/?uri=CELEX:32020R0217&from=EN.

Faizan, M., Faraz, A., Yusuf, M., Khan, S. T., & Hayat, S. (2018). Zinc oxide nanoparticle-mediated changes in photosynthetic efficiency and antioxidant system of tomato plants. *Photosynthetica, 56*(2), 678–686. https://doi.org/10.1007/s11099-017-0717-0

Ferl, R., & Paul, A. L. (2000). Genome organization and expression. In B. Buchanan, W. Gruissem, & R. Jones (Eds.), *Biochemistry and molecular biology of plants*. American Society of Plant Physiologists.

Gao, X., Avellan, A., Laughton, S., Vaidya, R., Rodrigues, S. M., Casman, E. A., & Lowry, G. V. (2018). CuO nanoparticle dissolution and toxicity to wheat (*Triticum aestivum*) in rhizosphere soil. *Environmental Science and Technology, 52*(5), 2888–2897. https://doi.org/10.1021/acs.est.7b05816

Ghosh, I., Mukherjee, A., & Mukherjee, A. (2017). In planta genotoxicity of nZVI: Influence of colloidal stability on uptake, DNA damage, oxidative stress and cell death. *Mutagenesis, 32*(3), 371–387. https://doi.org/10.1093/mutage/gex006

Ghosh, I., Sadhu, A., Moriyasu, Y., Bandyopadhyay, M., & Mukherjee, A. (2019). Manganese oxide nanoparticles induce genotoxicity and DNA hypomethylation in the moss *Physcomitrella patens*. *Mutation Research – Genetic Toxicology and Environmental Mutagenesis, 842*, 146–157. https://doi.org/10.1016/j.mrgentox.2018.12.006

Halliwell, B., Clement, M. V., & Long, L. H. (2000). Hydrogen peroxide in the human body. *FEBS Letters, 486*(1), 10–13. https://doi.org/10.1016/S0014-5793(00)02197-9

Hernandez-Viezcas, J. A., Castillo-Michel, H., Andrews, J. C., Cotte, M., Rico, C., Peralta-Videa, J. R., Ge, Y., Priester, J. H., Holden, P. A., & Gardea-Torresdey, J. L. (2013). In situ synchrotron X-ray fluorescence mapping and speciation of CeO_2 and ZnO nanoparticles in soil cultivated soybean (*Glycine max*). *ACS Nano, 7*(2), 1415–1423. https://doi.org/10.1021/nn305196q

Hernandez-Viezcas, J. A., Castillo-Michel, H., Peralta-Videa, J. R., & Gardea-Torresdey, J. L. (2016). Interactions between CeO_2 nanoparticles and the desert plant mesquite: A spectroscopy approach. *ACS Sustainable Chemistry and Engineering, 4*(3), 1187–1192. https://doi.org/10.1021/acssuschemeng.5b01251

He, W., Zhou, Y. T., Wamer, W. G., Boudreau, M. D., & Yin, J. J. (2012). Mechanisms of the pH dependent generation of hydroxyl radicals and oxygen induced by Ag nanoparticles. *Biomaterials, 33*(30), 7547–7555. https://doi.org/10.1016/j.biomaterials.2012.06.076

Jiang, X., Gray, P., Patel, M., Zheng, J., & Yin, J. J. (2020). Crossover between anti- and pro-oxidant activities of different manganese oxide nanoparticles and their biological implications. *Journal of Materials Chemistry B, 8*(6), 1191–1201. https://doi.org/10.1039/c9tb02524c

Kaveh, R., Li, Y. S., Ranjbar, S., Tehrani, R., Brueck, C. L., & Van Aken, B. (2013). Changes in *Arabidopsis thaliana* gene expression in response to silver nanoparticles and silver ions. *Environmental Science and Technology, 47*(18), 10637–10644. https://doi.org/10.1021/es402209w

Küpper, H., & Andresen, E. (2016). Mechanisms of metal toxicity in plants. *Metallomics, 8*(3), 269–285. https://doi.org/10.1039/c5mt00244c

Landa, P., Vankova, R., Andrlova, J., Hodek, J., Marsik, P., Storchova, H., White, J. C., & Vanek, T. (2012). Nanoparticle-specific changes in *Arabidopsis thaliana* gene expression after exposure to ZnO, TiO_2, and fullerene soot. *Journal of Hazardous Materials, 241*(242), 55–62. https://doi.org/10.1016/j.jhazmat.2012.08.059

Larue, C., Khodja, H., Herlin-Boime, N., Brisset, F., Flank, A. M., Fayard, B., Chaillou, S., & Carrière, M. (2011). Investigation of titanium dioxide nanoparticles toxicity and uptake by plants. In , *304. Journal of physics: Conference series.* Institute of Physics Publishing. https://doi.org/10.1088/1742-6596/304/1/012057. Issue 1.

Liu, Y., Ma, Y., Jiao, C., Liu, M., Luo, W., Dong, C., Fan, S., He, X., Yang, F., & Zhang, Z. (2021). Comparative toxicity of rod-shaped nano-CeO_2 and nano-$CePO_4$ to lettuce. *Metallomics, 13*(7). https://doi.org/10.1093/mtomcs/mfab033

Li, X., Yang, Y., Gao, B., Zhang, M., & Singer, A. C. (2015). Stimulation of peanut seedling development and growth by zero-valent iron nanoparticles at low concentrations. *PLoS One, 10*(4), e0122884. https://doi.org/10.1371/journal.pone.0122884

López-Moreno, M. L., De La Rosa, G., Hernández-Viezcas, J. A., Castillo-Michel, H., Botez, C. E., Peralta-Videa, J. R., & Gardea-Torresdey, J. L. (2010). Evidence of the differential biotransformation and genotoxicity of ZnO and CeO_2 nanoparticles on soybean (*Glycine max*) plants. *Environmental Science and Technology, 44*(19), 7315–7320. https://doi.org/10.1021/es903891g

Ma, Y., He, X., Zhang, P., Zhang, Z., Guo, Z., Tai, R., Xu, Z., Zhang, L., Ding, Y., Zhao, Y., & Chai, Z. (2011). Phytotoxicity and biotransformation of La_2O_3 nanoparticles in a terrestrial plant cucumber (*Cucumis sativus*). *Nanotoxicology, 5*(4), 743–753. https://doi.org/10.3109/17435390.2010.545487

Majumdar, S., Peralta-Videa, J. R., Bandyopadhyay, S., Castillo-Michel, H., Hernandez-Viezcas, J. A., Sahi, S., & Gardea-Torresdey, J. L. (2014). Exposure of cerium oxide nanoparticles to kidney bean shows disturbance in the plant defense mechanisms. *Journal of Hazardous Materials, 278*, 279–287. https://doi.org/10.1016/j.jhazmat.2014.06.009

Ma, Y., Zhang, P., Zhang, Z., He, X., Li, Y., Zhang, J., Zheng, L., Chu, S., Yang, K., Zhao, Y., & Chai, Z. (2015). Origin of the different phytotoxicity and biotransformation of cerium and lanthanum oxide nanoparticles in cucumber. *Nanotoxicology, 9*(2), 262–270. https://doi.org/10.3109/17435390.2014.921344

Ma, Y., Zhang, P., Zhang, Z., He, X., Zhang, J., Ding, Y., Zhang, J., Zheng, L., Guo, Z., Zhang, L., Chai, Z., & Zhao, Y. (2015). Where does the transformation of precipitated ceria nanoparticles in hydroponic plants take place? *Environmental Science & Technology, 49*(17), 10667–10674. https://doi.org/10.1021/acs.est.5b02761

Mehrian, S. K., & De Lima, R. (2016). Nanoparticles cyto and genotoxicity in plants: Mechanisms and abnormalities. *Environmental Nanotechnology, Monitoring and Management, 6*, 184–193. https://doi.org/10.1016/j.enmm.2016.08.003

Mitreva, N. I. (1989). Sunflower and maize transorganal tissue pH gradients and problems related to the ion transport mechanisms. *Plant and Soil, 113*(1), 144–146. https://doi.org/10.1007/BF02181935

Morrissey, J., & Guerinot, M. L. (2009). Iron uptake and transport in plants: The good, the bad, and the ionome. *Chemical Reviews, 109*(10), 4553–4567. https://doi.org/10.1021/cr900112r

Muccifora, S., Castillo-Michel, H., Barbieri, F., Bellani, L., Castiglione, M. R., Spanò, C., Pradas Del Real, A. E., Giorgetti, L., & Tassi, E. L. (2021). Synchrotron radiation spectroscopy and transmission electron microscopy techniques to evaluate TiO_2 nps incorporation, speciation, and impact on root cells ultrastructure of *Pisum sativum* l. Plants. *Nanomaterials, 11*(4). https://doi.org/10.3390/nano11040921

Ogunkunle, C. O., Bornmann, B., Wagner, R., Fatoba, P. O., Frahm, R., & Lützenkirchen-Hecht, D. (2019). Copper uptake, tissue partitioning and biotransformation evidence by XANES in cowpea (*Vigna unguiculata* L) grown in soil amended with nano-sized copper particles. *Environmental Nanotechnology, Monitoring and Management, 12*. https://doi.org/10.1016/j.enmm.2019.100231

Peng, C., Duan, D., Xu, C., Chen, Y., Sun, L., Zhang, H., Yuan, X., Zheng, L., Yang, Y., Yang, J., Zhen, X., Chen, Y., & Shi, J. (2015). Translocation and biotransformation of CuO nanoparticles in rice (*Oryza sativa* L.) plants. *Environmental Pollution, 197*, 99–107. https://doi.org/10.1016/j.envpol.2014.12.008

Peng, C., Xu, C., Liu, Q., Sun, L., Luo, Y., & Shi, J. (2017). Fate and transformation of CuO nanoparticles in the soil-rice system during the life cycle of rice plants. *Environmental Science and Technology, 51*(9), 4907–4917. https://doi.org/10.1021/acs.est.6b05882

Ramezani, M., Asghari, S., Gerami, M., Ramezani, F., & Karimi Abdolmaleki, M. (2020). Effect of silver nanoparticle treatment on the expression of key genes involved in glycosides biosynthetic pathway in *Stevia rebaudiana* B. Plant. *Sugar Tech, 22*(3), 518–527. https://doi.org/10.1007/s12355-019-00786-x

Ronzio, R. A. (2020). *Naturally occurring antioxidants*. Elsevier BV. https://doi.org/10.1016/b978-0-323-43044-9.00097-2. e12.

Ruiz-Torres, N., Flores-Naveda, A., Barriga-Castro, E. D., Camposeco-Montejo, N., Ramírez-Barrón, S., Borrego-Escalante, F., Niño-Medina, G., Hernández-Juárez, A., Garza-Alonso, C., Rodríguez-Salinas, P., & García-López, J. I. (2021). Zinc oxide nanoparticles and zinc sulfate impact physiological parameters and boosts lipid peroxidation in soil grown coriander plants (*Coriandrum sativum*). *Molecules, 26*(7). https://doi.org/10.3390/molecules26071998

Rui, Y., Zhang, P., Zhang, Y., Ma, Y., He, X., Gui, X., Li, Y., Zhang, J., Zheng, L., Chu, S., Guo, Z., Chai, Z., Zhao, Y., & Zhang, Z. (2015). Transformation of ceria nanoparticles in cucumber plants is influenced by phosphate. *Environmental Pollution, 198*, 8–14. https://doi.org/10.1016/j.envpol.2014.12.017

Saito, E., & Bielski, B. H. J. (1961). The electron paramagnetic resonance spectrum of the HO2 radical in aqueous solution. *Journal of the American Chemical Society, 83*(21), 4467–4468. https://doi.org/10.1021/ja01482a039

Sakihama, Y., Tamaki, R., Shimoji, H., Ichiba, T., Fukushi, Y., Tahara, S., & Yamasaki, H. (2003). Enzymatic nitration of phytophenolics: Evidence for peroxynitrite- independent nitration of plant secondary metabolites. *FEBS Letters, 553*(3), 377–380. https://doi.org/10.1016/S0014-5793(03)01059-7

Scherer, M. D., Sposito, J. C. V., Falco, W. F., Grisolia, A. B., Andrade, L. H. C., Lima, S. M., Machado, G., Nascimento, V. A., Gonçalves, D. A., Wender, H., Oliveira, S. L., & Caires, A. R. L. (2019). Cytotoxic and genotoxic effects of silver nanoparticles on meristematic cells of Allium cepa roots: A close analysis of particle size dependence. *Science of the Total Environment, 660*, 459–467. https://doi.org/10.1016/j.scitotenv.2018.12.444

Schwabe, F., Schulin, R., Rupper, P., Rotzetter, A., Stark, W., & Nowack, B. (2014). Dissolution and transformation of cerium oxide nanoparticles in plant growth media. *Journal of Nanoparticle Research, 16*(10). https://doi.org/10.1007/s11051-014-2668-8

Schwabe, F., Tanner, S., Schulin, R., Rotzetter, A., Stark, W., von Quadt, A., & Nowack, B. (2015). Dissolved cerium contributes to uptake of Ce in the presence of differently sized CeO_2-nanoparticles by three crop plants. *Metallomics, 7*(3), 466–477. https://doi.org/10.1039/c4mt00343h

Servin, A. D., Castillo-Michel, H., Hernandez-Viezcas, J. A., Diaz, B. C., Peralta-Videa, J. R., & Gardea-Torresdey, J. L. (2012). Synchrotron micro-XRF and micro-XANES confirmation of the uptake and translocation of TiO_2 nanoparticles in cucumber (*Cucumis sativus*) plants. *Environmental Science and Technology, 46*(14), 7637–7643. https://doi.org/10.1021/es300955b

Shi, J., Peng, C., Yang, Y., Yang, J., Zhang, H., Yuan, X., Chen, Y., & Hu, T. (2014). Phytotoxicity and accumulation of copper oxide nanoparticles to the Cu-tolerant plant *Elsholtzia splendens*. *Nanotoxicology, 8*(2), 179–188. https://doi.org/10.3109/17435390.2013.766768

Singh, K. R. B., Nayak, V., Sarkar, T., & Singh, R. P. (2020). Cerium oxide nanoparticles: Properties, biosynthesis and biomedical application. *RSC Advances, 10*(45), 27194–27214. https://doi.org/10.1039/d0ra04736h

Spielman-Sun, E., Lombi, E., Donner, E., Howard, D., Unrine, J. M., & Lowry, G. V. (2017). Impact of surface charge on cerium oxide nanoparticle uptake and translocation by wheat (*Triticum aestivum*). *Environmental Science and Technology, 51*(13), 7361–7368. https://doi.org/10.1021/acs.est.7b00813

Strock, C. F., & Lynch, J. P. (2020). Root secondary growth: An unexplored component of soil resource acquisition. *Annals of Botany, 126*(2), 205–218. https://doi.org/10.1093/aob/mcaa068

Tighe-Neira, R., Carmora, E., Recio, G., Nunes-Nesi, A., Reyes-Diaz, M., Alberdi, M., Rengel, Z., & Inostroza-Blancheteau, C. (2018). Metallic nanoparticles influence the structure and function of the photosynthetic apparatus in plants. *Plant Physiology and Biochemistry, 130*, 408–417. https://doi.org/10.1016/j.plaphy.2018.07.024

Torrent, L., Iglesias, M., Marguí, E., Hidalgo, M., Verdaguer, D., Llorens, L., Kodre, A., Kavčič, A., & Vogel-Mikuš, K. (2020). Uptake, translocation and ligand of silver in *Lactuca sativa* exposed to silver nanoparticles of different size, coatings and concentration. *Journal of Hazardous Materials, 384*, 121201. https://doi.org/10.1016/j.jhazmat.2019.121201

Tumburu, L., Andersen, C. P., Rygiewicz, P. T., & Reichman, J. R. (2017). Molecular and physiological responses to titanium dioxide and cerium oxide nanoparticles in *Arabidopsis*. *Environmental Toxicology and Chemistry, 36*(1), 71–82. https://doi.org/10.1002/etc.3500

Vaccari, D. A., Strom, P. F., & Alleman, J. E. (2006). *Environmental biology for engineers and scientists*.

Vinković Vrček, I., Pavičić, I., Crnković, T., Jurašin, D., Babič, M., Horák, D., Lovrić, M., Ferhatović, L., Ćurlin, M., & Gajović, S. (2015). Does surface coating of metallic nanoparticles modulate their interference with in vitro assays? *RSC Advances, 5*(87), 70787–70807. https://doi.org/10.1039/c5ra14100a

van der Vliet, A., Eiserich, J. P., O'Neill, C. A., Halliwell, B., & Cross, C. E. (1995). Tyrosine modification by reactive nitrogen species: A closer look. *Archives of Biochemistry and Biophysics, 319*(2), 341–349. https://doi.org/10.1006/abbi.1995.1303

Wang, Y. M., Kinraide, T. B., Wang, P., Hao, X. Z., & Zhou, D. M. (2014). Surface electrical potentials of root cell plasma membranes: Implications for ion interactions, rhizotoxicity, and uptake. *International Journal of Molecular Sciences, 15*(12), 22661–22677. https://doi.org/10.3390/ijms151222661

Wang, Z., Xie, X., Zhao, J., Liu, X., Feng, W., White, J. C., & Xing, B. (2012). Xylem- and phloem-based transport of CuO nanoparticles in maize (*Zea mays* L.). *Environmental Science and Technology, 46*(8), 4434–4441. https://doi.org/10.1021/es204212z

Wang, Z., Xu, L., Zhao, J., Wang, X., White, J. C., & Xing, B. (2016). CuO nanoparticle interaction with *Arabidopsis thaliana*: Toxicity, parent-progeny transfer, and gene expression. *Environmental Science and Technology, 50*(11), 6008–6016. https://doi.org/10.1021/acs.est.6b01017

Wu, H., Tito, N., & Giraldo, J. P. (2017). Anionic cerium oxide nanoparticles protect plant photosynthesis from abiotic stress by scavenging reactive oxygen species. *ACS Nano, 11*(11), 11283–11297. https://doi.org/10.1021/acsnano.7b05723

Yin, L., Cheng, Y., Espinasse, B., Colman, B. P., Auffan, M., Wiesner, M., Rose, J., Liu, J., & Bernhardt, E. S. (2011). More than the ions: The effects of silver nanoparticles on lolium multiflorum. *Environmental Science and Technology, 45*(6), 2360–2367. https://doi.org/10.1021/es103995x

Yruela, I. (2013). Transition metal in plant photosynthesis. *Metallomics, 5*(9), 1090–1109.

Zhang, W., Dan, Y., Shi, H., & Ma, X. (2017). Elucidating the mechanisms for plant uptake and in-planta speciation of cerium in radish (*Raphanus sativus* L.) treated with cerium oxide nanoparticles. *Journal of Environmental Chemical Engineering, 5*(1), 572–577. https://doi.org/10.1016/j.jece.2016.12.036

Zhang, P., Ma, Y., Liu, S., Wang, G., Zhang, J., He, X., Zhang, J., Rui, Y., & Zhang, Z. (2017). Phytotoxicity, uptake and transformation of nano-CeO_2 in sand cultured romaine lettuce. *Environmental Pollution, 220*, 1400–1408. https://doi.org/10.1016/j.envpol.2016.10.094

Zhang, P., Ma, Y., Zhang, Z., He, X., Guo, Z., Tai, R., Ding, Y., Zhao, Y., & Chai, Z. (2012). Comparative toxicity of nanoparticulate/bulk Yb_2O_3 and $YbCl_3$ to cucumber (*Cucumis sativus*). *Environmental Science & Technology, 46*(3), 1834–1841. https://doi.org/10.1021/es2027295

Zhang, P., Ma, Y., Zhang, Z., He, X., Zhang, J., Guo, Z., Tai, R., Zhao, Y., & Chai, Z. (2012). Biotransformation of ceria nanoparticles in cucumber plants. *ACS Nano, 6*(11), 9943–9950. https://doi.org/10.1021/nn303543n

Zhang, P., Xie, C., Ma, Y., He, X., Zhang, Z., Ding, Y., Zheng, L., & Zhang, J. (2017). Shape-dependent transformation and translocation of ceria nanoparticles in cucumber plants. *Environmental Science and Technology Letters, 4*(9), 380–385. https://doi.org/10.1021/acs.estlett.7b00359

Zhao, L., Peralta-Videa, J. R., Varela-Ramirez, A., Castillo-Michel, H., Li, C., Zhang, J., Aguilera, R. J., Keller, A. A., & Gardea-Torresdey, J. L. (2012). Effect of surface coating and organic matter on the uptake of CeO_2 NPs by corn plants grown in soil: Insight into the uptake mechanism. *Journal of Hazardous Materials*, 131–138. https://doi.org/10.1016/j.jhazmat.2012.05.008

CHAPTER

10

Biotransformation in leaves of foliar applied ENMs

Zoe C. Simon, and Edison Castro

Department of Chemistry, University of Pittsburgh, Pittsburgh, PA, United States

10.1 Introduction

As the world population continues to rapidly increase, so does the demand for food, and pressure has been put on the agricultural industry to increase crop production. To optimize available land to meet these growing demands, many research efforts have been focused on developing more productive and sustainable farming practices (Tilman et al., 2011). One way to maximize output in the agricultural sector is to use engineered nanomaterials (ENMs) in place of traditional fertilizers. ENMs can be sprayed on plant leaves during the growth period, and they enter the plant via the stomata, cuticle, trichomes, hydathodes, or necrotic spots (Avellan et al., 2021). The unique properties of ENMs make them a promising class of fertilizer because their particle size ranges between 1 and 100 nm, making them small enough to enter the plant through the leaf. Additionally, ENMs are highly tunable, as their size, shape, surface chemistry, and composition all play a critical role in determining their physical and chemical properties. Indeed, nanofertilizers define a large class of materials consisting of metal (Mehta et al., 2016), metalloid (Zahedi et al., 2019), metal-oxide (Raliya & Tarafdar, 2013; Rossi et al., 2019), and nonmetal structures (Rahmani et al., 2020; Rai et al., 2015). All of these structures have been synthesized at a small enough size regime for foliar application and uptake. So far, many studies have been conducted optimizing these synthetic handles to maximize active ingredient absorption. Moreover, other parameters such as nanoparticle ligands, surface area, magnetics, and photochemical properties can also be considered and tailored to specific plant needs.

Physicochemical Interactions of Engineered Nanoparticles and Plants
https://doi.org/10.1016/B978-0-323-90558-9.00003-6

Nanofertilizers can be classified into three groups: fertilizer inputs, where traditional fertilizer is reduced in size to nanoscale dimensions; fertilizer additives, where nanomaterials are added to traditional fertilizer formulations; and nanocoatings or nanohosts, where thin films or nanoporous materials control nutrient release (Mejías et al., n.d.; Rai et al., 2015). Of the patents filed for foliar applied nanomaterials up until 2015, the majority were either fertilizer inputs or fertilizer additives (Mastronardi et al., 2015). Because both methods involve direct use of nanoparticles, careful control of their properties is crucial to optimize nutrient delivery. One of the most important properties that gives ENMs an advantage over traditional fertilizers for foliar application is the nanoparticle size, which can be small enough to enter the plant leaf through the stomatal openings, which range in size but typically fall in the micron length scale (Wang, Tarafdar et al., 2013; Zhao et al., 2017). Because the largest nanomaterial is only 1 micron by definition, ENMs are well suited for entering the plant through open stomata, although opening depends on other environmental factors (Hall & Byrd, 2012). Taken together, foliar applied ENMs are advantageous in that the material can be tuned extensively to meet agricultural demands and also to maximize uptake. Many studies reveal that ENM-exposed crops exhibit adaptive processes in response to stress, production of antioxidant enzymes, regulation of genes related to cell division/extension, and membrane transport, among others. Currently, many studies are in development to understand the effects of crops produced in ENMs-contaminated environments; however, mechanistic information is still inconclusive. This chapter includes relevant information about the interaction and biotransformation of ENMs in plants after foliar application, including protection mechanisms, which allow the plants to eradicate ENMs.

10.2 Foliar application versus other methods

Compared to other uses of ENMs, foliar application offers several advantages. Foliar application involves direct adherence of the ENM to the plant leaf, which mitigates issues such as rainfall and soil-ENM interactions present in soil application methods (provided the foliar adhesion is robust enough to withstand rainfall) (Hong et al., 2021). Because of these issues arising in other methods, there is consequently lower ENM uptake. Lowered uptake results in higher concentrations needed to produce the desired outcome, which is both inefficient and can have negative downstream ecological consequences, such as eutrophication (Raliya et al., 2016). Additionally, nutrient uptake by plants can occur from a root to leaf or leaf to root mechanism, which allows foliar applied ENMs to be utilized by the plant in the same manner as traditional soil-based fertilizers (Grillo et al., 2021).

Foliar applied ENMs have also been used to alleviate impacts of environmental stressors, such as drought, soil salinity, and chilling. One study by Borišev et al. explored the effects of fullerenol nanoparticles (molecular formula $C_{60}(OH)_{24}$), a carbon-based ENM, on the mitigation of drought effects in sugar beets (Borišev et al., 2016). When fullerenol ENMs are applied foliarly, they are able to penetrate the leaf and translocate to the plant roots, where they can bind to water. In this study, a 4-day drought was simulated, and it was found that the fullerenol ENM's hygroscopic activity allowed them to act as an intercellular water reserve; thus, the ENM-treated plants were better able to survive drought stress, compared to the control. Foliar applied ENMs have also been used to improve resistance to high salinity. Again, using fullerenol nanoparticles, Shafiqu et al. studied the use of these ENMs to enhance physiobiochemical responses to salt stress (Shafiq et al., 2020). Wheat plants were grown in sand-filled pots (500 g) and 150 mM NaCl for 15 days, after which fullerenol nanoparticles were applied foliarly at various concentrations. After 2 weeks, they were able to correlate increased ENM application to increased shoot growth and chlorophyll content, among other benefits.

In some cases, it has been shown that foliar applied ENMs can be used to enhance photosynthesis. For example, Elsheery et al. found that several ENMs, including silicon dioxide (5–15 nm), were able to protect sugarcane plants from chilling, one of the plant's main abiotic stresses (Elsheery et al., 2020). They concluded that the beneficial effect of the foliar applied ENMs was due to improved photoprotection, which lead to increased chlorophyll and carotenoid content and resulted in enhanced photosynthesis. Another study by Alidoust et al. compared foliar application to root application of iron oxide nanoparticles (Fe_2O_3) in soybean plants (Alidoust & Isoda, 2013). They found that the use of these ENMs did not inhibit growth at any phase and that foliar application of the Fe_2O_3 during the eight-trifoliate leaf stage enhanced photosynthesis rates due to increased stomatal opening. Overall, they determined the ENMs applied foliarly had a more positive effect than the root application.

While some studies have shown foliar applied ENMs can have a beneficial effect on photosynthesis, others have shown they can have detrimental effects on photosynthetic capabilities. A recent study by Lu et al. explored the effects of foliar applied Fe_2O_3 in wheat plants, a nanofertilizer that has commonly been used and studied in seed or root exposure fertilization methods (Lu et al., 2020). They found that the ENMs were able to enter the wheat plant leaves though the stomata and translocate to the roots and stems. However, the Fe_2O_3 contributed to the overproduction of the hydroxide radical (•OH), and led to the degradation of chlorophyll, which negatively impacted the wheat plant's ability to perform photosynthesis. This, consequently, led to decreased biomass production. Taken together, these mixed results suggest the effects of foliar applied ENMs on a plant's photosynthetic capabilities are not yet fully understood, and may be plant species and nanomaterial dependent.

Additionally, there are other challenges unique to the foliar application method that must be considered. Unlike roots, plant leaves are coated with a waxy cuticle, which can inhibit adherence of water-soluble ENMs, thus impeding the uptake. However, appropriate synthetic measures can be taken to circumvent this issue and promote adherence. Studies have found that both organic and inorganic molecules can enter the plant through cuticular and stomatal pathways (Franke, 1967).

10.3 Foliar adhesion of ENMs

For the ENMs to enter stomata, they must first adhere to the foliar tissues. Strong adherence is necessary to minimize the effects of biocidal leaching, which can affect the surrounding environment and unintended organisms (Grillo et al., 2021). The waxy plant cuticle varies with different environmental stressors, such as temperature and humidity, though it primarily consists of long-chain hydrophobic compounds such as alkanes, aldehydes, primary and secondary alcohols, ketones, and esters (Yeats & Rose, 2013). Fig. 10.1 shows an overview of the uptake and translocation pathway of ENMs entering the plant through the foliage, which cannot occur without sufficient adhesion of the ENM to the cuticle surface. The hydrophobicity of the cuticle can limit adhesion of certain ENMs, though this limitation can be overcome by varying synthetic parameters of the ENM morphology, such as shape, to maximize nanoparticulate surface area. Increased surface area enhances adhesion, though the use of complex morphologies, such as disks, rods, and prisms, complicates large-scale nanofertilizer synthesis by making these syntheses more time and energy intensive. Due to these limiting factors, most foliar-applied ENMs are spherical, which increases synthetic efficiency though decreases cuticle adhesion.

In addition to morphology, rational ligand design could also be used to leverage ENM surface chemistry to enhance foliar adhesion. The active ingredients in the ENMs can be delivered through three primary methods, each containing complex surface chemistry (Smith & Gilbertson, 2018). The first is via encapsulations within a ligand shell, the second is via deposition onto a carrier substrate, and the third is by directly synthesizing the ENM to consist of the active ingredient. In each case, surface chemistry dictates ENM properties, as it controls the association

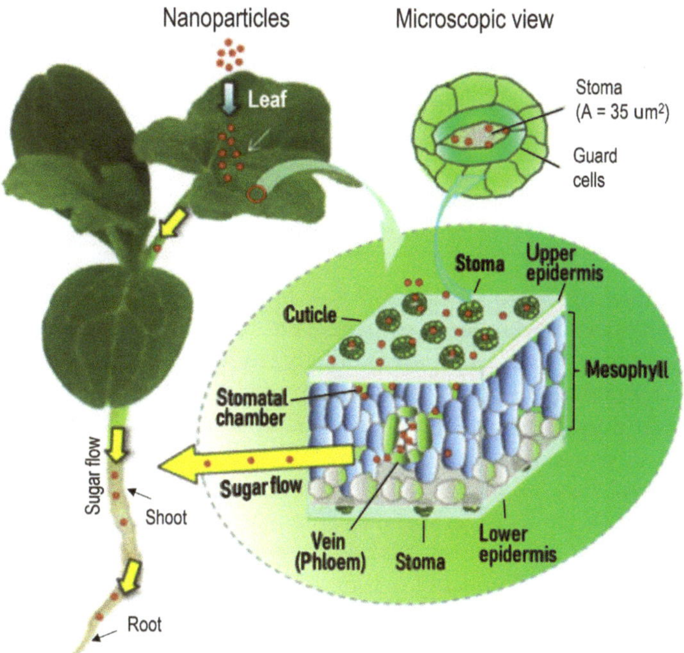

FIGURE 10.1 Foliar application, uptake, and translocation of ENMs in a watermelon plant. *Data from Wang, W. N., Tarafdar, J. C. & Biswas, P. (2013). Nanoparticle synthesis and delivery by an aerosol route for watermelon plant foliar uptake. Journal of Nanoparticle Research, 15*(1). *https://doi.org/10.1007/s11051-013-1417-8. Image reproduced with permission of Springer Nature.*

of the active ingredient to the ENM and the association of the ENM to the leaf. These ligands can range in size from small monoatomic ions (Millstone et al., 2008) to large polymeric surfactants (Smith et al., 2017). Many research efforts have explored the ways in which highly tunable ligand chemistry can be utilized to increase uptake in both cuticular and stomatal pathways (Avellan et al., 2019; Hu et al., 2020; Karny et al., 2018; Lowry et al., 2019; Yu et al., 2017).

One way ligand chemistry has been used to control foliar adhesion is by functionalizing ENM surfaces with molecules known to strongly interact with leaf tissues. An example of using molecules of this type was studied by Yu et al. using a tannic acid-based nanoparticle coating to improve adhesion (Yu et al., 2019). By modifying the surface of both abamectin and azoxystrobin nanopesticides, they were able to utilize the bioadhesive nature of tannic acid to improve adhesion of the ENMs to the leaf surface, which in turn enhanced foliar retention. Another study by Avellan et al. compared two ligand coatings, citrate and polyvinyl pyrrolidone (PVP), on various sized gold nanoparticles applied on wheat leaves (Avellan et al., 2019). They found that the PVP capped particles exhibited higher foliar adhesion, compared to the citrate capped ENMs, due to its relatively higher hydrophobicity, leading to increased interaction with the leaf surface. Both studies show that the ENM ligand shell is a crucial component of the material to optimize adhesive properties.

10.4 Uptake

The uptake of the foliar applied ENMs into the leaf is critical. Many ENMs' physical and chemical properties can influence this process including charge, size, and hydrophilicity/hydrophobicity. For example, ENM's charge must be considered, as the zeta (ζ) potential influences leaf adhesion and uptake (Hu et al., 2020). While it has been well established that the ζ potential of a nanomaterial affects its penetration into plasma membranes (Wong et al., 2016), a recent study demonstrated that is also an important factor in foliar uptake. This was demonstrated by studying the ζ potential of gold nanoparticles ranging from −31.1 to −69.2 mV (at 10 mg/L in 1mM KCl and pH 6.8) in wheat leaves, in which uptake follows a cuticular mechanism (Avellan et al., 2019). These researchers found that a higher percentage of the nanoparticles with a lower ζ potential reached the root system. In another study, the ζ potential of hydrophilic ENMs was tuned from −52.3 mV to +36.6 mV in carbon dots (CDs, 0.5 and 5 mg/mL), cerium oxide (CeO_2, 0.5 mg/mL), and silica (SiO_2, 0.5 mg/mL) and tested in both dicotyledon and monocotyledon plants (cotton and maize, respectively). In this study, they found that a higher positive charge (>15 mV) led to higher uptake in guard cells (100%), extracellular space (90.3%), and chloroplasts (55.8%). They hypothesized that this relationship is due to the negative charge of both the plant cell walls and transmembrane electrical potential of the cell membrane.

Another important factor to consider with respect to foliar uptake is the size of the ENM, especially with regards to the mechanism of uptake. A study focused on foliar applied TiO_2 ENMs in lettuce plants determined that the particles entered the leaf through both the cuticular and stomatal pathways (Larue, Castillo-Michel, Sobanska, Trcera, et al., 2014). Interestingly, this applied to both the nano- and micro-particles tested in the study. The authors hypothesized that particles greater than 100 nm were able to enter the leaf through both pathways, even though average cuticular pore radii typically only range in size from 0.45 to 1.18 nm (Schönherr, 2006). They suggested the reason why larger materials may be able to enter through the cuticle is due to a disruption of the waxy layer. Another study, which applied aerosolized Fe_2O_3, TiO_2, MgO, and ZnO ENMs to the leave of watermelon plants, found that the particles entered the leaf through the stomata via gas exchange (Wang, Tarafdar, et al.,

2013). They determined that when ENMs were greater than 100 nm in size, obstruction of the stomata occurred, leading to detrimental effects of limited gas exchange.

Additionally, the hydrophobicity/hydrophilicity is an impactful property of ENMs because it can influence the uptake pathway (Avellan et al., 2021). It is hypothesized that hydrophilic nanomaterials can enter the plant leaf through polar pores such as hydathodes, which are pores that secrete droplets of water through the epidermis or leaf margin, though limited studies have shown the possibility of uptake through this pathway (Hong et al., 2016). Hydrophilic ENMs are more commonly studied entering the leaf through stomata, although pathways through the cuticle are also known, though stomata and stomatal opening are largely dependent on plant species and environmental factors, all of which will affect this uptake pathway. Hydrophobic ENMs on the other hand are much more likely to enter the leaf through the waxy cuticle, and these nanoparticles have been observed to enter the leaf even when no stomata are present (Avellan et al., 2019).

10.5 Defense mechanisms

Given the diverse array of synthetically accessible ENMs, combined with the various exposure routes and a high potential to accumulate in plants, the literature shows that ENMs can have positive, neutral, and/or negative effects on plant growth, reproduction, and development. The development of mechanistic insights is still needed to explain these effects; yet they are at an early stage and need to be explored in greater detail (Mittal et al., 2020; Shang et al., 2019). As reported by Gardea-Torresdey et al. (2014), Jackson et al. (2013), Miralles et al. (2012), and Ma et al. (2010) there is a lack of examples in the literature studying the toxicity of ENMs after uptake. However, Calabrese reported an assessment of the toxicological bases of the hermetic dose-response relationship, in which he observed hormesis in the presence of ENMs, i.e., stimulation due to activation of repair mechanisms at low doses (Calabrese, 2005). In another study, Wang, Koo, et al. (2013) found that growth and transpiration increased at low doses of Ag-ENMs. The observed phytostimulatory effects preclude generalizations about the toxicity of Ag-ENMs at low levels, which encourage the need for further research in this field.

Several studies have clearly demonstrated that ENMs can traverse plant cell walls and localize in the cytosol or within cell organelles. At the same time, ENMs can also be transported from roots to aboveground tissues via xylem, followed by transfer to the phloem and transfer back to the roots (Nhan et al., 2015; Wang et al., 2012). This transport mechanism does not keep the ENMs concentration in equilibrium within the plant, thus allowing accumulation of ENMs in certain regions more than in others. This fact allowed Ma et al. (2017) to study Ce speciation in the xylem and phloem of Ce-ENMs–treated cucumber seedlings under hydroponic conditions, and a clear biotransformation of Ce-ENMs was observed. In this study around 15% of Ce(IV) was reduced to Ce(III) in the roots and around 20% was reduced to Ce(III) in the shoots.

The Ag-ENMs uptake in the crop species *Lactuca Sativa* after foliar exposure and their possible biotransformation and phytotoxic effects were studied by Larue, Castillo-Michel, Sobanska, Cécillon, et al. (2014). It was found that Ag-ENMs were oxidized and Ag^+ ions formed Ag complexes with thiols. Although no sign of phytotoxicity was observed, Ag was effectively trapped on lettuce leaves and a thorough washing was not effective to remove the Ag content. It has also been observed that in rice roots treated with Ag-ENMs most of the Ag speciation was in the form of both Ag-thiol complexes and Ag_2S (Larue et al., 2016; del Real et al., 2017). As reported by Ferretti et al. (2009) the thiol residues bind to the Ag released from the functionalized Ag-ENMs as a defense mechanism to inhibit the cytotoxicity to other plants. Wang et al. (2016) also studied the stability of captured Ag-ENMs and found that more than 79% of the total Ag-ENMs was transformed to Ag_2S, which indicates a pronounced reduction of the Ag-uptake by the plants. Similarly, when Ag-ENMs were treated anaerobically more than 90% of the Ag-ENMs were converted to Ag_2S (Li et al., 2017). Although many studies have shown efficient biotransformation in plants, these studies have been carried out at high-dose conditions (around 4000 mg/L) and mostly using hydroponics to avoid possible confounding environmental factors.

Once ENMs enter the plant cells, they interact with organelles and intracellular components such as chloroplasts and mitochondria, and depending on the interaction potential, chemical or physical alterations can result. Independent of the nature of the ENMs, stress can be induced and generate excess reactive oxygen species (ROS), which affects cell organelles and structures, proteins, carbohydrates, lipids, secondary metabolites, and DNA in plants (Xingmao Ma et al., 2016; Miralles et al., 2012; Singh et al., 2021). It has been reported that high levels of ROS are potentially harmful to plant cells; however, their presence positively impacts stress perception and protection due to secondary metabolite production (Suzuki & Mittler, 2006). To date, the effects of ENM exposure on plants is highly dependent on many factors which need to be taken into consideration, such as the plant age, species, growth conditions and medium, exposure time, ENM concentration, and the type of ENM to which the plant has been exposed to (Boršev et al., 2016; Poddar et al., 2020; Swift et al., 2019).

10.6 Photosynthesis related processes

Most living species depend directly or indirectly on photosynthesis as a source of energy. In this chemical process, light-sensitive chloroplasts from photosynthetic cells convert solar energy into chemical energy (glucose, triose phosphate, and maltose). Photosynthetic reactions include light absorption, light-induced charge separation, electron transfer between photosystems, (PS) II and PS I, water photolysis, oxygen evolution (water oxidation), and CO_2 assimilation. It has been reported that ENMs have important effects in this complex mechanism, and thus photosynthetic efficiency, photochemical fluorescence, and even quantum yields can be affected (Conway et al., 2015; Gomez-Garay et al., 2014; Hu et al., 2014). An increase in the production of antioxidants is reflected as a reduction of ROS levels, which is the result of the synergistic effect of ENMs acting as essential nutrients and enzymatic cofactors (Hou et al., 2018). In Fig. 10.2 a mechanism is depicted in which ENMs applied foliarly mitigate abiotic stress on crops. Here, the ENMs increase both ROS and antioxidant capacity affecting $^1O_2$, O , O_2 - and also H_2O_2 and OH radicals in larger quantities. It is believed that a Fenton reaction takes place first followed by a Haber-Weiss reaction, all of which is mediated by the transition metal in the ENM. $^1O_2$ and O radicals are dissipated by the action of α-tocopherol and β-carotene, SOD enzymes catalyze the transformation of O_2—radical to H_2O_2, whereas CAT enzymes produce H_2O and O_2 from H_2O_2. H_2O_2 is also transformed into H_2O by both enzymes GPX and APX involving NADPH as the source of energy (Pérez-Labrada et al., 2020).

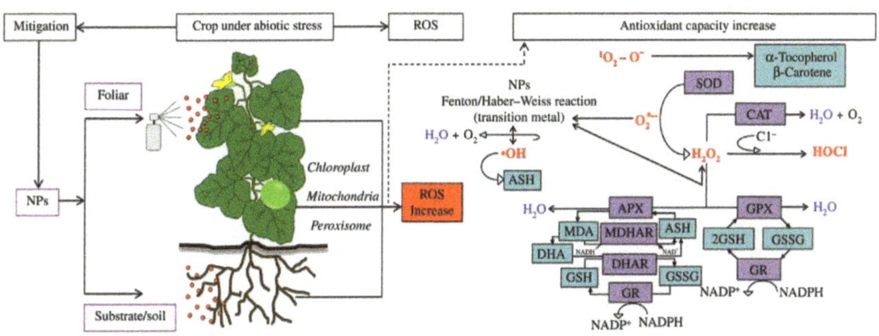

FIGURE 10.2 Increase in antioxidant capacity induced by NPs mitigates abiotic stress in plants. *Data from Pérez-Labrada, F., Hernández-Hernández, H., López-Pérez, M. C., González-Morales, S., Benavides-Mendoza, A. & Juárez-Maldonado, A. (2020).* Nanoparticles in plants: morphophysiological, biochemical, and molecular responses *(pp. 289–322). Elsevier BV. https://doi.org/10.1016/b978-0-12-818204-8.00016-3. Image reprinted from plant life under changing environment/nanoparticles in plants: morphophysiological, biochemical, and molecular responses, 289–322, Copyright (2020), with permission from Elsevier.*

Since chlorophyll *a* is more sensitive to photodegradation than other pigments, it can be used as a reliable biomarker of ENMs toxicity, which was observed by Rico et al. (2013). When rice (*Oryza sativa*) was treated with CeO_2 ENMs, no apparent symptoms of toxicity were observed, but the chlorophyll content was drastically affected. This effect was also observed in other species of plants and with other types of ENMs. For example, when plants were treated with Fe_3O_4 and $CoFe_2O_3$ ENMs, they did not induce toxic effects, but the chlorophyll content decreased compared to the untreated control (Ursache-Oprisan et al., 2011).

On the other hand, many studies have shown a positive impact of ENMs on both photosynthesis and biosynthesis mechanisms. Qi et al. (2013) reported that TiO_2-ENMs improved the leaf photosynthetic rates on tomato plants under mild heat stress. Additionally, this study also reported that TiO_2-ENMs significantly decreased the minimum chlorophyll fluorescence and relative electron transport in leaves. It was also found that spinach growth was improved when treated with TiO_2-ENMs. In this report ENMs played a crucial role in intensifying the absorption of inorganic nutrients, enhancing the degradation of organic substances, and inducing quenching by oxygen free radicals produced during photosynthesis (Yang et al., 2007). The improvement of spinach growth was attributed to the generation of N_2 fixation by TiO_2-ENMs.

Another study measured the amount of photosynthetic pigments in ZnO-ENM-treated plants (chlorophyll *a*, chlorophyll *b*, and total chlorophyll content) by recording the fluorescence of chlorophyll *a* for 30 days after cultivation and concluded they were significantly enhanced. They also reported that the number of active reaction centers per chlorophyll molecule were increased on ENM-treated plants. Additionally, an increase in the absorption, efficiency of excitation energy trapping, and electron transport from active reaction centers were observed (Rai-Kalal & Jajoo, 2021). ZnO-ENMs have been demonstrated to have a positive impact to regulate plant tolerance to multiple environmental stresses as well (Faizan, Bhat, Chen, et al., 2021; Faizan, Bhat, Hessini, et al., 2021; Pejam et al., 2021). Foliar application of ZnO-ENMs at different concentrations (10, 50, and 100 mg/L) in the presence and absence of NaCl (150 mM) indicated that foliar spray of ZnO-ENMs significantly increased shoot length, biomass, leaf area, chlorophyll content, and photosynthetic activities of tomato plants in the presence/absence of salt stress (Faizan, Bhat, Chen, et al., 2021; Faizan, Bhat, Hessini, et al., 2021).

Another important aspect to consider for minimizing the nanotoxicity is the coating material used to stabilize the ENMs. For example, Au-ENMs coated with small organic molecules like citrate or inorganic compounds like carbonate have shown no harmful impacts on the photosynthetic process of *Chlamydomonas Reinhardtii* (Behra et al., 2015). In contrast, when a similar investigation was carried out species using

Au-ENMs coated with mannose, a drastic decrease in the photosynthetic activity was observed (Perreault et al., 2012). This effect has been associated with the ability of Au-ENMs to lock the electron cascade on the photosynthetic I path (Ghosh & Chattopadhyay, 2015). Currently, in vitro studies using Au-ENMs are exhibiting positive impacts by enhancing PS II thus impacting overall photosynthesis (Torres et al., 2018). To date, despite the tremendous efforts to understand detailed mechanisms of ENMs' biotransformations in plants, they are still not well understood, and many questions remain open.

References

Alidoust, D., & Isoda, A. (2013). Effect of γFe_2O_3 nanoparticles on photosynthetic characteristic of soybean (*Glycine max* (L.) Merr.): Foliar spray versus soil amendment. *Acta Physiologiae Plantarum, 35*(12), 3365–3375. https://doi.org/10.1007/s11738-013-1369-8

Avellan, A., Yun, J., Morais, B. P., Clement, E. T., Rodrigues, S. M., & Lowry, G. V. (2021). Critical review: Role of inorganic nanoparticle properties on their foliar uptake and in planta translocation. *Environmental Science and Technology.* https://doi.org/10.1021/acs.est.1c00178

Avellan, A., Yun, J., Zhang, Y., Spielman-Sun, E., Unrine, J. M., Thieme, J., Li, J., Lombi, E., Bland, G., & Lowry, G. V. (2019). Nanoparticle size and coating chemistry control foliar uptake pathways, translocation, and leaf-to-rhizosphere transport in wheat. *ACS Nano, 13*(5), 5291–5305. https://doi.org/10.1021/acsnano.8b09781

Behra, R., Wagner, B., Sgier, L., & Kistler, D. (2015). Colloidal stability and toxicity of gold nanoparticles and gold chloride on *Chlamydomonas reinhardtii. Aquatic Geochemistry, 21*(2–4), 331–342. https://doi.org/10.1007/s10498-015-9255-1

Borišev, M., Borišev, I., Župunski, M., Arsenov, D., Pajević, S., Ćurčić, Ž., Vasin, J., Djordjevic, A., & Ng, C. (2016). Drought impact is alleviated in sugar beets (*Beta vulgaris* L.) by foliar application of fullerenol nanoparticles. *PLoS One, 11*(11), e0166248. https://doi.org/10.1371/journal.pone.0166248

Calabrese, E. J. (2005). Paradigm lost, paradigm found: The re-emergence of hormesis as a fundamental dose response model in the toxicological sciences. *Environmental Pollution, 138*(3), 378–411. https://doi.org/10.1016/j.envpol.2004.10.001

Conway, J. R., Beaulieu, A. L., Beaulieu, N. L., Mazer, S. J., & Keller, A. A. (2015). Environmental stresses increase photosynthetic disruption by metal oxide nanomaterials in a soil-grown plant. *ACS Nano, 9*(12), 11737–11749. https://doi.org/10.1021/acsnano.5b03091

Elsheery, N. I., Sunoj, V. S. J., Wen, Y., Zhu, J. J., Muralidharan, G., & Cao, K. F. (2020). Foliar application of nanoparticles mitigates the chilling effect on photosynthesis and photoprotection in sugarcane. *Plant Physiology and Biochemistry, 149*, 50–60. https://doi.org/10.1016/j.plaphy.2020.01.035

Faizan, M., Bhat, J. A., Chen, C., Alyemeni, M. N., Wijaya, L., Ahmad, P., & Yu, F. (2021). Zinc oxide nanoparticles (ZnO-NPs) induce salt tolerance by improving the antioxidant system and photosynthetic machinery in tomato. *Plant Physiology and Biochemistry, 161*, 122–130. https://doi.org/10.1016/j.plaphy.2021.02.002

Faizan, M., Bhat, J. A., Hessini, K., Yu, F., & Ahmad, P. (2021). Zinc oxide nanoparticles alleviates the adverse effects of cadmium stress on Oryza sativa via modulation of the photosynthesis and antioxidant defense system. *Ecotoxicology and Environmental Safety, 220*, 112401. https://doi.org/10.1016/j.ecoenv.2021.112401

Ferretti, M., Destro, T., Tosatto, Rascio, a R. N., & Masi, A. (2009). Glutamyl transferase in the cell wall participates in extracellular glutathione salvage from the root apoplast. *New Phytologist, 181*, 115–1126.

Franke, W. (1967). Mechanisms of foliar penetration of solutions. *Annual Review of Plant Physiology*, 281–300. https://doi.org/10.1146/annurev, 18.060167.001433.

Gardea-Torresdey, J. L., Rico, C. M., & White, J. C. (2014). Trophic transfer, transformation, and impact of engineered nanomaterials in terrestrial environments. *Environmental Science and Technology, 48*(5), 2526–2540. https://doi.org/10.1021/es4050665

Ghosh, D., & Chattopadhyay, N. (2015). Gold and silver nanoparticles based superquenching of fluorescence: A review. *Journal of Luminescence, 160*, 223–232. https://doi.org/10.1016/j.jlumin.2014.12.018

Gomez-Garay, A., Pintos, B., Manzanera, J. A., Lobo, C., Villalobos, N., & Martín, L. (2014). Uptake of CeO_2 nanoparticles and its effect on growth of *Medicago arborea* in vitro plantlets. *Biological Trace Element Research, 161*(1), 143–150. https://doi.org/10.1007/s12011-014-0089-2

Grillo, R., Mattos, B. D., Antunes, D. R., Forini, M. M. L., Monikh, F. A., & Rojas, O. J. (2021). Foliage adhesion and interactions with particulate delivery systems for plant nanobionics and intelligent agriculture. *Nano Today, 37*. https://doi.org/10.1016/j.nantod.2021.101078

Hall, D. W., & Byrd, J. (2012). *Forensic botany: A practical guide.*

Hong, J., Wang, L., Sun, Y., Zhao, L., Niu, G., Tan, W., Rico, C. M., Peralta-Videa, J. R., & Gardea-Torresdey, J. L. (2016). Foliar applied nanoscale and microscale CeO_2 and CuO alter cucumber (*Cucumis sativus*) fruit quality. *Science of the Total Environment, 563–564*, 904–911. https://doi.org/10.1016/j.scitotenv.2015.08.029

Hong, J., Wang, C., Wagner, D. C., Gardea-Torresdey, J. L., He, F., & Rico, C. M. (2021). Foliar application of nanoparticles: Mechanisms of absorption, transfer, and multiple impacts. *Environmental Science: Nano, 8*(5), 1196–1210. https://doi.org/10.1039/d0en01129k

Hou, J., Wu, Y., Li, X., Wei, B., Li, S., & Wang, X. (2018). Toxic effects of different types of zinc oxide nanoparticles on algae, plants, invertebrates, vertebrates and microorganisms. *Chemosphere, 193*, 852–860. https://doi.org/10.1016/j.chemosphere.2017.11.077

Hu, P., An, J., Faulkner, M. M., Wu, H., Li, Z., Tian, X., & Giraldo, J. P. (2020). Nanoparticle charge and size control foliar delivery efficiency to plant cells and organelles. *ACS Nano, 14*(7), 7970–7986. https://doi.org/10.1021/acsnano.9b09178

Hu, X., Lu, K., Mu, L., Kang, J., & Zhou, Q. (2014). Interactions between graphene oxide and plant cells: Regulation of cell morphology, uptake, organelle damage, oxidative effects and metabolic disorders. *Carbon, 80*(1), 665–676. https://doi.org/10.1016/j.carbon.2014.09.010

Jackson, P., Jacobsen, N. R., Baun, A., Birkedal, R., Kühnel, D., Jensen, K. A., Vogel, U., & Wallin, H. (2013). Bioaccumulation and ecotoxicity of carbon nanotubes. *Chemistry Central Journal, 7*(1). https://doi.org/10.1186/1752-153X-7-154

Karny, A., Zinger, A., Kajal, A., Shainsky-Roitman, J., & Schroeder, A. (2018). Therapeutic nanoparticles penetrate leaves and deliver nutrients to agricultural crops. *Scientific Reports, 8*(1). https://doi.org/10.1038/s41598-018-25197-y

Larue, C., Castillo-Michel, H., Sobanska, S., Cécillon, L., Bureau, S., Barthès, V., Ouerdane, L., Carrière, M., & Sarret, G. (2014). Foliar exposure of the crop Lactuca sativa to silver nanoparticles: Evidence for internalization and changes in Ag speciation. *Journal of Hazardous Materials, 264*, 98–106. https://doi.org/10.1016/j.jhazmat.2013.10.053

Larue, C., Castillo-Michel, H., Sobanska, S., Trcera, N., Sorieul, S., Cécillon, L., Ouerdane, L., Legros, S., & Sarret, G. (2014). Fate of pristine TiO_2 nanoparticles and aged paint-containing TiO_2 nanoparticles in lettuce crop after foliar exposure. *Journal of Hazardous Materials, 273*, 17–26. https://doi.org/10.1016/j.jhazmat.2014.03.014

Larue, C., Castillo-Michel, H., Stein, R. J., Fayard, B., Pouyet, E., Villanova, J., Magnin, V., del Real, A. E. P., Trcera, N., Legros, S., Sorieul, S., & Sarret, G. (2016). Innovative combination of spectroscopic techniques to reveal nanoparticle fate in a crop plant. *Spectrochimica Acta – Part B Atomic Spectroscopy, 119*, 17–24. https://doi.org/10.1016/j.sab.2016.03.005

Li, M., Wang, P., Dang, F., & Zhou, D. M. (2017). The transformation and fate of silver nanoparticles in paddy soil: Effects of soil organic matter and redox conditions. *Environmental Science: Nano, 4*(4), 919–928. https://doi.org/10.1039/c6en00682e

Lowry, G. V., Avellan, A., & Gilbertson, L. M. (2019). Opportunities and challenges for nanotechnology in the agri-tech revolution. *Nature Nanotechnology, 14*(6), 517–522. https://doi.org/10.1038/s41565-019-0461-7

Lu, K., Shen, D., Liu, X., Dong, S., Jing, X., Wu, W., Tong, Y., Gao, S., & Mao, L. (2020). Uptake of iron oxide nanoparticles inhibits the photosynthesis of the wheat after foliar exposure. *Chemosphere, 259*, 127445. https://doi.org/10.1016/j.chemosphere.2020.127445

Ma, X., Geiser-Lee, J., Deng, Y., & Kolmakov, A. (2010). Interactions between engineered nanoparticles (ENPs) and plants: Phytotoxicity, uptake and accumulation. *Science of the Total Environment, 408*(16), 3053–3061. https://doi.org/10.1016/j.scitotenv.2010.03.031

Ma, Y., He, X., Zhang, P., Zhang, Z., Ding, Y., Zhang, J., Wang, G., Xie, C., Luo, W., Zhang, J., Zhang, L., Chai, Z., & Yang, K. (2017). IXylem and phloem based transport of CeO2 nanoparticles in hydroponic cucumber. *Environmental Science & Technology, 51*(9), 5215–5221. https://doi.org/https://doi.org/10.1021/acs.est.6b05998

Mastronardi, E., Tsae, P., Zhang, X., Monreal, C., & DeRosa, M. C. (2015). Strategic role of nanotechnology in fertilizers: Potential and limitations. In *Nanotechnologies in food and agriculture* (pp. 25–67). Springer International Publishing. https://doi.org/10.1007/978-3-319-14024-7_2

Ma, X., Wang, Q., Rossi, L., & Zhang, W. (2016). Cerium oxide nanoparticles and bulk cerium oxide leading to different physiological and biochemical responses in Brassica rapa. *Environmental Science & Technology, 50*(13), 6793–6802. https://doi.org/10.1021/acs.est.5b04111

Mehta, C. M., Srivastava, R., Arora, S., & Sharma, A. K. (2016). Impact assessment of silver nanoparticles on plant growth and soil bacterial diversity. *Biotech, 3*, 380. https://doi.org/10.1007/s13205-016-0567-7

Mejías, J. H., Salazar, F. J., Pérez, L., Hube, S., Rodriguez, M., & Alfaro, M. A. (n.d.). *Nanofertilizers: A cutting-edge approach to increase nitrogen use efficiency in grasslands.*

Millstone, J. E., Wei, W., Jones, M. R., Yoo, H., & Mirkin, C. A. (2008). Iodide ions control seed-mediated growth of anisotropic gold Nanoparticles. *Nano Letters, 8*(8), 2526–2529. https://doi.org/10.1021/nl8016253

Miralles, P., Church, T. L., & Harris, A. T. (2012). Toxicity, uptake, and translocation of engineered nanomaterials in vascular plants. *Environmental Science and Technology, 46*(17), 9224–9239. https://doi.org/10.1021/es202995d

Mittal, D., Kaur, G., Singh, P., Yadav, K., & Ali, S. A. (2020). Nanoparticle-based sustainable agriculture and food science: Recent advances and future outlook. *Frontiers in Nanotechnology,* (10).

Nhan, L. V., Ma, C., Rui, Y., Liu, S., Li, X., Xing, B., & Liu, L. (2015). Phytotoxic mechanism of nanoparticles: Destruction of chloroplasts and vascular bundles and alteration of nutrient absorption. *Scientific Reports,* (1).

Pejam, F., Ardebili, Z. O., Ladan-Moghadam, A., & Danaee, E. (2021). Zinc oxide nanoparticles mediated substantial physiological and molecular changes in tomato. *PLoS One,* (3).

Pérez-Labrada, F., Hernández-Hernández, H., López-Pérez, M. C., González-Morales, S., Benavides-Mendoza, A., & Juárez-Maldonado, A. (2020). *Nanoparticles in plants: morpho-physiological, biochemical, and molecular responses* (pp. 289–322). Elsevier BV. https://doi.org/10.1016/b978-0-12-818204-8.00016-3

Perreault, F., Bogdan, N., Morin, M., Claverie, J., & Popovic, R. (2012). Interaction of gold nanoglycodendrimers with algal cells (Chlamydomonas reinhardtii) and their effect on physiological processes. *Nanotoxicology, 6*(2), 109–120. https://doi.org/10.3109/17435390.2011.562325

Poddar, K., Sarkar, D., & Sarkar, A. (2020). Nanoparticles on photosynthesis of plants: Effects and role. In *Nanotechnology in the life sciences* (pp. 273–287). Springer Science and Business Media B.V. https://doi.org/10.1007/978-3-030-39246-8_13

Qi, M., Liu, Y., & Li, T. (2013). Nano-TiO_2 improve the photosynthesis of tomato leaves under mild heat stress. *Biological Trace Element Research, 156*(1–3), 323–328. https://doi.org/10.1007/s12011-013-9833-2

Rahmani, N., Radjabian, T., & Soltani, B. M. (2020). Impacts of foliar exposure to multi-walled carbon nanotubes on physiological and molecular traits of Salvia verticillata L., as a medicinal plant. *Plant Physiology and Biochemistry, 150*, 27–38. https://doi.org/10.1016/j.plaphy.2020.02.022

Rai-Kalal, P., & Jajoo, A. (2021). Priming with zinc oxide nanoparticles improve germination and photosynthetic performance in wheat. *Plant Physiology and Biochemistry, 160*, 341–351. https://doi.org/10.1016/j.plaphy.2021.01.032

Rai, M., Ribeiro, C., Mattoso, L., & Duran, N. (2015). Nanotechnologies in food and agriculture. In *Nanotechnologies in food and agriculture* (pp. 1–347). Springer International Publishing. https://doi.org/10.1007/978-3-319-14024-7

Raliya, R., Franke, C., Chavalmane, S., Nair, R., Reed, N., & Biswas, P. (2016). Quantitative understanding of nanoparticle uptake in watermelon plants. *Frontiers in Plant Science*, 1288.

Raliya, R., & Tarafdar, J. C. (2013). ZnO nanoparticle biosynthesis and its effect on phosphorous-mobilizing enzyme secretion and gum contents in clusterbean (*Cyamopsis tetragonoloba* L.). *Agricultural Research, 2*(1), 48–57. https://doi.org/10.1007/s40003-012-0049-z

del Real, P., Vidal, A. E., Carrière, V., Castillo-Michel, M., Levard, H., Chaurand, C., & Sarret, P. (2017). Silver nanoparticles and wheat roots: A complex interplay. *Environmental Science & Technology*, (10), 5774–5782.

Rico, C. M., Hong, J., Morales, M. I., Zhao, L., Barrios, A. C., Zhang, J. Y., Peralta-Videa, J. R., & Gardea-Torresdey, J. L. (2013). Effect of cerium oxide nanoparticles on rice: A study involving the antioxidant defense system and in vivo fluorescence imaging. *Environmental Science and Technology, 47*(11), 5635–5642. https://doi.org/10.1021/es401032m

Rossi, L., Fedenia, L. N., Sharifan, H., Ma, X., & Lombardini, L. (2019). Effects of foliar application of zinc sulfate and zinc nanoparticles in coffee (*Coffea arabica* L.) plants. *Plant Physiology and Biochemistry, 135*, 160–166. https://doi.org/10.1016/j.plaphy.2018.12.005

Schönherr, J. (2006). Characterization of aqueous pores in plant cuticles and permeation of ionic solutes. *Journal of Experimental Botany, 57*(11), 2471–2491. https://doi.org/10.1093/jxb/erj217

Shafiq, F., Iqbal, M., Ashraf, M. A., & Ali, M. (2020). Foliar applied fullerol differentially improves salt tolerance in wheat through ion compartmentalization, osmotic adjustments and regulation of enzymatic antioxidants. *Physiology and Molecular Biology of Plants, 26*(3), 475–487. https://doi.org/10.1007/s12298-020-00761-x

Shang, Y., Hasan, M. K., Ahammed, G. J., Li, M., Yin, H., & Zhou, J. (2019). Applications of nanotechnology in plant growth and crop protection: A review. *Molecules*, (14).

Singh, A., Tiwari, S., Pandey, J., Lata, C., & Singh, I. K. (2021). Role of nanoparticles in crop improvement and abiotic stress management. *Journal of Biotechnology, 337*, 57–70. https://doi.org/10.1016/j.jbiotec.2021.06.022

Smith, A. M., & Gilbertson, L. M. (2018). Rational ligand design to improve agrochemical delivery efficiency and advance agriculture sustainability. *ACS Sustainable Chemistry and Engineering, 6*(11), 13599–13610. https://doi.org/10.1021/acssuschemeng.8b03457

Smith, A. M., Johnston, K. A., Crawford, S. E., Marbella, L. E., & Millstone, J. E. (2017). Ligand density quantification on colloidal inorganic nanoparticles. *Analyst, 142*(1), 11–29. https://doi.org/10.1039/c6an02206e

Suzuki, N., & Mittler, R. (2006). Reactive oxygen species and temperature stresses: A delicate balance between signaling and destruction. *Physiologia Plantarum, 126*(1), 45–51. https://doi.org/10.1111/j.0031-9317.2005.00582.x

Swift, T. A., Oliver, T. A. A., Carmen Galan, M., & Whitney, H. M. (2019). Functional nanomaterials to augment photosynthesis: Evidence and considerations for their responsible use in agricultural applications. *Interface Focus, 9*(1). https://doi.org/10.1098/rsfs.2018.0048

Tilman, D., Balzer, C., Hill, J., & Befort, B. L. (2011). Global food demand and the sustainable intensification of agriculture. *Proceedings of the National Academy of Sciences of the United States of America, 108*(50), 20260–20264. https://doi.org/10.1073/pnas.1116437108

Torres, R., Diz, V. E., & Lagorio, M. G. (2018). Effects of gold nanoparticles on the photophysical and photosynthetic parameters of leaves and chloroplasts. *Photochemical and Photobiological Sciences, 17*(4), 505–516. https://doi.org/10.1039/c8pp00067k

Ursache-Oprisan, M., Focanici, E., Creanga, D., & Caltun, O. (2011). Sunflower chlorophyll levels after magnetic nanoparticle supply. *African Journal of Biotechnology, 10*(36), 7092–7098. http://www.academicjournals.org/AJB/PDF/pdf2011/18Jul/Ursache-Oprisan%20et%20al.pdf.

Wang, T., Bai, J., Jiang, X., & Nienhaus, G. U. (2012). Cellular uptake of nanoparticles by membrane penetration: A study combining confocal microscopy with FTIR spectroelectrochemistry. *ACS Nano, 6*(2), 1251–1259. https://doi.org/10.1021/nn203892h

Wang, J., Koo, Y., Alexander, A., Yang, Y., Westerhof, S., Zhang, Q., Schnoor, J. L., Colvin, V. L., Braam, J., & Alvarez, P. J. J. (2013). Phytostimulation of poplars and Arabidopsis exposed to silver nanoparticles and Ag+ at sublethal concentrations. *Environmental Science and Technology, 47*(10), 5442–5449. https://doi.org/10.1021/es4004334

Wang, P., Menzies, N. W., Dennis, P. G., Guo, J., Forstner, C., Sekine, R., Lombi, E., Kappen, P., Bertsch, P. M., & Kopittke, P. M. (2016). Silver nanoparticles entering soils via the wastewater-sludge-soil pathway pose low risk to plants but elevated Cl concentrations increase Ag bioavailability. *Environmental Science and Technology, 50*(15), 8274–8281. https://doi.org/10.1021/acs.est.6b01180

Wang, W. N., Tarafdar, J. C., & Biswas, P. (2013). Nanoparticle synthesis and delivery by an aerosol route for watermelon plant foliar uptake. *Journal of Nanoparticle Research, 15*(1). https://doi.org/10.1007/s11051-013-1417-8

Wong, M. H., Misra, R. P., Giraldo, J. P., Kwak, S. Y., Son, Y., Landry, M. P., Swan, J. W., Blankschtein, D., & Strano, M. S. (2016). Lipid exchange envelope penetration (LEEP) of nanoparticles for plant engineering: A universal localization mechanism. *Nano Letters, 16*(2), 1161–1172. https://doi.org/10.1021/acs.nanolett.5b04467

Yang, F., Liu, C., Gao, F., Su, M., Wu, X., Zheng, L., Hong, F., & Yang, P. (2007). The improvement of spinach growth by nano-anatase TiO2 treatment is related to nitrogen photoreduction. *Biological Trace Element Research, 119*(1), 77–88. https://doi.org/10.1007/s12011-007-0046-4

Yeats, T. H., & Rose, J. K. C. (2013). The formation and function of plant cuticles. *Plant Physiology, 163*(1), 5–20. https://doi.org/10.1104/pp.113.222737

Yu, M., Sun, C., Xue, Y., Liu, C., Qiu, D., Cui, B., Zhang, Y., Cui, H., & Zeng, Z. (2019). Tannic acid-based nanopesticides coating with highly improved foliage adhesion to enhance foliar retention. *RSC Advances, 9*(46), 27096–27104. https://doi.org/10.1039/c9ra05843e

Yu, M., Yao, J., Liang, J., Zeng, Z., Cui, B., Zhao, X., Sun, C., Wang, Y., Liu, G., & Cui, H. (2017). Development of functionalized abamectin poly(lactic acid) nanoparticles with regulatable adhesion to enhance foliar retention. *RSC Advances, 7*(19), 11271–11280. https://doi.org/10.1039/c6ra27345a

Zahedi, S. M., Abdelrahman, M., Hosseini, M. S., Hoveizeh, N. F., & Tran, L. S. P. (2019). Alleviation of the effect of salinity on growth and yield of strawberry by foliar spray of selenium-nanoparticles. *Environmental Pollution, 253*, 246–258. https://doi.org/10.1016/j.envpol.2019.04.078

Zhao, L., Huang, Y., Adeleye, A. S., & Keller, A. A. (2017). Metabolomics reveals Cu $(OH)_2$ nanopesticide-activated anti-oxidative pathways and decreased beneficial antioxidants in spinach leaves. *Environmental Science & Technology,* (17), 10184–10194.

CHAPTER

11

Potential toxicity and bioavailability of ENMs and their products in plant tissues

Rene Loredo-Portales[1], *Luis J. Castillo-Pérez*[2], *Angel Josabad Alonso-Castro*[3], *and Candy Carranza-Álvarez*[4]

[1] Departamento de Ciencias Ambientales, División de Ciencias de la Vida, Universidad de Guanajuato, Irapuato, Guanajuato, México; [2] Programa Multidisciplinario de Posgrado en Ciencias Ambientales, Universidad Autónoma de San Luis Potosí, San Luis Potosí, San Luis Potosí, México; [3] Departamento de Farmacia, División de Ciencias Naturales y Exactas, Universidad de Guanajuato, Guanajuato, Guanajuato, México; [4] Facultad de Estudios Profesionales Zona Huasteca, Universidad Autónoma de San Luis Potosí, San Luis Potosí, México

11.1 Introduction

Engineered nanomaterials (ENMs) are manufactured materials with nanometric dimensions (approximately 1–100 nm, 1 nm = 10^{-9} m) that have a plethora of applications. The various applications of ENMs have resulted in a global increase in their production, and consequently, in their release in the environment. These materials have special physicochemical properties that can cause negative environmental impacts. The environmental effects that these ENMs cause are related to their final deposits in the biosphere (soil, water, plant, and organisms), in consequence affecting the health of ecosystems (Jeevanandam et al., 2018).

Several studies have demonstrated that ENMs, particularly those containing heavy metals, have effects at the cellular, subcellular, and protein levels in different plant species (Lanone & Boczkowski, 2006;

Physicochemical Interactions of Engineered Nanoparticles and Plants
https://doi.org/10.1016/B978-0-323-90558-9.00005-X

Winkler, 2016; Yu et al., 2020). Furthermore, some ENMs have the potential to be transported in a whole organism, penetrating cell membranes. They can even be stored in mitochondria producing lethal effects (Unfried et al., 2007; Baker et al., 2014). Recent reports indicate that ENMs can cause oxidative stress due to the high number of reactive oxygen species (ROS) they produce, which damage cellular biomolecules such as nucleic acids, lipids, and proteins (Mendoza & Brown, 2019). This is generated by changes in the oxidation state of the elements in ENMs that enter into the plant cell. However, the toxicity of heavy metals coming from ENMs is not limited only to the production of these oxidizing compounds, since it has also been reported that they generate interactions with enzymes, chlorophyll biosynthesis, and the electron transport system (Rai et al., 2016).

Since plants are a potential route of exposure to ENMs for higher organisms including humans, the knowledge of the interactions between plants and ENMs can provide relevant information on the effects that nanotechnology may have on ecosystems and the environment in general. However, there are gaps in the knowledge about the interactions occurring between plants and ENMs. These include specific data on plant toxicity and organisms exposure, transformation, degradation, dissolution, bioaccumulation, and bioavailability, among others (Santiago-Martín et al., 2016; Kranjc & Drobne, 2019).

11.2 ENMs toxicity in plants

ENMs may constitute a prominent source of soil contamination. For instance, around 200 tons per year of Cu-ENMs are deposited in soils (Keller & Lazareva, 2014) through agrochemicals, nanofertilizers, nanopesticides, seed treatment preparations, agrofilms, hydroponic solutions, and in some soil and water remediation treatments (Santiago-Martín et al., 2016). Once ENMs reach soil, they can be taken up and bioaccumulated by plants at different rates depending on several factors including plant species and soil properties. Additionally, the coexposure of ENMs with other pollutants may affect their uptake, translocation, and phytotoxicity (Kranjc & Drobne, 2019). ENMs can alter soil bacterial communities, which also affect soil-plant-microbe interactions (Dimkpa et al., 2012). In addition, nanophytotoxicity is influenced by the localization of the nanomaterial in plant tissues (Montes et al., 2017), and the type of plant species (Moll et al., 2016). In plant tissues, the principal mechanisms by which ENMs can cause phytotoxicity are: (1) induction of ROS production, (2) alterations in DNA structure, and (3) reduction of photosynthesis.

Oxidative stress starts the overproduction of ROS: singlet oxygen, superoxide, glutathione peroxidase, hydrogen peroxide, and hydroxyl radicals (Nair et al., 2010). Oxidative stress induces membrane electrolyte leakage and the disruption of the cell membrane that leads to lipid peroxidation due to the increase in malonaldehyde. This mainly causes inhibition of seed germination and reduction of plant growth (Nair et al., 2010). The plant's response to oxidative stress is the activation of enzymatic (i.e., superoxide dismutase and catalase) and nonenzymatic systems (salicylic acid, indole-3-acetic acid, abscisic acid, carotenoids, and ascorbate) (Song et al., 2013). In the case of chronic exposure to ENMs, plant cell damage and subsequent death will occur if the antioxidant response cannot neutralize the overproduction of reactive oxygen species.

DNA damage is another mechanism by which ENMs induce phytotoxicity (Speranza et al., 2013). Since ENMs are in the nanometric scale, this material may even cross the nuclear pores and establish interactions with DNA, potentially causing mitotic aberrations, mutations, genomic instability, and micronuclei formation. All these effects may also result in plant growth inhibition (Rico et al., 2011; Speranza et al., 2013).

Finally, the reduction in photosynthesis rate is another mechanism by which ENMs exert phytotoxicity (Miralles et al., 2012). This mechanism involves the interference with the electron transport chain of chloroplast, the inhibition of protochlorophyllide reductase, stomatal closure, and decrease in chlorophyll content (Miralles et al., 2012). In addition to the chemical direct interaction, at physiological level ENMs induce phytotoxicity by decreasing water conductivity and nutrient uptake, which induce a decrease in shoot biomass (Wang et al., 2011). ENMs can also block pores in the apoplastic stream, which decreases biomass accumulation (Van Aken, 2015).

The phytotoxicity of ENMs is also associated to the concentration, particle shape, zeta potential, dimensions, particle dissolution, chemical structure, bioavailability of heavy metals ions, exposure time, and reactivity (Musante & White, 2012; Qiu & Smolders, 2017). These parameters can be used to predict ENMs-induced phytotoxicity (Musante & White, 2012). However, since ENMs can change size, shape, and surface chemistry over time, their toxic effects can constantly vary. These changes in particle size and speciation may significantly affect the bioavailability of ENMs and their trophic transfer, which are also dependent on the type, exposure conditions, and producer/consumer species (Servin et al., 2017). Regarding ENMs size, small ENMs (5–20 nm) can easily be transported through plasmodesmata into the cell wall of the plant and enter in other plant tissues (Lucas & Lee, 2004). In some cases, ENMs are more toxic than free heavy metal ions (Navarro et al., 2008). This suggests that ENMs have intrinsic toxic properties.

On the other hand, it has been proven that phytotoxicity induced by ENMs can be minimized by the addition of humic acids into the soil solution; this causes ion complexation with the subsequent increase in biomass production and transpiration rate (Musante & White, 2012). Reports indicate that nitric oxide, in the form of sodium nitroprusside (SNP), can decrease the AgNPs-induced phytotoxicity by protecting the integrity of the plasma membrane and increasing the thickness of the cell wall, thus reducing the accumulation of ENMs in the root and shoots (Tripathi et al., 2017). As general considerations on the actual evidence, most of the plants used to assess phytotoxicity are crop species, and the studies are mainly focused on seed germination and root elongation. Additionally, most of the studies are implemented with high dose and single solute exposures over short periods of time (Servin et al., 2017). Consequently, standardized guidelines to evaluate ENMs toxicity in plants should be implemented.

Many molecular mechanisms by which ENMs induce toxicity remain to be elucidated. An important issue that requires further investigation is the effect of ENMs on the in-situ production of secondary metabolites, which are important in the adaptation against biotic and abiotic stresses (Yu et al., 2020). For example, an increase in the antimalarial agent artemisinin was observed in *Artemisia annua* hairy roots after 3 days of exposure to AgNPs. Similarly, in *Trigonella foenum-graecum* seedlings, 5 days of exposure to AgNPs resulted in the increased production of diosgenin, a precursor in the synthesis of progesterone (Jasim et al., 2017). On the contrary, the concentration of stearic acid in *Helianthus annuus* decreased after 30 days of exposure to AgNPs (Batool et al., 2021).

11.3 ENMs biotransformation in plant tissues

As a novel contaminant, the environmental significance and biological effects of ENMs in plants have attracted significant attention (Reddy et al., 2016). Furthermore, ENMs may remain suspended in the air for several weeks (Duncan, 2015), to finally sediment in soils and plant surfaces. Additionally, ENMs are intentionally deposited in soils as mentioned before. Due to this, plants are one of the primary receptors of ENMs, which can acquire and integrate into the plant system directly by individual cells and cell nuclei. This represents the main route of exposure through the food chain (Jeevanandam et al., 2018; Yu et al., 2020) (Fig. 11.1).

Several environmental factors including the pH, ionic strength, presence and type of dissolved organic matter, and biotic interactions can influence the properties of ENMs and then their transformation

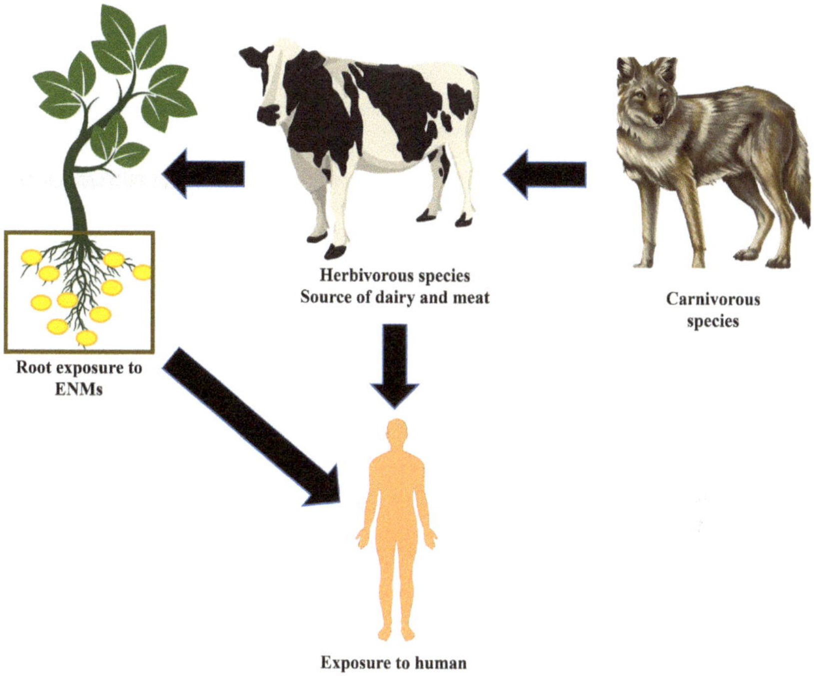

FIGURE 11.1 ENMs biotransformation in plants and transfer of ENMs within the terrestrial food chain.

(Zhu et al., 2013; Liu et al., 2020). However, the biotransformation that ENMs experiment in plant tissues are due to their physicochemical properties, specially to their high dynamism and reactivity. Furthermore, due to these characteristics, they may bioaccumulate and affect plant growth and development (Liu et al., 2020). Biotransformation is the result of a wide variety of processes, including aggregation, agglomeration, redox reactions, dissolution, exchange of surface residues, and reactions with biomacromolecules. These dynamic transformations in turn affect the transport, fate, and toxicity of nanoparticles in the environment. Thus, understanding and characterizing these transformations is critical (Zhu et al., 2013; Liu et al., 2020). For example, metal-based NPs, such as Cu, Zn, and Ag, can release metal ions and biotransform when reacting with organic and inorganic substances. Other examples of effects by ENMs biotransformation are shown in Table 11.1.

Manufactured ENMs enter living systems through both intentional and unintentional releases including solid/liquid waste streams, air emissions from manufacturing facilities, and wastewater treatment plants (Maurer-Jones et al., 2013; Winkler, 2016). Biosolids from wastewater

TABLE 11.1 Effects of ENMs on plants in the past decade.

ENMs	Plant used	Plant section used	Effect caused	References
ZnO	*Oryza sativa*	Roots	There is no reduction in the percent seed germination but have detrimental effects on rice roots at early seedling stage.	Boonyanitipong et al. (2011)
	Prosopis juliflora-velutina	Roots, stems, and leaves	Increased CAT enzyme in roots, stems, and leaves, while APOX enzyme increased only in stems and leaves.	Hernández-Viezcas et al. (2011)
	Allium cepa	Roots	There were clastogenic, genotoxic, and cytotoxic effects	Kumari et al. (2011)
	Allium sativum	Roots	Several types of mitotic aberrations were induced, mainly chromosomal stickiness, bridging, breaks, and lags. Furthermore, the total percentage of abnormal cells increased with increasing ZnO NP concentration.	Shaymurat et al. (2012)
	Fagopyrum esculentum	Roots	Reactive oxygen species were produced.	Lee et al. (2012)
	Pisum sativum	Whole plant	Root elongation was observed but no effects were observed in the stem. Moreover, chlorophyll in leaves decreased.	Mukherjee et al. (2014)

TABLE 11.1 Effects of ENMs on plants in the past decade.—cont'd

ENMs	Plant used	Plant section used	Effect caused	References
	Brassica pekinensis	Seeds	Germination was not affected, but root elongation and the formation of new shoots were inhibited.	Xiang et al. (2015)
TiO_2	*Triticum* spp.	Roots and leaves	There was a greater elongation of the roots, but it did not affect the germination, evapotranspiration, and biomass of the plant.	Larue et al. (2012)
	Foeniculum vulgare	Seeds	Application of 40 ppm nanosized TiO_2 treatment improved mean germination time.	Feizi et al. (2013)
	Petroselinum crispum	Whole plant	There was an increase in germination, root and shoot length, fresh weight, and chlorophyll content of seedlings.	Dehkourdi and Mosavi (2013)
	Solanum lycopersicum	Leaves	Nano-TiO_2 particles play a positive role in promoting photosynthesis in tomato leaves under mild heat stress.	Qi et al. (2013)
	Cucumis sativus	Leaves and fruits	The μ-XRF and μ-XANES techniques showed root-to-fruit translocation of TiO_2 in cucumber without biotransformation.	Servin et al. (2013)

Continued

TABLE 11.1 Effects of ENMs on plants in the past decade.—cont'd

ENMs	Plant used	Plant section used	Effect caused	References
	Triticum aestivum	Whole plant	Germination indices, germination rate, and weighted germination index were affected by TiO_2 nanoparticles treatments.	Mahmoodzadeh and Aghili (2014)
CeO_2	*Oryza sativa*	Grains	In the treated grains, the antioxidant compounds, Fe, S, prolamin, glutelin, lauric and valeric acids, and starch were reduced.	Rico et al. (2013)
	Triticum aestivum	Whole plant	Plant growth, shoot biomass and grain yield were improved. Furthermore, the roots also increased.	Rico et al. (2014)
	Cucumis sativus	Fruits	The stress levels were modified in leaves without apparent signs of toxicity. Moreover, the elements P, K, ca, Mg, S, Fe, Zn, Cu, and Al were increased.	Zhao et al. (2014)
	Hordeum vulgare	Whole plant	The stress levels were modified in leaves without apparent signs of toxicity. Moreover, the elements P, K, ca, Mg, S, Fe, Zn, Cu, and Al were increased.	Rico et al. (2015)

TABLE 11.1 Effects of ENMs on plants in the past decade.—cont'd

ENMs	Plant used	Plant section used	Effect caused	References
Ag	*Triticum aestivum*	Whole plant	The Ag NPs reduced the length of shoots and roots. Furthermore, they increased branching in the roots, thereby affecting plant biomass.	Dimkpa et al. (2013)
	Oryza sativa	Whole plant	Germination level of the seeds and the subsequent growth of the seedlings that germinated decreased with increasing size and concentrations of AgNP.	Thuesombat et al. (2014)
	Pennisetum glaucum	Seeds	The seeds showed better germination, but the seedling growth was affected by exposure to concentrations of AgNPs.	Parveen and Srinath (2015)
CuO	*Oryza sativa*	Seedlings	The germination percentage was reduced, and oxidative stress occurred.	Shaw and Hossain (2013)
	Vigna radiata	Seedlings	Decrease in elongation and root weight was observed. Furthermore, the total chlorophyll content was also reduced.	Nair and Chung (2015)
Si	*Zea mays*	Seeds	An increase in the percentage of germination, dry weight and a greater amount of nutrients was observed.	Suriyaprabha et al. (2012)

Continued

TABLE 11.1 Effects of ENMs on plants in the past decade.—cont'd

ENMs	Plant used	Plant section used	Effect caused	References
	Lycopersicum esculentum	Seeds	An increase in the germination percentage, the mean germination time, the seed germination index, the seed vigor index, the fresh weight, and the dry weight of the seedling was observed.	Siddiqui and Al-Whaibi (2014)
	Zea mays	Seeds	A favorable effect was observed on the beneficial bacterial population and the nutritive value of the soil.	Karunakaran et al. (2013)

treatment facilities, nanopesticides, nanofertilizers, paints, fabrics, personal health care, and accidental spill are also sources of ENMs contamination (Aslani et al., 2014). Due to their unique surface properties, ENMs have a high affinity with the surface membrane of living cells, which facilitate their absorption and toxicity. However, it is important to highlight that the toxicity of ENMs is conditioned by factors such as concentration, size, surface curvature, species-dependent toxicity, and surface load (Maharramov et al., 2019; Liu et al., 2020). One advantage of ENMs is the stimulation of plant development when they are added at low concentrations. For example, nanoparticles of Fe, Ag, Au, and CeO_2 contribute to the reduction of ROS and stimulate defense mechanisms by an adaptive response. Concerning negative effects, ENMs smaller than 5 nm are capable of penetrating the cell wall through its pores. After that, they travel to the endodermis through the apoplast to be transported through the xylem and then return to the roots through the phloem. Larger ENMs (20–160 nm) enter the plant through the cell wall, combined with other molecules or using symbiotic microorganisms. ENMs can reorganize inside plants. For example, they can reaggregate within the plant cell or undergo reduction reactions if they interact with terpenoids, flavones, ketones, aldehydes, amines, or carboxylic acids. These reactions are interesting because the toxicity will depend on the properties acquired by the new chemical species formed (Liu et al., 2020).

11.4 Bioavailability of ENMs in plant tissues

ENMs bioavailability has been defined as "a relative measure of that fraction of the total ambient ENMs that an organism takes up when encountering or processing environmental media, summed across all possible sources, including water and food" (Luoma & Rainbow, 2008). Bioavailability is important, mainly, for ENMs bioaccumulation (Luoma et al., 2014). Additionally, as mentioned above, the transformation of ENMs is crucial for their bioavailability, which can occur at different levels including the main source, the plant itself, and the soils. This makes biotransformation a complex phenomenon difficult to study and describe because, at the end, it is controlled by a wide range of factors. Details about different ENMs biotransformation are given in Table 11.2.

TABLE 11.2 ENMs transformations at the soil-plant system.

Level	Type of transformations	Transforming agents
Soil	Aggregation, dissolution, reduction, oxidation, sulfidation, chlorination, and phosphorylation	Organic matter, ionic strength, pH, redox potential, inorganic ligands, and microorganisms
Rhizosphere	Reduction, oxidation, complexation, dissolution	Root exudates, organic acids, phenols
Stem	Aggregation, oxidation, reduction, and dissolution	Plant sap components; reducing sugars, inorganic ions, amino acids, proteins, organic acids and hormones, free radicals
Leaf	Oxidation and reduction	Flavonoids, glucose and fructose, free radicals
Phyllosphere	Reduction, oxidation, complexation, dissolution	Water exudates and microorganisms

References: (Zhang et al., 2020; Jahan et al., 2017).

Soils can induce physical and chemical transformations including aggregation, dissolution, reduction, oxidation, sulfidation, chlorination, and phosphorylation. For example, organic matter can induce aggregation by forming a coating that changes the surface chemistry, in consequence affecting the subsequent behavior and fate of a particular ENM because it produces electrostatic stabilization, flocculation, and aggregation (Zhang et al., 2020). This has a direct consequence on their transport and stability. These processes are controlled by different environmental factors including pH, salt concentration, type of electrolyte, isoelectric point, organic acids, and humic acids. All these parameters produce a huge impact on the aggregation, dissolution, and further release of metal ions and in consequence their bioavailability (Jahan et al., 2017). However, the most important transformation occurs in the rhizosphere. The transformation of ENMs in this region is driven by the chemistry of the rhizosphere, mainly by the root exudates (secondary metabolites). For example, organic acids and pectin in root exudates may form stable complexes with heavy metals like Pb^{2+}, Cu^{2+}, and Cd^{2+} (Zhang et al., 2020; Jahan et al., 2017). On the other hand, excretion of organic acids in the rhizosphere may reduce pH, which increases metal dissolution and in consequence the bioavailability of metal-based ENMs. Besides the dissolution, reduction of ENMs occur in the rhizosphere and is produced by substances like phenols and ascorbic acid. For example, the reduction of CeO_2 results in the release of Ce^{3+}, while Cu-based ENMs release Cu^{2+}, Cu^{+}, and Cu^{0} species in presence of root exudates. However, the exact composition of root exudates is species-dependent. This kind of transformation can also occur in the phyllosphere (lead surface) through water exudates and microbial communities (Zhang et al., 2020).

Finally, transformations of ENMs within the plants have been poorly differentiated from those occurring prior to the uptake of ENM. In general, it is suggested that the transformation inside plants is ENM-dependent. However, plant sap components, such as reducing sugars and enzymes, play a key role in the transformation of ENMs. In xylem sap, aggregation can occur due to the presence of inorganic ions. On the other hand, phloem is richer than xylem in inorganic ions, sugar, amino acids, proteins, organic acids, and hormones. Thus, physical and chemical transformations may occur. While organic acids may influence ENMs dissolution, reducing sugars such as glucose and fructose may induce NM formation from ions. The last has a high dependence on flow, which is higher in xylem than in phloem. Transformation in plant leaves can be also influenced by different reducing agents including flavonoids, glucose, and fructose. Additionally, oxidation can be possible through free radicals generated by various physiological processes in plants. Further studies on ENMs transformation inside the cell and intracellular spaces are still not completely understood (Zhang et al., 2020). However, the latest reports proved that the transport and toxicological impact of ENMs are solely attributed to the dissolved ion concentration rather than the nanomaterial itself or its aggregated form (Jahan et al., 2017). This strongly supports the idea that bioavailability is a crucial parameter that needs to be measured when toxicity and exposure to ENMs are studied.

11.5 Conclusions

The behavior and toxicity of ENMs in plants and further exposed organisms can be affected by environmental factors such as high ionic strength, which leads to a rapid aggregation of the particles, or by pH that induces dissolution. Aggregation and dissolution (that is strongly related to ENMs bioavailability) play an important role in ENMs phytotoxicity. The interaction of ENMs with organic molecules such as humic acid, fulvic acid, citric acid, and extracellular polymeric substances, improve their stability and reduce their toxicity. In plants, the toxic effects of ENMs occur at the cellular and molecular level. Moreover, different physiological processes including photosynthesis can be interrupted. It is also known that oxidative stress affects unsaturated fatty acids of cell membranes through peroxidation processes.

It is important to note that some ENMs can damage genes and alter gene expression in plants, affecting different cellular functions. For example, different stages of cell division can be affected by the presence of ENMs. Additionally, ENMs' surface affinity for some cell membrane components facilitates their absorption and uptake and in consequence, their phytotoxicity. Transformations outside and inside plants control the ENMs bioavailability. In soils, bioavailability is controlled by environmental factors and soil biota interactions, among others. In plant tissues, these transformations are controlled by the type of ENMs and potential changes in their physicochemical properties in each compartment. ENMs transformations include aggregation, reduction, oxidation, solubilization, and complexation, among others. These transformations are influenced by several processes including chlorination, sulfidation, and phosphorylation, in addition to others.

It is well known that chemical speciation plays an important role in toxicity. Speciation may influence chemical changes in the ENMs surface and liberation of metal ions. In consequence, ENMs behavior and fate inside the plants are also influenced. Speciation is also related to bioavailability, which directly affects the way ENMs will be transported in living organisms as well as their toxicological characteristics. Thus, this parameter must be taken into account when performing research on nanophytotoxicity.

References

Aslani, F., Bagheri, S., Muhd Julkapli, N., Juraimi, A. S., Hashemi, F. S. G., & Baghdadi, A. (2014). Effects of engineered nanomaterials on plants growth: An overview. *Scientific World Journal, 2014*. https://doi.org/10.1155/2014/641759

Baker, T. J., Tyler, C. R., & Galloway, T. S. (2014). Impacts of metal and metal oxide nanoparticles on marine organisms. *Environmental Pollution, 186*, 257–271. https://doi.org/10.1016/j.envpol.2013.11.014

Batool, S. U., Javed, B., Zehra, S. S., Mashwani, Z.-R., Raja, N. I., Khan, T., ALHaithloul, H. A. S., Alghanem, S. M., Al-Mushhin, A. A. M., Hashem, M., & Alamri, S. (2021). Exogenous applications of bio-fabricated silver nanoparticles to improve biochemical, antioxidant, fatty acid and secondary metabolite contents of sunflower. *Nanomaterials, 11*(7), 1750. https://doi.org/10.3390/nano11071750

Boonyanitipong, P., Kositsup, B., Kumar, P., Bamah, S., & Dutta, J. (2011). Toxicit y of ZnO and TiO_2 nanoparticles on germinating rice seed *O,yza saliva L. International Journal of Bioscience, Biochemistry and Bioinformatics, 1*(4), 282–285. https://doi.org/10.7763/IJBBB.2011.V1.53

Dehkourdi, E. H., & Mosavi, M. (2013). Effect of anatase nanoparticles (TiO_2) on parsley seed germination (*Petroselinum crispum*) in vitro. *Biol Trace Elem Res, 155*(2), 283–286. https://doi.org/10.1007/s12011-013-9788-3

Dimkpa, C. O., McLean, J. E., Martineau, N., Britt, D. W., Haverkamp, R., & Anderson, A. J. (2013). Silver Nanoparticles Disrupt Wheat (*Triticum aestivum L.*) Growth in a Sand Matrix. *Environ Sci Technol, 47*(2), 1082–1090. https://doi.org/10.1021/es302973y

Dimkpa, C. O., Zeng, J., McLean, J. E., Britt, D. W., Zhan, J., & Anderson, A. J. (2012). Production of indole-3-acetic acid via the indole-3-acetamide pathway in the plant-beneficial bacterium pseudomonas chlororaphis O6 is inhibited by ZnO nanoparticles but enhanced by CuO nanoparticles. *Applied and Environmental Microbiology, 78*(5), 1404–1410. https://doi.org/10.1128/AEM.07424-11

Duncan, T. V. (2015). Release of engineered nanomaterials from polymer nanocomposites: The effect of matrix degradation. *ACS Applied Materials and Interfaces, 7*(1), 20–39. https://doi.org/10.1021/am5062757

Feizi, H., kamali, M., Jafari, L., & Moghaddam, P. R. (2013). Phytotoxicity and stimulatory impacts of nanosized and bulk titanium dioxide on fennel (*Foeniculum vulgare Mill*). *Chemosphere, 91*(4), 506–511. https://doi.org/10.1016/j.chemosphere.2012.12.012

Hernández-Viezcas, J. A., Castillo-Michel, H., Servin, A. D., Peralta-Videa, J. R., & Gardea-Torresdey, J. L. (2011). Spectroscopic verification of zinc absorption and distribution in the desert plant Prosopis juliflora-velutina (velvet mesquite) treated with ZnO nanoparticles. *Chem Eng J, 170*(1), 346–352. https://doi.org/10.1016/j.cej.2010.12.021

Jahan, S., Yusoff, I. B., Alias, Y. B., & Bakar, A. F. B. A. (2017). Reviews of the toxicity behavior of five potential engineered nanomaterials (ENMs) into the aquatic ecosystem. *Toxicology Reports, 4*, 211–220. https://doi.org/10.1016/j.toxrep.2017.04.001

Jasim, B., Thomas, R., Mathew, J., & Radhakrishnan, E. K. (2017). Plant growth and diosgenin enhancement effect of silver nanoparticles in Fenugreek (*Trigonella foenum-graecum* L.). *Saudi Pharmaceutical Journal, 25*(3), 443–447. https://doi.org/10.1016/j.jsps.2016.09.012

Jeevanandam, J., Barhoum, A., Chan, Y. S., Dufresne, A., & Danquah, M. K. (2018). Review on nanoparticles and nanostructured materials: History, sources, toxicity and regulations. *Beilstein Journal of Nanotechnology, 9*(1), 1050–1074. https://doi.org/10.3762/bjnano.9.98

Karunakaran, G., Suriyaprabha, R., Manivasakan, P., Yuvakkumar, R., Rajendran, V., Prabu, P., & Kannan, N. (2013). Effect of nanosilica and silicon sources on plant growth promoting rhizobacteria, soil nutrients and maize seed germination. *IET Nanobiotechnol, 7*(3), 70–77. https://doi.org/10.1049/iet-nbt.2012.0048

Keller, A. A., & Lazareva, A. (2014). Predicted releases of engineered nanomaterials: From global to regional to local. *Environmental Science & Technology Letters, 1*(1), 65–70. https://doi.org/10.1021/ez400106t

Kranjc, E., & Drobne, D. (2019). Nanomaterials in plants: A review of hazard and applications in the agri-food sector. *Nanomaterials, 9*, 1–33. https://doi.org/10.3390/nano9081094 (1094).

Kumari, M., Khan, S. S., Pakrashi, S., Mukherjee, A., & Chandraserkaran, N. (2011). Cytogenetic and genotoxic effects of zinc oxide nanoparticles on root cells of *Allium cepa*. *J Hazard Mater, 190*(1-3), 613–621. https://doi.org/10.1016/j.jhazmat.2011.03.095

Lanone, S., & Boczkowski, J. (2006). Biomedical applications and potential health risks of nanomaterials: Molecular mechanisms. *Current Molecular Medicine, 6*(6), 651–663. https://doi.org/10.2174/156652406778195026

Larue, C., Laurette, J., Herlin-Boime, N., Khodja, H., Fayard, B., Flank, A. M., Brisset, F., & Carriere, M. (2012). Accumulation, translocation and impact of TiO_2 nanoparticles in wheat (*Triticum aestivum spp.*): Influence of diameter and crystal phase. *Sci Total Environ, 431*(1), 197–208. https://doi.org/10.1016/j.scitotenv.2012.04.073

Lee, S., Kim, S., Kim, S., & Lee, I. (2012). Assessment of phytotoxicity of ZnO NPs on a medicinal plant, *Fagopyrum esculentum*. *Environ Sci Pollut Res Int, 20*(2), 848–854. https://doi.org/10.1007/s11356-012-1069-8

Liu, Y., Pan, B., Li, H., Lang, D., Zhao, Q., Zhang, D., Wu, M., Steinberg, C. E. W., & Xing, B. (2020). Can the properties of engineered nanoparticles be indicative of their functions and effects in plants? *Ecotoxicology and Environmental Safety, 205*. https://doi.org/10.1016/j.ecoenv.2020.111128

Lucas, W. J., & Lee, J. Y. (2004). Plasmodesmata as a supracellular control network in plants. *Nature Reviews Molecular Cell Biology, 5*(9), 712–726. https://doi.org/10.1038/nrm1470

Luoma, S. N., Khan, F. R., Croteau, M.-N., Lead, J. R., & Valsami-Jones, E. (2014). Chapter 5 - bioavailability and bioaccumulation of metal-based engineered nanomaterials in aquatic environments: Concepts and processes. In *Nanoscience and the environment* (Vol. 7, pp. 157–193). Elsevier. https://doi.org/10.1016/B978-0-08-099408-6.00005-0

Luoma, S. N., & Rainbow, P. S. (2008). Selenium: Dietary exposure, trophic transfer and food web effects. In S. N. Luoma, & P. S. Rainbow (Eds.), *Metal contamination in aquatic environments: Science and lateral management*. Cambridge University Press. https://doi.org/10.1111/j.1365-2427.2009.02238.x

Maharramov, A. M., Hasanova, U. A., Suleymanova, I. A., Osmanova, G. E., & Hajiyeva, N. E. (2019). The engineered nanoparticles in food chain: Potential toxicity and effects. *SN Applied Sciences, 1*(11). https://doi.org/10.1007/s42452-019-1412-5

Mahmoodzadeh, H., & Aghili, R. (2014). Effect on Germination and Early Growth Characteristics in Wheat Plants (*Triticumaestivum L.*) Seeds Exposed to TiO2 Nanoparticles. *Journal of Chemical Health Risks, 4*(1), 29–36. https://doi.org/10.22034/jchr.2018.544054

Maurer-Jones, M. A., Gunsolus, I. L., Murphy, C. J., & Haynes, C. L. (2013). Toxicity of engineered nanoparticles in the environment. *Analytical Chemistry, 85*(6), 3036–3049. https://doi.org/10.1021/ac303636s

Mendoza, R. P., & Brown, J. M. (2019). Engineered nanomaterials and oxidative stress: Current understanding and future challenges. *Current Opinion in Toxicology, 13*, 74–80. https://doi.org/10.1016/j.cotox.2018.09.001

Miralles, P., Church, T. L., & Harris, A. T. (2012). Toxicity, uptake, and translocation of engineered nanomaterials in vascular plants. *Environmental Science and Technology, 46*(17), 9224–9239. https://doi.org/10.1021/es202995d

Moll, J., Gogos, A., Bucheli, T. D., Widmer, F., & van der Heijden, M. G. A. (2016). Effect of nanoparticles on red clover and its symbiotic microorganisms. *Journal of Nanobiotechnology, 14*(1). https://doi.org/10.1186/s12951-016-0188-7

Montes, A., Bisson, M. A., Gardella, J. A., & Aga, D. S. (2017). Uptake and transformations of engineered nanomaterials: Critical responses observed in terrestrial plants and the model plant *Arabidopsis thaliana*. *Science of the Total Environment, 607–608*, 1497–1516. https://doi.org/10.1016/j.scitotenv.2017.06.190

Mukherjee, A., Peralta-Videa, J. R., Bandyopadhyay, S., Rico, M. C., Zhao, L., & Gardea-Torresdey, J. L. (2014). Physiological effects of nanoparticulate ZnO in green peas (*Pisum sativum L.*) cultivated in soil. *Metallomics, 6*(1), 132–138. https://doi.org/10.1039/c3mt00064h

Musante, C., & White, J. C. (2012). Toxicity of silver and copper to Cucurbita pepo: Differential effects of nano and bulk-size particles. *Environmental Toxicology, 27*(9), 510–517. https://doi.org/10.1002/tox.20667

Nair, P. M. G., & Chung, M. (2015). Physiological and molecular level studies on the toxicity of silver nanoparticles in germinating seedlings of mung bean (*Vigna radiata L.*). *Acta Physiologiae Plantarum, 37*(1), 1–11. https://doi.org/10.1007/s11738-014-1719-1

Nair, R., Varghese, S. H., Nair, B. G., Maekawa, T., Yoshida, Y., & Kumar, D. S. (2010). Nanoparticulate material delivery to plants. *Plant Science, 179*(3), 154–163. https://doi.org/10.1016/j.plantsci.2010.04.012

Navarro, E., Piccapietra, F., Wagner, B., Marconi, F., Kaegi, R., Odzak, N., Sigg, L., & Behra, R. (2008). Toxicity of silver nanoparticles to *Chlamydomonas reinhardtii*. *Environmental Science and Technology, 42*(23), 8959–8964. https://doi.org/10.1021/es801785m

Parveen, A., & Srinath, R. (2015). Effect of Nanosilver on Seed Germination and Seedling Growth in *Pennisetum glaucum*. *Journal of Cluster Science, 26*(1), 693–701. https://doi.org/10.1007/s10876-014-0728-y

Qi, M., Liu, Y., & Li, T. (2013). Nano-TiO(2) improve the photosynthesis of tomato leaves under mild heat stress. *Biol Trace Elem Res, 156*(1-3), 323–328. https://doi.org/10.1007/s12011-013-9833-2

Qiu, H., & Smolders, E. (2017). Nanospecific phytotoxicity of CuO nanoparticles in soils disappeared when bioavailability factors were considered. *Environmental Science and Technology, 51*(20), 11976–11985. https://doi.org/10.1021/acs.est.7b01892

Rai, R., Agrawal, M., & Agrawal, S. B. (2016). Impact of heavy metals on physiological processes of plants: With special reference to photosynthetic system. In *Plant responses to xenobiotics* (pp. 127–140). Singapore: Springer. https://doi.org/10.1007/978-981-10-2860-1_6

Reddy, P. V. L., Hernandez-Viezcas, J. A., Peralta-Videa, J. R., & Gardea-Torresdey, J. L. (2016). Lessons learned: Are engineered nanomaterials toxic to terrestrial plants? *Science of the Total Environment, 568*, 470–479. https://doi.org/10.1016/j.scitotenv.2016.06.042

Rico, C. M., Barrios, A. C., Tan, W., Rubenecia, R., Lee, S. C., Varela-Ramirez, A., Peralta-Videa, J. R., & Gardea-Torresdey, J. L. (2015). Physiological and biochemical response of soil-grown barley (*Hordeum vulgare L.*) to cerium oxide nanoparticles. *Environ Sci Pollut Res Int, 22*(14), 10551–10558. https://doi.org/10.1007/s11356-015-4243-y

Rico, C. M., Lee, S. C., Rubenecia, R., Mukherjee, A., Hong, J., Peralta-Videa, J. R., & Gardea-Torresdey, J. L. (2014). Cerium Oxide Nanoparticles Impact Yield and Modify Nutritional Parameters in Wheat (*Triticum aestivum L.*). *J Agric Food Chem, 62*(40), 9669–9675. https://doi.org/10.1021/jf503526r

Rico, C. M., Majumdar, S., Duarte-Gardea, M., Peralta-Videa, J. R., & Gardea-Torresdey, J. L. (2011). Interaction of nanoparticles with edible plants and their possible implications in the food chain. *Journal of Agricultural and Food Chemistry, 59*(8), 3485–3498. https://doi.org/10.1021/jf104517j

Rico, C. M., Morales, M. I., Barrios, A. C., McCreary, R., Hong, J., Lee, W. Y., Nunez, J., Peralta-Videa, J. R., & Gardea-Torresdey, J. L. (2013). Effect of Cerium Oxide Nanoparticles on the Quality of Rice (*Oryza sativa L.*) Grains. *J Agric Food Chem, 61*(47), 11278–11285. https://doi.org/10.1021/jf404046v

Santiago-Martín, A., Constanti, B., Guesdon, G., Kagambega, N., Raymond, S., & Cloutier, R. G. (2016). Bioavailability of engineered nanoparticles in soil systems. *Journal of Hazardous, Toxic, and Radioactive Waste, 20*(1), B4015001. https://doi.org/10.1061/(ASCE)HZ.2153-5515.0000263

Servin, A. D., Morales, M. I., Castillo-Michel, H., Hernández-Viezcas, J. A., Munoz, B., Zhao, L., Nunez, J. E., Peralta-Videa, J. R., & Gardea-Torresdey, J. L. (2013). Synchrotron Verification of TiO_2 Accumulation in Cucumber Fruit: A Possible Pathway of TiO_2 Nanoparticle Transfer from Soil into the Food Chain. *Environ Sci Technol, 47*(20), 11592–11598.

Servin, A. D., Pagano, L., Castillo-Michel, H., De la Torre-Roche, R., Hawthorne, J., Hernandez-Viezcas, J. A., Loredo-Portales, R., Majumdar, S., Gardea-Torresday, J., Dhankher, O. P., & White, J. C. (2017). Weathering in soil increases nanoparticle CuO bioaccumulation within a terrestrial food chain. *Null, 11*(1), 98–111. https://doi.org/10.1080/17435390.2016.1277274

Shaw, A. K., & Hossain, Z. (2013). Impact of nano-CuO stress on rice (*Oryza sativa L.*) seedlings. *Chemosphere, 93*(6), 906–915. https://doi.org/10.1016/j.chemosphere.2013.05.044

Shaymurat, T., Gu, J., Xu, G., Yang, Z., Zhao, Q., Liu, Y., & Liu, Y. (2012). Phytotoxic and genotoxic effects of ZnO nanoparticles on garlic (*Allium sativum* L.): A morphological study. *Nanotoxicology, 6*(3), 241–248. https://doi.org/10.3109/17435390.2011.570462

Siddiqui, M. H., & Al-Whaibi, M. H. (2014). Role of nano-SiO_2 in germination of tomato (*Lycopersicum esculentum* seeds Mill.). *Saudi Journal of Biological Sciences, 21*(1), 13–17. https://doi.org/10.1016/j.sjbs.2013.04.005

Song, U., Jun, H., Waldman, B., Roh, J., Kim, Y., Yi, J., & Lee, E. J. (2013). Functional analyses of nanoparticle toxicity: A comparative study of the effects of TiO_2 and Ag on tomatoes (*Lycopersicon esculentum*). *Ecotoxicology and Environmental Safety, 93*, 60–67. https://doi.org/10.1016/j.ecoenv.2013.03.033

Speranza, A., Crinelli, R., Scoccianti, V., Taddei, A. R., Iacobucci, M., Bhattacharya, P., & Ke, P. C. (2013). In vitro toxicity of silver nanoparticles to kiwifruit pollen exhibits peculiar traits beyond the cause of silver ion release. *Environmental Pollution, 179*, 258–267. https://doi.org/10.1016/j.envpol.2013.04.021

Suriyaprabha, R., Karunakaran, G., Yuvakkumar, R., Rajendran, V., & Kannan, N. (2012). Silica Nanoparticles for Increased Silica Availability in Maize (*Zea mays. L*) Seeds Under Hydroponic Conditions. *Current Nanoscience, 8*(6), 902–908. https://doi.org/10.2174/157341312803989033

Thuesombat, P., Hannongbua, S., Akasit, S., & Chadchawan, S. (2014). Effect of silver nanoparticles on rice (*Oryza sativa L.* cv. KDML 105) seed germination and seedling growth. *Ecotoxicology and Environmental Safety, 104*(1), 302–309. https://doi.org/10.1016/j.ecoenv.2014.03.022

Tripathi, D. K., Singh, S., Singh, S., Srivastava, P. K., Singh, V. P., Singh, S., Prasad, S. M., Singh, P. K., Dubey, N. K., Pandey, A. C., & Chauhan, D. K. (2017). Nitric oxide alleviates silver nanoparticles (AgNps)-induced phytotoxicity in *Pisum sativum* seedlings. *Plant Physiology and Biochemistry, 110*, 167–177. https://doi.org/10.1016/j.plaphy.2016.06.015

Unfried, K., Albrecht, C., Klotz, L. O., Von Mikecz, A., Grether-Beck, S., & Schins, R. P. F. (2007). Cellular responses to nanoparticles: Target structures and mechanisms. *Nanotoxicology, 1*(1), 52–71. https://doi.org/10.1080/00222930701314932

Van Aken, B. (2015). Gene expression changes in plants and microorganisms exposed to nanomaterials. *Current Opinion in Biotechnology, 33*, 206–219. https://doi.org/10.1016/j.copbio.2015.03.005

Wang, S., Kurepa, J., & Smalle, J. A. (2011). Ultra-small TiO2 nanoparticles disrupt microtubular networks in *Arabidopsis thaliana*. *Plant, Cell and Environment, 34*(5), 811–820. https://doi.org/10.1111/j.1365-3040.2011.02284.x

Winkler, D. A. (2016). Recent advances, and unresolved issues, in the application of computational modelling to the prediction of the biological effects of nanomaterials. *Toxicology and Applied Pharmacology, 299*, 96–100. https://doi.org/10.1016/j.taap.2015.12.016

Xiang, L., Zhao, H. M., Li, Y. W., Huang, X. P., Wu, X. L., Zhai, T., Yuan, Y., Cai, Q. Y., & Mo, C. H. (2015). Effects of the size and morphology of zinc oxide nanoparticles on the germination of Chinese cabbage seeds. *Environ Sci Pollut Res Int, 22*(14), 10452–10462. https://doi.org/10.1007/s11356-015-4172-9

Yu, J., Loh, X. J., Luo, Y., Ge, S., Fan, X., & Ruan, J. (2020). Insights into the epigenetic effects of nanomaterials on cells. *Biomaterials Science, 8*(3), 763–775. https://doi.org/10.1039/c9bm01526d

Zhang, P., Guo, Z., Zhang, Z., Fu, H., White, J. C., & Lynch, I. (2020). Nanomaterial transformation in the soil–plant system: Implications for food safety and application in agriculture. *Small, 16*(21).

Zhao, L., Peralta-Videa, J. R., Rico, C. M., Hernández-Viezcas, J. A., Sun, Y., Niu, G., Servin, A., Nunez, J. E., Duarrte-Gardea, M., & Gardea-Toresdey, J. L. (2014). CeO_2 and ZnO nanoparticles change the nutritional qualities of cucumber (*Cucumis sativus*). *J Agric Food Chem*, 2(62), 2752–2759. https://doi.org/10.1021/jf405476u

Zhu, M., Nie, G., Meng, H., Xia, T., Nel, A., & Zhao, Y. (2013). Physicochemical properties determine nanomaterial cellular uptake, transport, and fate. *Accounts of Chemical Research, 46*(3), 622–631. https://doi.org/10.1021/ar300031y

CHAPTER

12

Accumulation of engineered nanomaterials by plants: environmental implications

Nubia Zuverza-Mena[1], *Carlos Tamez*[1], *Jaya Borgatta*[2], *Tania M. Guardado-Alvarez*[1], and *Jason C. White*[1]

[1] The Connecticut Agricultural Experiment Station, New Haven, CT, United States; [2] Corteva Agriscience, Indianapolis, IN, United States

12.1 Introduction

Estimates indicate that more than 260,000 metric tons of engineered nanomaterials (ENMs) were generated in 2010, with most being metal-based ENMs (~60%–90%) going to landfills (Keller et al., 2013; Miseljic & Olsen, 2014). ENMs have been entering the air, soil, and water at increasing concentrations, potentially coming into contact with biota. Plants can also be intentionally exposed to ENMs from agrichemicals that contain these particles as part of fertilizer or pesticide formulations (Adeleye et al., 2014; Avellan et al., 2021; Kah et al., 2019). Notably, most work done with plants containing metal/metal oxide ENMs are found largely in research settings, where ENMs are studied as amendments to mitigate abiotic stress (Das & Das, 2019; Zhou et al., 2021), as a source of nutrients, pesticides, or fertilizers (Fatima et al., 2021; Kah et al., 2019), and to evaluate overall ENMs' hazard to plants, including toxicity (Zuverza-Mena et al., 2017). Indirectly, plants can also become contaminated with nanomaterials from phytoremediation practices. Although the main purpose may not be the removal of nanomaterials, ENMs are being researched as additives or synergistic materials to aid in the removal or to mitigate the toxicity of other contaminants (Song et al., 2019). These ENMs end up being absorbed and accumulated in plants. Moreover, plants can

Physicochemical Interactions of Engineered Nanoparticles and Plants
https://doi.org/10.1016/B978-0-323-90558-9.00004-8

also generate nanoparticles (NPs), both as a result of intentional biosynthesis approaches or an innate detoxification process where plants accumulate ions with subsequent reduction to NPs (Singh, Dutta, et al., 2018). The investigated inorganic nanomaterials fall into three categories:

Essential inorganic nutrients. Essential elements are crucial for plant growth and metabolism and include macronutrients, which are required at relatively high quantities (in addition to C, H, and O), including N, P, K, Ca, Mg, and S; and micronutrients that are required in lower amounts, including Fe, Mn, Zn, Cu, B, Mo, Cl, Ni, and Co (Lipa, 2008).

Nonessential elements. Those are elements that confer benefits for plant development, but are not required for successful completion of their life cycle, such as the metalloid Si (Pavlovic et al., 2021).

Others. These include nonessential metal elements (e.g., Li, Na), precious metals (e.g., Ag, Au), rare earth elements (e.g., Ce, La, Ti), and heavy metals (Pb, Cd) that can have positive or negative impacts on plants.

While some nanoscale inorganics are present as elemental metals, most are exposed to plants in the form of oxides, hydroxides, or as compound chemicals made out of more than two elements (e.g., CuO, CeO_2, $CoFe_2O_4$, $NiFe_2O_4$) (Sharma et al., 2017).

In addition to the direct effects of ENMs in plants, this chapter focuses on the transformations that metal and metal oxide ENMs undergo within plant tissues. The environmental impact associated with the decomposition of plants contaminated with ENMs will be mediated by the physical and chemical properties of the plant-derived ENMs, the amount of ENMs that are contained in the plants' biomass, and the characteristics of the recipient system.

12.2 Fate of plant-derived ENMs

In order to understand, predict, or speculate about the fate of plant-derived ENMs, it is critical to know how much and in what manner these ENMs are leaving a decaying plant and entering the next environmental compartment. It is challenging to assess the amount of nanoparticles that a plant can release and how these will be entering another matrix. The difficulty comes from the fact that ENMs that enter the plant can be biotransformed within the tissues and thus, the material leaving the plant may no longer be nanoparticles, but ions or other type of compounds. Moreover, plants have the ability to generate nanoparticles (e.g., converting ions into nanoparticles to mitigate their toxicity) and thus, plants could potentially have more nanomaterials to give than what they initially contained. The fate of ENMs is largely influenced by the amount released and the mechanisms, two key factors discussed in the following subsections.

12.2.1 Fate of plant-derived ENMs—release

Engineered nanomaterials can make their way into plants, and plants can also make nanoparticles within their tissues (Khan et al., 2019). We expect an increase of ENM-loaded plants from ENM-containing sludge applied to fields (The fate of engineered nanomaterials in Sewage treatment plants and agricultural applications, 2016), from common agricultural practices as commercial nanoagrichemicals (Kah et al., 2019), and from other less studied sources such as the aerial deposition and absorption of ultrafine particulate matter (PM) pollutants onto plant surfaces (Ha et al., 2021). However, we speculate that currently most of the plants containing ENMs are a result of research practices rather than other activities, and that the amounts of ENMs from ENM-loaded plants generated from research are minimal compared to the ENMs from consumer products entering the waste stream. Ultimately, the final disposition of the ENM-containing plants will dictate where the nanomaterials reside. The majority of ENMs in plants from research activities are feeding into landfills, followed by potentially significant discharges to soil, with less amounts to water and air. Notably, this assumption is necessary given that most of the research is being done under contained conditions in laboratories, greenhouses, growth rooms, and environmental chambers, with a small fraction of studies carried out in the field. The growth media are varied and include agar, hydroponics, or pots with soil or soillike systems (perlite, sand, vermiculite and their mixtures) (Kranjc & Drobne, 2019). Excluding hydroponic media, the plants and growing matrix will most likely be handled as common solid waste and be discarded in landfills or to waste treatment facilities. The management of ENMs-containing waste varies by country, as does their waste regulations and management activities. For example, in some countries, materials containing ENMs will go to common waste and directly to the landfill or soil, while in other places products with ENMs would be considered hazardous and will go to waste incineration facilities. There are numerous gaps regarding the fate of ENMs in waste streams (Part et al., 2018). A fraction of experimental plants are grown in the field, in which case the plants' biomass will most likely remain in the agricultural soil, making this soil a second fate compartment for plant-derived ENMs where availability could be a significant issue. As earthworms and other invertebrates consume decaying plant matter, ENM availability may be significantly higher than soil, leading to greater accumulation and potential toxicity/trophic transfer. Note that crops might also be directly impacted with nano-enabled agrichemicals. Not all marketable products will disclose if they contain ENMs as it is not a regulatory requirement in every country, which challenges our generalized assumptions because this could result in the underestimation of the amount of plant-derived ENMs entering the soil.

12.2.2 Fate of plant-derived ENMs—mechanisms

When a plant perishes, it becomes a food source for detritivores. The biodegradation process in plants is derived from the activity of microorganisms (mostly bacteria and fungi) breaking down organic matter (Focht, 2020). The minerals and any ENMs present are prone to erosion, dissolution, aggregation, and other transformations that depend on their surrounding environmental conditions, as well as the interactions with compounds present in the site where ENMs are deposited. To understand or predict what will happen when plant-derived ENMs are released, it is necessary to know the quantity of plant-derived ENMs entering a waste stream, landfill, or recipient soil system; the properties of those ENMs; and the subsequent receptors those ENMs will encounter. The impact of other elements that are present in the environmental compartment and the potential interactions with other living organisms (trophic transfer) are topics outside the scope of this chapter. Measuring the ENMs utilized in studies is feasible because the concentrations and details of exposure are often described. The initial ENM quantities can provide a point of reference, but measuring the actual amounts of ENMs that plants release can be difficult. Despite the extensive studies on ENMs and plants, very few report the physicochemical changes of the materials involved (Tolaymat et al., 2017). Table 12.1 summarizes the characteristics and transformations of ENMs within plants, such as size, crystallinity, elemental speciation, coordination, surface area, surface charge, particle coating, and shape; these topics are explored in the following sections.

12.3 Transformations in plants

The transformations that ENMs undergo within plants has to be considered when assessing the effects of ENMs-laden plants biodegradation. The form in which the ENMs leave the decaying biomass might be different from the pristine form in which ENMs were applied to plants. Many agricultural soils are deficient in vital nutrients and this state of nutrient deficiency leads to growers supplementing nutrient content through soil amendments. Current estimates on the efficiency of applied N and P fertilizers on cropland is less than 10% (Kah et al., 2019). Likewise, micronutrients including Cu, Zn, and Fe become unavailable after soil application. As overapplication of mineral fertilizers is an environmental concern (i.e., eutrophication), nanoscale-based amendments have

TABLE 12.1 ENMs transformations in plants.

	Form supplied						Form inside plant	
Rowhead	Nanoformulation	Size (nm)	Application mode	Amount added/conc. Used	Plant species	Tissue accumulated	Degrades? Transforms?	Citations
Ti	TiO_2	27 ± 4	Hydroponic	0–4000 mg/L	*Cucumis sativus*	Leaves, roots, trichomes, xylem, phloem	Not stated	Servin et al. (2012)
	TiO_2	27 ± 4	Soil	0–750 mg/L	*Cucumis sativus*	Fruit	Not stated	Servin et al. (2013)
Ce	CeO_2	6.9 ± 0.4	Soil	0, 500, 1000 mg/L	*Lactuca sativa*	Roots, shoots	Trophically transferred to snails, in digestive gut 85% is reduced, in lettuce only 4% is reduced	Ma et al. (2018)
	CeO_2	<10	Soil	1000 µg/g	*Cucurbita pepo*	Leaf, stem, flower, cricket	Not stated	Hawthorne et al. (2014)
	CeO_2	100	Suspension	20 mg/L applied weekly for 9 mo	Mesocosm, *Egeria densa*	3 ± 0.5 mg interior, 6 ± 3 mg exterior	10% reduced to Ce(III)	Geitner et al. (2018)

Continued

TABLE 12.1 ENMs transformations in plants.—cont'd

	Form supplied						Form inside plant	
Rowhead	**Nanoformulation**	**Size (nm)**	**Application mode**	**Amount added/conc. Used**	**Plant species**	**Tissue accumulated**	**Degrades? Transforms?**	**Citations**
	CeO_2	20	Suspension	20 mg/L applied weekly for 9 mo	Mesocosm, *Egeria densa*	23 ± 2 mg interior 9.5 ± 0.9 exterior	57% reduced to Ce(III)	Geitner et al. (2018)
	CeO_2	67 ± 8 × 8 ± 1	Soil	250 mg/kg	Barley, *Hordeum vulgare*	Adsorbed to roots	3%–12% overall reduced to Ce(III), hotspots at roots show up to 98% reduction at root	Rico et al. (2018)
	CeO_2	25.2 ± 2.3	Hydroponic	200, 2000 mg/L	*Cucumis sativus*	Xylem, phloem, shoots, roots	In root 15% reduced to Ce(III), in shoot 200 mg/L 18% reduced and 2000 mg/L 8% reduced. Needle like CePO4 observed.	Ma et al. (2017)

	CeO_2	25.2 ± 2.4	Hydroponic	2000 mg/L	*Cucumis sativus*	Shoot, root, intracellular space, root surface	Shoot in the presence of P: 81% CeO_2 and 21% $Ce(CH_3OO)_3$; root in the presence of P: 74.4% CeO_2, 2.1% $Ce(CH_3OO)_3$, 1.3% $Ce_2(C_2O_3)_3$, 20% $CePO_4$; shoot no P: 32.5% CeO_2, 67.5% $Ce(CH_3OO)_3$ root no P: 74.1% CeO_2, 4.5% $Ce_2(C_2O_4)_3$, 19.1% $CePO_4$	Rui et al. (2015)
	CeO_2	7.1 ± 0.4	Agar	0–2000 mg/L	*Lactuca sativa* Linn. var. angustata	Root	78.3% CeO_2, 21.7% $Ce(CH_3OO)_3$	Cui et al. (2014)
	CeO_2	9 and 23	Hydroponic	100 mg/L	*Cucurbita maxima, Triticum aestivum, Helianthus annuus*	Root, shoot	Not stated	Schwabe et al. (2015)

Continued

TABLE 12.1 ENMs transformations in plants.—cont'd

	Form supplied						Form inside plant	
Rowhead	**Nanoformulation**	**Size (nm)**	**Application mode**	**Amount added/conc. Used**	**Plant species**	**Tissue accumulated**	**Degrades? Transforms?**	**Citations**
La	La_2O_3	15–30	Nutrient medium	400 mg/L	*Pfaffia glomerata*	Root, shoot, veins	Not stated	Neves et al. (2019)
	La_2O_3	23 ± 4	Hydroponic	0.2–2000 mg/L	*Cucumis sativus*	Root epidermis, peripheral endodermis, leaf veins, vascular bundle	80% converted to La-carboxylate or $LaPO_4$	Ma et al. (2015)
	La_2O_3	10–100	Soil	500 mg/kg	*Lactuca sativa*	Root, shoot, trophic transfer	Not stated	De et al. (2015)
Yb	Yb_2O_3	12 ± 2	Hydroponic	0–2000 mg/L	*Cucumis sativus*	Roots, shoots, intercellular spaces, middle lamella,	$YbPO_4$	Zhang et al. (2012)
Ag	Ag triangular	47 ± 7	Seedling root/agar	0–100 μM	*Arabidopsis thaliana*	Not stated	Not stated	Shi et al. (2014)
	Ag spherical	8 ± 2	Seedling root/agar	0–100 μM	*Arabidopsis thaliana*	Not stated	Not stated	Shi et al. (2014)
	Ag decahedral	45 ± 5	Seedling root/agar	0–100 μM	*Arabidopsis thaliana*	Not stated	Not stated	Shi et al. (2014)

	Ag-PVP	50 ± 20	Microcosm	2 mg/L	*Potamogeton diversifolius, Egeria densa*	Not stated	No transformation	Bone et al. (2012)
	Ag-gum Arabic	12 ± 9	Microcosm	2 mg/L	*Potamogeton diversifolius, Egeria densa*	Not stated	Inorganic complexes and Ag-thiols	Bone et al. (2012)
	Ag	13 ± 3	Hydroponic	10 mg/L	*Landoltia punctata*	Root	Ag_2S Ag-thiol	Stegemeier et al. (2017)
	PVP-Ag	8 ± 2	Hydroponic	50 μg/L	*Oryza sativa*	Root, shoot	Ag^+, AgCl	Yang et al. (2019)
Au	Au	15, 25, 50	Hydroponic	15,25, 50 mg/L	*Populus deltoides × nigra*	Phloem and xylem cells, shoots and roots	No transformation	Zhai et al. (2014)
Cu	CuO	20–40	Seedling/ hydroponics	100 mg/L	*Zea mays*	Redistributed root-shoot-root	Reduces from Cu(II) to Cu(I)	Wang et al. (2012)
	CuO	31 ± 4	Soil	0–500 mg/kg	*Oryza sativa*	Root exodermis	Not stated	Peng et al. (2020)
	CuO	<50	Sand	500 mg/kg	*Triticum aestivum*	Shoots	CuO and Cu^{2+}	Dimkpa et al. (2013)
	CuO	43 ± 9	Soil	0–1000 mg/kg	*Oryza sativa*	Roots, stems, leaves, chaff, grain,	CuO, Cu-citrate, Cu-acetate. Cu (II) reduced to Cu(I) as Cu-cysteine, Cu-acetate, and Cu_2O	Peng et al. (2017)

Continued

TABLE 12.1 ENMs transformations in plants.—cont'd

	Form supplied						Form inside plant	
Rowhead	Nanoformulation	Size (nm)	Application mode	Amount added/conc. Used	Plant species	Tissue accumulated	Degrades? Transforms?	Citations
	CuO	30–50	Soil	0–10,000 mg/kg	*Hordeum sativa*	Roots, stems, grains	70% CuO, ~25% Cu_2O; stems and grains not specified	Burachevskaya et al. (2021)
	CuO	40	Hydroponic	100 mg/L	*Oryza sativa*	Root exodermis, endodermis; root, leaves	CuO; Cu-cysteine, Cu-citrate, Cu-phosphate. Reduction of Cu(II) to Cu(I).	Peng et al. (2015)
	Cu	25	Soil	0–1000 mg/kg	*Vigna unguiculata*	Roots	Cu(I) and Cu(II) (dominate)	Ogunkunle et al. (2019)
	CuO	43 ± 9	Hydroponic	0–2000 mg/kg	*Elsholtzia splendens*	Roots, stems, leaves	CuO (major); Cu-alginate, Cu-cysteine, Cu-oxalate (minor)	Shi et al. (2014)
	CuO	<50	Aquatic	0–200 mg/L	*Eichhornia crassipes*	Roots and leaves	CuO (major), Cu_2S, Cu-acetate, Cu-phosphate, Cu-citrate, Cu_2O	Zhao et al. (2017)

	CuO	23–37	Hydroponic	0–100 mg/L	*Oryza sativa*	Roots and shoots	CuO, Cu-acetate, Cu-oxalate. Cu(II) reduced to Cu(I) as Cu-acetate	Liu et al. (2019)
Fe	Fe_3O_4	20–2000	Growth medium	500 mg/L	*Cucurbita maxima*	Roots and leaves	Fe_3O_4 NPs	Zhu et al. (2008)
	ZVI	344–366	Hydroponic	0–1000 mg/L	*Typha latifolia*	No accumulation of NPs	Not stated	(Ma et al., 2013)
			Hydroponic	0–1000 mg/L	Populous deltoids x *P. nigra*	Roots	Not stated	
	ZVI	20–60	Germination	0–1000 mg/kg	*Oryza sativa*	Roots	Not stated	Wang et al. (2016)
Zn	ZnO	<100	Solution culture	25 mg/L	*Vigna unguiculata*	Roots; shoots	ZnO NPs; Zn-citrate, Zn-histidine, Zn-phytate	Wang et al. (2013)
			Soil	500 mg/kg	*Vigna unguiculata*	Roots, shoots	Zn-citrate, Zn-histidine, Zn-phytate	
	ZnO	30 ± 5	Hydroponic	0–100 mg/L	*Zea mays*	Roots, shoots	Zn-phosphate complexes	Lv et al. (2015)

Continued

TABLE 12.1 ENMs transformations in plants.—cont'd

	Form supplied						Form inside plant	
Rowhead	**Nanoformulation**	**Size (nm)**	**Application mode**	**Amount added/conc. Used**	**Plant species**	**Tissue accumulated**	**Degrades? Transforms?**	**Citations**
	ZnO	<100	Sand	500 mg/kg	*Triticum aestivum*	Shoots	Zn-phosphate	Dimkpa et al. (2013)
	ZnO	20 ± 5	Hydroponic	0–1000 mg/L	*Lolium perenne*	Root endodermis, and stele	ZnO NPs	Lin et al. (2008)
	ZnO	25 ± 4	Soil	0–500 mg/kg	*Oryza sativa*	Root cortex	Not stated	Peng et al. (2020)
	ZnO	10	Germination	0–4000 mg/L	*Prosopis juliflora-velutina*	Roots and leaves	Zn(II) ions	Hernandez-Viezcas et al. (2011)
	ZnO	20 ± 5	Soil	2 g/L	*Triticum aestivum*	Grain aleurone and crease	70%–80% Zn-phosphate complexes	Zhang et al. (2018)
Si	SiO_2	~50	Foliar	1500 mg/L	*Citrullus lanatus*	Leaf	Silicic acid	Kang et al. (2021)

begun to gain popularity to fill this deficiency and help reduce application rates (Fatima et al., 2021; Kah et al., 2019). Of these, the most reported with respect to transformation processes in plant tissues are Cu, Zn, and Fe. The translocation, accumulation, and final chemical form of the applied ENMs are highly dependent on a variety of factors including particle size, shape, and composition; plant species; growth conditions; and environmental factors. For instance, maize exposed to CuO ENMs via root tissues will transport CuO ENMs to aerial tissues through xylem, and conversely ENMs applied to shoot tissues will translocate to root tissues through phloem (Wang et al., 2012). It is likely that similar transport mechanisms occur in other species; evidence for the presence of CuO ENMs in plant tissues has been reported in wheat, barley, and rice (Burachevskaya et al., 2021; Dimkpa et al., 2013; Peng et al., 2015, 2017, 2020). In both barley and rice, the presence of CuO ENMs has been confirmed in root tissues using X-ray absorbance near-edge spectroscopy (XANES). When rice was grown hydroponically and exposed to CuO ENMs at 100 mg/L, ENMs were detected in plant roots, stems, and leaves. However, as plants matured the percent content of Cu in the leaf tissue in the form of CuO increased (Peng et al., 2015). Further, CuO nanoparticles were also found in the stem, leaf, and chaff of soil-grown rice plants exposed to 500 mg/kg (Peng et al., 2017). When Cu distribution in the rice grains was analyzed, grains from unexposed plants contained an even distribution of Cu across the endosperm, but rice grains from plants treated with CuO ENMs accumulated Cu in the pericarp and aleurone. Additional analyses on the uptake of CuO ENMs in rice root tissues found that treated plants showed an abundance of Cu in the root exodermis, while control plants contained Cu in the root stele and cortex. This suggests that the Casparian strip could limit infiltration of CuO ENMs into the interior of root tissues, although previous studies have shown the presence of CuO ENMs in aerial tissues (Peng et al., 2020).

However, metals applied as ENMs are not always translocated throughout plant tissues in the same form they were applied. Numerous studies have found that ENMs in plant tissues are transformed—either into ionic form, complexed with other ions or ligands, or as reduced species. To elucidate the chemical form of the translocated ENMs, plant tissues containing the metal ENMs can be compared to model compounds using advanced X-ray spectroscopy techniques. Studies conducted on sand-grown wheat revealed that shoots from nano-CuO treated plants contained CuO as the major form of Cu, with approximately 40% Cu^{2+} (Dimkpa et al., 2013). However, when plants were exposed to micron-sized CuO, both CuO and Cu^{2+} forms were present in nearly equal amounts. Usually, Cu^{2+} will associate with available molecules to form stable complexes. In the case of hydroponically grown rice, Cu has been identified in root and leaf tissues ionically bound to cysteine, citrate, or phosphate compounds (Peng et al., 2015). Further, reduced Cu_2O was detected in the root endodermis and central cylinder, as well as in mature leaves. These results were reinforced with data from soil grown rice (Peng et al., 2017). Plants treated with 500 mg/kg possessed roots, stems, and leaves where the majority of detected Cu was bound to citrate (Cu^{2+}) or reduced to Cu(I)-cystine or Cu(I)-acetate. Interestingly, only root tissues contained Cu in the form of Cu_2O. Regarding Cu content in the edible rice grains, 42% was in the form of Cu_2O, 25% as Cu-acetate, 22% as Cu(I)-acetate, while none was present in the form of CuO. There has been published evidence that coexposure of rice to both nano-CuO and As (ionic) can alter the rice's transformation behavior toward Cu. Rice germinated for 18 days exposed to nano-CuO at 100 mg/L then transferred to a soil-medium for 131 days contained nearly 40% Cu in the form of CuO and 60% in the form of Cu(I)-acetate (Liu et al., 2019). However, when seedlings were transferred to soil spiked with 10 mg/kg As, roots contained a different Cu content profile: 22% CuO, 23% Cu-oxalate, 26% Cu-acetate, and 29% Cu(I)-acetate. The change in Cu components in shoot tissues was significant, shoots not exposed to As contained 58% of the Cu as CuO, which decreased to 16% CuO in plants exposed to As. Additionally, Cu-acetate changed from 10% to 59%, and 32% to no detectable Cu(I)-acetate. Exposure of cowpea to Cu metal ENMs revealed the presence of both Cu(I) and Cu(II); however, Cu(II) was the dominant form present in root tissues (Ogunkunle et al., 2019). CuO was also the dominant form (71%) of Cu detected in barley roots, with a large fraction of Cu_2O (Burachevskaya et al., 2021). XANES spectra obtained from *Elsholtzia splendens* grown in solution containing CuO ENMs demonstrated that Cu translocated from root to stem and leaf tissues, having 76%, 35%, and 44% CuO, respectively (Shi et al., 2014). Additionally, transformed Cu in the form of Cu^{2+} was detected as Cu-alginate, Cu-cysteine, and Cu-oxalate, following previous reports that Cu(II) preferentially complexes with oxygen (Printz et al., 2016).

Further experiments conducted using tobacco bright yellow-2 (BY-2) cells and CuO ENMs demonstrated the accumulation of CuO, Cu_2O, and Cu_2S inside the cell wall (Dai et al., 2019). The authors hypothesized that CuO transformation begins upon CuO contact with the plant cell wall. Spectral analysis determined that the dominant species of Cu present in both the cell wall and within the cells were consistent with CuO and Cu_2O. Additional testing revealed that the mechanism of Cu(II) reduction to Cu(I) occurred through reactions with galacturonic acid containing sugars and with cysteine. However, it is important to note that this mechanism may be plant species dependent. The final fate for the majority of Cu ENMs released into the environment, via plant exposure or otherwise, will be soil and aquatic sediments. As they accumulate in the soil/sediment, aggregation dominates, trapping Cu in the soil to potentially toxic levels (Keller et al., 2017).

Typically, ionic Zn^{2+} is the form bioavailable to plants and the most common amendments to Zn-deficient soils are ZnO or $ZnSO_4$ (Fatima et al., 2021). However, these bulk-sized amendments are inefficient, leading to interest in nanoscale-based Zn fertilizers. Although pathways for the uptake of ZnO ENMs have been proposed (Singh, Singh, et al., 2018), it is important to note that ZnO nanoparticles are often quickly transformed to Zn ions (either in the soil or at the root surface). For example, roots from cowpea grown in a solution of 25 mg Zn/L were found to contain intact ZnO NPs, while the roots from plants grown in soil (500 mg/kg Zn) contained Zn bound to citrate (59%), histidine (25%), or phytate (16%) (Wang et al., 2013). Under both growth conditions, only organic-bound Zn was detected in shoot tissues. Similarly, hydroponically grown ryegrass treated with ZnO ENMs contained ZnO in the root endodermis and stele (Lin & Xing, 2008). Maize grown under hydroponic conditions accumulated 24% of Zn as ZnO in roots, with the remaining complexed with phosphates (Lv et al., 2015). Zinc phosphates were also the dominant form (70%–83%) in shoot tissues. Additionally, wheat grown in sand treated with ZnO ENMs showed that the accumulated Zn in shoot tissues was bound to phosphate (Dimkpa et al., 2013). Further, Zn speciation work with soybean showed that Zn is preferentially transported in leaves as Zn-citrate or Zn-malate, compared to the applied $ZnSO_4$, while there was no detectable transport of ZnO ENMs (Gomes et al., 2019).

There is less information regarding the transformation of iron-based NMs in plants. Pumpkin plants treated with Fe_3O_4 ENMs and grown to maturity were found to contain Fe_3O_4 inside root and leaf tissues (Zhu et al., 2008). Other studies investigating the uptake and translocation of nano zero valent iron (nZVI) found that iron accumulated in the roots of rice, broad leaf cattail, and poplar (Ma et al., 2013; Wang et al., 2016). Although the final oxidation state of Fe was not determined, scanning/transmission electron microscopy (STEM) imaging coupled with energy dispersive X-ray (EDX) spectroscopy confirmed that there was no internalization of nZVI in the roots of treated cattails or poplar (Ma et al., 2013). However, the formation of a coating of iron plaque over the roots of treated plants was observed, and this barrier was hypothesized to impede nZVI uptake. In rice plants, severe damage to the root cortex was found in plants treated with nZVI, but Fe was unable to cross the Casparian strip (Wang et al., 2016). Electron microscopy did provide evidence for the presence of agglomerates of Fe, but the exact form of the Fe could not be determined.

CeO_2 ENMs have been developed for applications in catalysts (Huang et al., 2005), polishing agents (Cheng et al., 2020), diesel fuel additives

(Cassee et al., 2011), and sensors (Khan et al., 2011). The widespread use of cerium-based ENMs has led to significant environmental exposure to ceria. CeO_2 ENMs are being supplied to agricultural systems to assess their toxicity as well as to investigate positive impacts such as benefits to plant health (Prakash et al., 2021). Nanoceria can disrupt the nutritional balance of plants (Barrios et al., 2017; Rico et al., 2017), but can also promote growth and enhance stress tolerance by neutralizing free radicals from reactive oxygen species (ROSs) (Collin et al., 2014; Prakash et al., 2021; Rico et al., 2018). Plants are also intentionally exposed to cerium-based nanomaterials as a useful tool for understanding uptake and transformations. Cerium is a rare-earth metal that is not commonly found in biological tissues, which helps to ascertain NP location using elemental analysis techniques. However, elemental analysis may be confounded for plants grown in soils in regions where Ce is present due to industrial activities or simply to native background levels (Thomas et al., 2014).

CeO_2 ENMs were added to an aquatic mesocosm to understand the distribution and chemical transformations of ENMs in the environment. Elemental analysis showed that smaller particles interacted significantly more than larger particles with the tissues of the aquatic plant *Egeria densa*. For the small particles (4 ± 1 nm), the internal Ce concentration was also twice as high as the externally adsorbed Ce concentration. The opposite trend was observed for the larger particles (190 ± 60 nm), with higher Ce concentrations in the external fraction. X-ray absorption near edge structure (XANES) measurements showed that the NP size also influenced the speciation of Ce inside plants; after 9 months in the mesocosm, small particles showed that 57% of the Ce identified in the *E. densa* tissue was reduced to Ce(III) while large NPs showed only a 10% reduction of Ce. The size of the NP influenced partitioning in the environment, with larger particles observed in deeper layers of the sediment and small particles observed in the upper floc and aquatic organisms (Geitner et al., 2018). A study was performed applying CeO_2 to the soil prior to lettuce growth. The majority of CeO_2 NPs accumulated in the roots, with a low translocation factor from the root to the foliage, $0.5 \pm 0.2\%$ for a 1000 mg/kg treatment. Trophic transfer of Ce was observed in snails fed on the lettuce and the majority of Ce was excreted. In plant tissues, Ce was mostly in the form of Ce(IV) consistent with CeO_2; however, in the snail digestive gut, 80%–90% of the cerium was reduced to Ce(III), the reduction of Ce in the snail gut was attributed to the presence of digestive enzymes and the low pH (Ma et al., 2018). The influence of particle size on trophic transfer was also evaluated for zucchini plants fed to crickets, and the crickets were subsequently fed to wolf spiders. The NPs led to significantly more Ce accumulation in the shoot, root, and flower when compared to bulk particles. Furthermore, the crickets fed with zucchini exposed to nanomaterials had excrement that contained more Ce than the crickets that were fed with plants exposed to larger particles. The findings of this study indicate that nanoscale materials can be more readily transferred through trophic levels than bulk particles of the same material (Hawthorne et al., 2014). In contrast, a study was performed to look at the trophic transfer of nanoscale and bulk CeO_2 particles from the soil to kidney bean (*Phaseolus vulgaris* var. red hawk) plants. The plants were then fed to beetles (*Epilachna varivestis*) that were consumed by an insect species (*Podisus maculiventris*). The study showed that uptake, translocation, and accumulation of particles did not depend on the particle size; however, nanoscale materials showed a greater biomagnification (Majumdar et al., 2016). For more insights into trophic transfer please refer to Zuverza-Mena and White (2018).

ENMs can be applied to the foliage or to the soil with a subsequent root exposure; in either pathway, NPs or the related ions could be internalized and subsequently transported to other plant tissues. For example, a split root hydroponic experiment with cucumber (*Cucumis sativus*) was performed to examine the transport of CeO_2 ENMs through the plant vasculature (Ma et al., 2017). The results showed that CeO_2 ENMs accumulated by the root can be transported to the shoot tissues, with subsequent transport back to the unexposed roots. However, very little Ce is released from the unexposed root to the hydroponic solution. Using transmission electron microscopy (TEM), NPs were identified in the shoots, unexposed roots, and xylem sap, indicating long range transport through the plant vasculature. Clusters of nano-CeO_2 were also observed in the exposed root tissue; interestingly, there were also electron dense needlelike NPs that were of different morphology from the initial material and were identified as $CePO_4$. However, in spite of this transformation of ceria at the exposed root, the translocated NPs were still the original CeO_2 (Ma et al., 2017). Similarly, another study with CeO_2 ENMs exposure to cucumber demonstrated the presence of needlelike $CePO_4$ in the root, as confirmed by scanning transmission X-ray microscopy (STXM) and XANES. Here, CeO_2 was applied in the presence or the absence of phosphate solution. For the CeO_2 ENMs applied without phosphate, the shoot Ce was 32.5% CeO_2 and 67.5% $Ce(CH_3COO)_3$, while the nano-CeO_2 applied in the presence of phosphate showed that 81.3% was CeO_2 and 21% $Ce(CH_3COO)_3$; these results indicate that the presence of phosphate reduces the transformation of the CeO_2 ENMs and may impact NP translocation. In the roots, in the presence and absence of phosphate, Ce was identified as ~74% CeO_2 and ~20%, with small amounts $Ce_2(C_2O_4)_2$ and $Ce(CH_3COO)_3$ (Rui et al., 2015). Similarly, asparagus lettuce (*Lactuca sativa* Linn. var. angustata) seeds were germinated in a solution containing CeO_2; about 20% of total Ce in the roots was transformed to Ce(III), and features in the XANES spectra matched $CePO_4$, $Ce(CH_3COO)_3$ and

$Ce_2(C_2O_4)_3$ (Cui et al., 2014). The uptake, translocation, and transformation of NPs can be influenced by processes such as dissolution and reprecipitation. To investigate transport processes, CeO_2NPs were alloyed with Zr and applied to hydroponically grown wheat, pumpkin, and sunflower plants. The comparison of the Zr-alloyed CeO_2 to CeO_2 NPs showed that transport from root to foliage is possible for both NPs and ions. In sunflower leaves, 60%–82% of the Ce was transported as ions, and 0%–86% of Ce was transported as ions in pumpkin leaves. When comparing the different plant species, wheat plants did not show detectable amounts of Ce in the shoot tissues, while the pumpkin and sunflower plants did transport CeO_2 (Schwabe et al., 2015). Existing literature shows that once in the plant, the CeO_2 NPs can be transformed to $CePO_4$, $Ce(CH_3COO)_3$ and $Ce_2(C_2O_4)_3$, corresponding to a reduction from Ce(IV) to Ce(III). These studies indicate that nanoceria readily transforms in plant systems and a variety of different forms of nanoscale and non-nanoscale forms of Ce may impact the environment upon plant disposal.

La_2O_3 and Yb_2O_3ENMs have been developed for use as optical materials (Méndez et al., 2010; Zheng et al., 2014), and as such, several studies involving plant exposure have been conducted. For example, the uptake of La_2O_3 ENMs by Spreng plants (*Pfaffia glomerata*) was investigated using laser ablation inductively coupled plasma mass spectroscopy (LA-ICP-MS) and micro X-ray fluorescence (μ-XRF) measurements. The findings demonstrate that La_2O_3 NPs were accumulated by the root and translocated to the plant stem and leaves. In the leaves, the main vein had a La concentration of 350 μg/g, the secondary veins had a La concentration of 100–200 μg/g, and the leaf blade had a La concentration of 50 μg/g (Neves et al., 2019). In cucumber (C. *sativa*), La_2O_3 ENMs were applied at concentrations between 0.2 and 2000 mg/L in a hydroponic study. The nanoparticles were taken up through the root and were accumulated in the vasculature and root epidermis. XANES measurements showed that 80% of the NPs were transformed to La-PO_4 and La-carboxylate (Ma et al., 2015). De La Torre Roche et al. (2015) found that La_2O_3 ENMs applied to the soil of lettuce (*Lactuca sativa*) plants at a concentration of 500 mg/kg showed the accumulation of La in the plant tissue with a concentration of 14.5 mg/kg in the leaves and 243 mg/kg in the roots. These results indicate that La_2O_3 can be taken up by plants and transferred through the food chain to consumers; however, whether the transferred La is in ionic, or nanoparticle form is not known (De La Torre Roche et al., 2015). Moreover, most of the La remained on the roots, which might result in a higher availability for organisms that reside in the soil once the plant decays.

Yb_2O_3 ENMs, bulk and $YbCl_3$ treatments were applied to foliage of cucumber plants at concentrations ranging from 0.32 to 2000 mg/L. The nanoparticle exposure lasted 2 weeks; TEM images showed the presence of dense deposits of needlelike NPs in the intracellular spaces and middle lamella, and importantly, the morphology of the deposits differed from the original morphology of the ENMs as shown in Fig. 12.1. Energy dispersive X-ray spectroscopy (EDS) analysis indicated the presence of Yb and P for both the particle and ionic treatments; subsequent near-edge X-ray absorption fine structure spectroscopy (NEXAFS) and STXM analysis were used to confirm that the compound observed was $YbPO_4$. The precipitates also appeared in the cytoplasm for the NP treatment indicating that the nanoparticles were transformed in all observed regions of the plant (Zhang et al., 2012).

Silver ENMs have been developed for antimicrobial, catalytic, optical applications; pesticides; and food packaging additives (Abou El-Nour et al., 2010; Donia & Carbone, 2019; Mishra & Singh, 2015) and have been used to create antimicrobial textiles, all of which may lead to widespread environmental exposure (Radetić, 2013). To understand uptake and transformations, Yang et al. (2019) performed a hydroponic study with rice that demonstrated that majority of Ag ENMs interacted directly with the root surface. Nanoscale Ag was identified in the root, while the majority of the fraction in the shoot was in the form of ionic Ag, indicating that the NPs dissolved or were otherwise oxidized in the plant tissues (Yang et al., 2019). In another hydroponic study, 10 mg/L of silver NPs interacted with duckweed (*Landoltia punctata*). XANES measurements showed that the majority of the silver was transformed to Ag_2S or Ag-thiols (Stegemeier et al., 2017). Ultimately, these studies indicate that Ag ENMs can be internalized in plants and rapidly transformed to ionic forms. Furthermore, complex biota can mitigate NP toxicity to other organisms.

Gold NPs have been developed for use in drug delivery; medical treatments; optical, catalysis, and plasmonic applications (Sardar et al., 2009). Furthermore, the low background concentrations of Au in the environment makes it a good candidate for elemental analysis and synchrotron spectroscopy. The uptake and translocation of Au NPs was studied with poplar (*Populus deltoides* × nigra, DN34) plants under hydroponic conditions at a concentration of 20 mg/L. TEM images showed the presence of Au NPs in both the roots and leaves of exposed plants. The particles were identified in both phloem and xylem tissues. In leaf tissues, the Au NPs were primarily observed in xylem cells and xylem cell walls. The Au NPs were more abundant in the root tissues than the leaves. TEM images demonstrated that the plasmodesmata are a primary route of NP transport. NPs were accumulated by the roots, penetrating the root caps and root hair cell walls, with subsequent translocation through the xylem and phloem. The NPs were observed in cell membranes, mitochondria, chloroplasts, and other organelles. Interestingly, the NPs were shown to

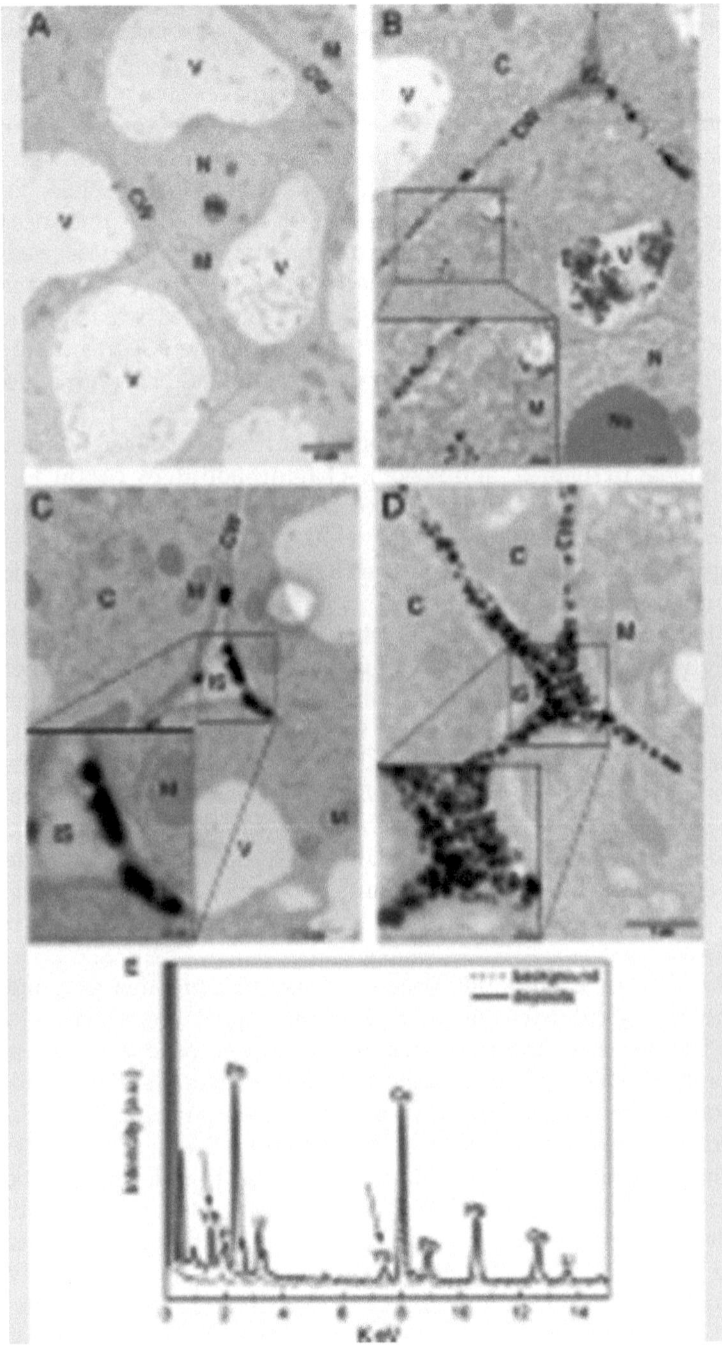

FIGURE 12.1 TEM cross-section images of cucumber root cells under the control (A), 2000 mg/L nm-Yb_2O_3 (B), 2000 mg/L bulk Yb_2O_3 (C), 200 mg/L $YbCl_3$ (D), and representative EDS spectra of background and dense deposits in TEM images (E). The insets are higher magnification of the rectangle areas. Cells walls (CW), nucleus (N) with nucleolus (Nu), intercellular space (IS), mitochondria (M), and vacuole (V). *From Zhang, P., Ma, Y., Zhang, Z., He, X., Guo, Z., Tai, R., Ding, Y., Zhao, Y. & Chai, Z. (2012). Comparative toxicity of nanoparticulate/bulk Yb_2O_3 and $YbCl_3$ to cucumber* (Cucumis sativus*).* Environmental Science and Technology, 46*(3), 1834–1841. https://doi.org/10.1021/es2027295. Copyright 2021 American Chemical Society.*

aggregate, which may have inhibited the transport through the plasmodesmata (Zhai et al., 2014).

The fate of nanomaterials may also be influenced by their crystal structure, which has been shown to influence the partitioning of TiO_2 NPs in plant tissues. TiO_2 ENMs have applications as a whitening powder pigment in food and paints, photovoltaic material, and catalyst (Dudefoi et al., 2017; Jagadale et al., 2008; Schlexer et al., 2018). TiO_2 can exist in several different crystallographic polymorphs: rutile, brookite, and anatase. In polymorphs, the stoichiometry is the same but the arrangement of atoms in space can differ, which can lead to crystal phases with different stabilities and surface energies (Reyes-Coronado et al., 2008). Servin et al. (2012) investigated TiO_2 ENMs from 50 to −4000 mg/L that were applied to roots in a hydroponic system; after 15 days, nanoparticles were identified in the xylem and phloem transport tissues of *C. sativus* using micro-XANES. A high Ti concentration was identified in the phloem walls. The Ti identified in the root cortex/epidermis and xylem was a mixture of crystal phases, composed of 81% anatase and 19% rutile TiO_2. The spectrum collected from the phloem possessed only the rutile phase, indicating the preferential transport of this crystal phase. Similarly, μ-XRF measurements of leaves showed rutile TiO_2 in the mesophyll tissues, epidermis, and trichomes (Servin et al., 2012). In addition, XANES was used to identify TiO_2 NPs in both rutile and anatase crystalline phases in cucumber fruit, indicating that the plants accumulated TiO_2ENMs from the soil and transported them to the edible tissues. This finding raises concern of potential ENM exposure via trophic transfer through the food chain. The European Food Safety Authority deemed the use of nano-TiO_2 food additives as not safe. Although toxic effects are not conclusive, nanotitania can bioaccumulate in the body and could potentially cause DNA damage ("Titanium dioxide: E171 no longer considered safe when used as a food additive," 2021).

Among the metalloids, silicon is the most common on earth and is the second most abundant element in the earth's crust after oxygen (Silicon, 2021). Silicon is not required for the survival of most plants, although many species experience direct benefit upon exposure, including better adaptation to different environmental abiotic and biotic stressors (Buchman et al., 2019; Kang et al., 2021). Additionally, silicon is known to be used by plants of the Equisetaceae family to strengthen their cell walls; this species cannot survive in a nutrient solution lacking silicon and as such, is considered essential for this family of plants (Epstein, 1994). Silicon content in plants was observed to vary from 0.1% to 10%, which was attributed to different mechanisms of silicon uptake in different plant species (Liang et al., 2007).

Silica nanoparticles are the main type of Si-based material that has been studied in plants and agriculture. Silica ENMs are biocompatible and have been widely used in fertilizers and as nanocarriers of chemicals. The toxicity of Si-nanoparticles in plants remains a topic of debate. Some studies show no toxicity of Si-NPs in plants while others show negative effects on plants (Bao-shan et al., 2004; Siddiqui et al., 2018). Porous silica ENMs are widely investigated because of their ease to be functionalized and thus, their versatility as smart sensors and chemical carriers (Guardado-Alvarez et al., 2016). Silicic acid (a compound of silicon, oxygen, and hydrogen) is the bioavailable form in which plants take up Si, and Si has demonstrated to have positive effects on plant health. Kang et al. (2021) exposed watermelon to silica ENMs of varying dissolution rates and demonstrated that the most rapidly dissolving NPs had the greatest impact on plants, although straight silicic acid had no benefit. Si-NPs overall seem to have a positive effect on plants; however, their size and other physicochemical properties have a large impact in their possible toxicity at higher doses. From various studies it has been observed that Si-based ENMs do transform within plants. Because silica nanoparticles tend to dissolve in aqueous environments (including inside plants), it is to be expected that after the plant dies silicon will eventually become incorporated into the soil in the form of silicic acid (Fig. 12.2). Depending on the surface area and/or other surface characteristics of the nanoparticles, the rate of silica nanoparticle dissolution will be different and therefore the transformation pre- and post-plant exposure. It is therefore unlikely that Si ENMs will be found in their pristine form once the ENM-exposed plant enters the soil or a waste stream compartment.

A multitude of factors can influence the bioavailability of translocated ENMs, which can therefore affect their biotransformation. One such event occurs during ENM surface interaction with plant biomolecules, such as xylem proteins (Borgatta et al., 2021; Wheeler et al., 2021). Interactions

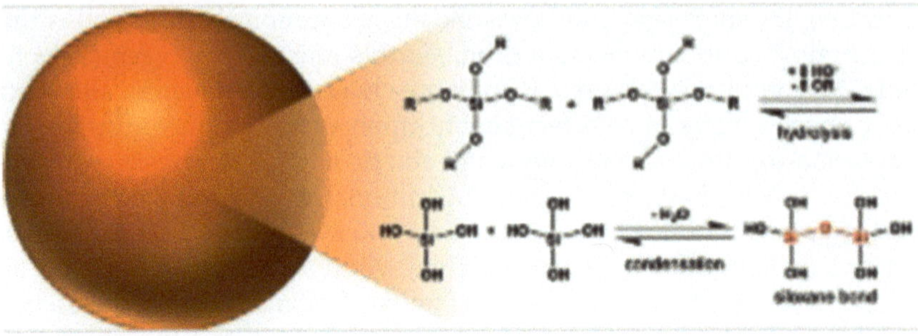

FIGURE 12.2 Hydrolysis condensation of silica nanoparticles. *Adapted with permission from Kang, H., Elmer, W., Shen, Y., Zuverza-Mena, N., Ma, C., Botella, P., White, J. C. & Haynes, C. L. (2021). Silica nanoparticle dissolution rate controls the suppression of fusarium wilt of watermelon* (Citrullus lanatus). Environmental Science and Technology. *https://doi.org/10.1021/acs.est.0c07126 Copyright 2021 American Chemical Society.*

with these proteins form a coating surrounding the ENM particles, often referred to as a corona. The resulting corona influences the behavior of the ENM by altering particle surface chemistry and morphology (Francia et al., 2019). The composition of the corona, as well as the dynamics associated with its formation, is dependent on the content of the biological matrix in which it formed, leading to unpredictable changes in dissolution and transformation processes.

12.4 Transformations in soil

Although there are many reports of ENMs biotransformation in studies conducted in a laboratory or greenhouse setting, some experiments have been carried out under field conditions. This being the case, there is certainly the opportunity for nanoparticle-containing plant material to enter the soil environment through loss of foliage, decaying plant material, seeds from previously exposed plants, or through root exudation. Though a portion of the applied ENMs will reside in plants in previously transformed states, there is the potential for any ENMs present to transform once released into soil. The inherent surface chemistry typical of soil organic matter (SOM) allows for NP stabilization via surface binding of metal-based NMs, either through ligand exchanges, hydrophilic interactions, or attachment to biomolecules (Amde et al., 2017). However, dissolution is a readily occurring transformation process for NPs in soil. For example, ZnO NMs amended into oxisol or ultisol soils at 500 mg/kg were no longer detectable as ZnO by XANES after an hour of incubation (Wang et al., 2013). In general, as NP size decreases, the dissolution rate increases; however, as noted above, high organic matter can have a stabilization effect, due to either chelation and/or surface coatings (Chen, 2018). The dissolution of Zn and Cu-based ENMs has been shown to increase in soil with low pH. Both ZnO and CuO NMs were shown to release up to 20 and 5.5 times more exchangeable metal ions in a soil of pH 4 as compared to 6.5 or 9, respectively. Furthermore, once present in the soil, ENMs can transform through oxidation or reduction reactions, as well as bind with available inorganic sulfides, phosphates, or other organic compounds (Chen, 2018). For example, synthetic root exudates have been shown to increase the dissolution of Cu NPs and facilitate the formation of Cu salts with organic compounds (Huang et al., 2017). Bone et al. (2012) conducted a microcosm study with silver nitrate, silver ENMs, polyvinylpyrrolidone (PVP) coated silver ENMs, and gum Arabic coated silver ENMs. ENMs were added at a concentration of 2 mg/L directly to the water column of a microcosm that contained sediment, aquatic plants (*Potamogeton diversifolius* and *Egeria densa*), *Daphnia magna*, and zebra fish (*Danio rerio*). The study was performed under the conditions of water only, water + sediment, water + plants, and water +

plants + sediment. For PVP-coated particles, the presence of sediment and plants reduced the toxicity of the NPs toward daphnia from 74 ± 16% for the water-only condition to 10 ± 4%. For zebra fish, the addition of plants significantly reduced toxicity from 83 ± 17% in water to 8 ± 5%. The sediment and plants remove the NPs from the water column, effectively reducing the available dose and decreasing overall toxicity. The authors found that plants and soil reduced the toxicity of gum Arabic coated NPs as well. ENMs accumulate in plants, soils and sediments, which reduces the concentration of ENMs in the water and limits the exposure to aquatic organisms. However, gum Arabic particles were more stable and had a higher toxicity when compared to the PVP-coated particles showing 100% toxicity in water; this is important as it suggests that NP properties such as coating can directly impact toxicity. The authors also hypothesized that the NPs stimulated exudate release from plants that limited toxicity (Bone et al., 2012).

Additionally, the weathering of ENMs after deposition into the soil can influence their transformation. For example, Cu found in secondary roots of lettuce shifted to 94% Cu_2O in plants grown in soil where CuO ENMs were weathered for 70 days, compared to 100% CuO in plants grown in unweathered soil (see Fig. 12.3) (Servin et al., 2017). Of course, these processes dramatically influence the initial availability of ENMs to plants when applied through the soil/roots, making tracking transformation reactions rather difficult.

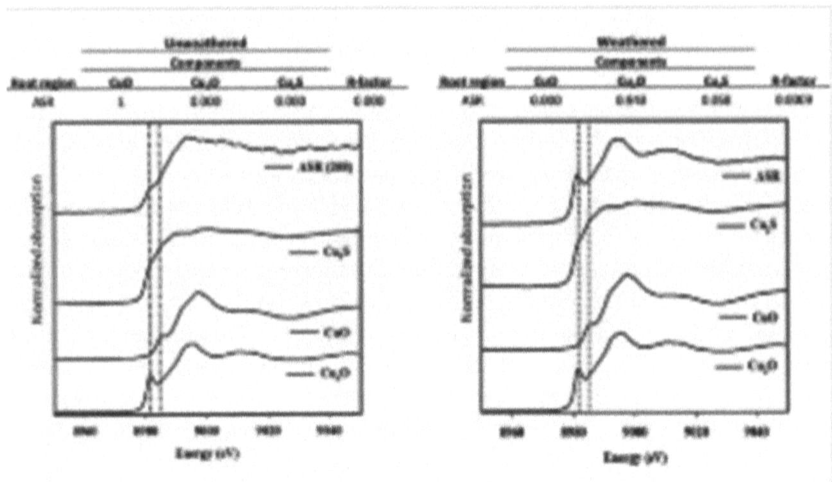

FIGURE 12.3 XANES spectra showing Cu species present aggregates of secondary roots (ASR) in lettuce, with or without soil weathering of CuO NMs for 70 days. *Graphic adapted from Servin, A. D., Pagano, L., Castillo-Michel, H., De la Torre-Roche, R., Hawthorne, J., Hernandez-Viezcas, J. A., Loredo-Portales, R., Majumdar, S., Gardea-Torresday, J., Dhankher, O. P., & White, J. C. (2017). Weathering in soil increases nanoparticle CuO bioaccumulation within a terrestrial food chain.* Nanotoxicology, 11*(1), 98–111. https://doi.org/10.1080/17435390.2016.1277274, reproduced with permission from Taylor & Francis.*

12.5 Conclusions

It is difficult to assess the implications of ENMs entering the environment after a contaminated plant drops, leaves or biodegrades. Many studies have shown the presence of ENMs in plants, but few studies discuss the entrance of nanomaterials into plants, their transformations, and the ENMs form/ fate at the end of a plant's life cycle. Prediction studies to assess risk become more challenging as advanced and custom-made ENMs are developed. The amount of ENMs entering the environment from plant sources is likely small, although not negligible, in comparison to other sources. For example, a 2013 study estimated that 40 tons of nano-TiO_2 entered the environment, with 90% coming from the paint industry. The sources for metal-based ENMs contaminated plants correspond to those used in agriculture as nutrients, fertilizers, pesticides, and to other ENMs that correspond to new formulations, novel materials in different industries that are external to agricultural settings such as paint additives in catalyst nanomaterials, including TiO_2 and CeO_2, among others. Important knowledge gaps that require future investigation are the implications of ENMs interacting with chemicals present in landfills, the actual amounts of ENMs-contaminated plants entering the environment, and other materials that are out of the scope of this chapter, such as natural and synthetic polymers, organic ENMs, and emerging ENMs that are yet to be developed.

References

Abou El-Nour, K. M. M., Eftaiha, A., Al-Warthan, A., & Ammar, R. A. A. (2010). Synthesis and applications of silver nanoparticles. *Arabian Journal of Chemistry, 3*(3), 135–140. https://doi.org/10.1016/j.arabjc.2010.04.008

Adeleye, A. S., Conway, J. R., Perez, T., Rutten, P., & Keller, A. A. (2014). Influence of extracellular polymeric substances on the long-term fate, dissolution, and speciation of copper-based nanoparticles. *Environmental Science and Technology, 48*(21), 12561–12568. https://doi.org/10.1021/es5033426

Amde, M., Liu, J.f., Tan, Z. Q., & Bekana, D. (2017). Transformation and bioavailability of metal oxide nanoparticles in aquatic and terrestrial environments. A review. *Environmental Pollution, 230*, 250–267. https://doi.org/10.1016/j.envpol.2017.06.064

Avellan, A., Yun, J., Morais, B. P., Clement, E. T., Rodrigues, S. M., & Lowry, G. V. (2021). Critical review: Role of inorganic nanoparticle properties on their foliar uptake and in planta translocation. *Environmental Science & Technology, 55*(20), 13417–13431. https://doi.org/10.1021/acs.est.1c00178

Bao-shan, L., shao-qi, D., Chun-hui, L., Li-jun, F., Shu-chun, Q., & Min, Y. (2004). Effect of TMS (nanostructured silicon dioxide) on growth of Changbai larch seedlings. *Journal of Forestry Research*, 138–140. https://doi.org/10.1007/bf02856749

Barrios, A. C., Medina-Velo, I. A., Zuverza-Mena, N., Dominguez, O. E., Peralta-Videa, J. R., & Gardea-Torresdey, J. L. (2017). Nutritional quality assessment of tomato fruits after exposure to uncoated and citric acid coated cerium oxide nanoparticles, bulk cerium oxide, cerium acetate and citric acid. *Plant Physiology and Biochemistry, 110*, 100–107. https://doi.org/10.1016/j.plaphy.2016.04.017

Bone, A. J., Colman, B. P., Gondikas, A. P., Newton, K. M., Harrold, K. H., Cory, R. M., Unrine, J. M., Klaine, S. J., Matson, C. W., & Di Giulio, R. T. (2012). Biotic and abiotic interactions in aquatic microcosms determine fate and toxicity of Ag nanoparticles: Part 2-toxicity and Ag speciation. *Environmental Science and Technology, 46*(13), 6925–6933. https://doi.org/10.1021/es204683m

Borgatta, J. R., Lochbaum, C. A., Elmer, W. H., White, J. C., Pedersen, J. A., & Hamers, R. J. (2021). Biomolecular corona formation on CuO nanoparticles in plant xylem fluid. *Environmental Science: Nano, 8*(4), 1067–1080. https://doi.org/10.1039/d1en00140j

Buchman, J. T., Elmer, W. H., Ma, C., Landy, K. M., White, J. C., & Haynes, C. L. (2019). Chitosan-coated mesoporous silica nanoparticle treatment of *Citrullus lanatus* (watermelon): Enhanced fungal disease suppression and modulated expression of stress-related genes. *ACS Sustainable Chemistry and Engineering, 7*(24), 19649–19659. https://doi.org/10.1021/acssuschemeng.9b04800

Burachevskaya, M., Minkina, T., Mandzhieva, S., Bauer, T., Nevidomskaya, D., Shuvaeva, V., Sushkova, S., Kizilkaya, R., Gülser, C., & Rajput, V. (2021). Transformation of copper oxide and copper oxide nanoparticles in the soil and their accumulation by *Hordeum sativum*. *Environmental Geochemistry and Health, 43*(4), 1655–1672. https://doi.org/10.1007/s10653-021-00857-7

Cassee, F. R., Van Balen, E. C., Singh, C., Green, D., Muijser, H., Weinstein, J., & Dreher, K. (2011). Exposure, health and ecological effects review of engineered nanoscale cerium and cerium oxide associated with its use as a fuel additive. *Critical Reviews in Toxicology, 41*(3), 213–229. https://doi.org/10.3109/10408444.2010.529105

Chen, H. (2018). Metal based nanoparticles in agricultural system: Behavior, transport, and interaction with plants. *Chemical Speciation and Bioavailability, 30*(1), 123–134. https://doi.org/10.1080/09542299.2018.1520050

Cheng, J., Huang, S., Li, Y., Wang, T., Xie, L., & Lu, X. (2020). RE (La, Nd and Yb) doped CeO_2 abrasive particles for chemical mechanical polishing of dielectric materials: Experimental and computational analysis. *Applied Surface Science, 506*, 144668. https://doi.org/10.1016/j.apsusc.2019.144668

Collin, B., Auffan, M., Johnson, A. C., Kaur, I., Keller, A. A., Lazareva, A., Lead, J. R., Ma, X., Merrifield, R. C., Svendsen, C., White, J. C., & Unrine, J. M. (2014). Environmental release, fate and ecotoxicological effects of manufactured ceria nanomaterials. *Environmental Science: Nano, 1*(6), 533–548. https://doi.org/10.1039/c4en00149d

Cui, D., Zhang, P., Ma, Y., He, X., Li, Y., Zhang, J., Zhao, Y., & Zhang, Z. (2014). Effect of cerium oxide nanoparticles on asparagus lettuce cultured in an agar medium. *Environmental Science: Nano, 1*(5), 459–465. https://doi.org/10.1039/C4EN00025K

Dai, Y., Zhao, J., Liu, X., Yu, X., Jiang, Z., Bu, Y., Xu, Z., Wang, Z., Zhu, X., & Xing, B. (2019). Transformation and species identification of CuO nanoparticles in plant cells (*Nicotiana tabacum*). *Environmental Science: Nano, 6*(9), 2724–2735. https://doi.org/10.1039/c9en00781d

Das, A., & Das, B. (2019). *Nanotechnology a potential tool to mitigate abiotic stress in crop plants.* IntechOpen. https://doi.org/10.5772/intechopen.83562

De La Torre Roche, R., Servin, A., Hawthorne, J., Xing, B., Newman, L. A., Ma, X., Chen, G., & White, J. C. (2015). Terrestrial trophic transfer of bulk and nanoparticle La_2O_3 does not depend on particle size. *Environmental Science and Technology, 49*(19), 11866–11874. https://doi.org/10.1021/acs.est.5b02583

Dimkpa, C. O., Latta, D. E., McLean, J. E., Britt, D. W., Boyanov, M. I., & Anderson, A. J. (2013). Fate of CuO and ZnO nano- and microparticles in the plant environment. *Environmental Science and Technology, 47*(9), 4734–4742. https://doi.org/10.1021/es304736y

Donia, D. T., & Carbone, M. (2019). Fate of the nanoparticles in environmental cycles. *International Journal of Environmental Science and Technology, 16*(1), 583–600. https://doi.org/10.1007/s13762-018-1960-z

Dudefoi, W., Moniz, K., Allen-Vercoe, E., Ropers, M. H., & Walker, V. K. (2017). Impact of food grade and nano-TiO_2 particles on a human intestinal community. *Food and Chemical Toxicology, 106*, 242–249. https://doi.org/10.1016/j.fct.2017.05.050

Epstein, E. (1994). The anomaly of silicon in plant biology. *Proceedings of the National Academy of Sciences of the United States of America, 91*(1), 11–17. https://doi.org/10.1073/pnas.91.1.11

Fatima, F., Hashim, A., & Anees, S. (2021). Efficacy of nanoparticles as nanofertilizer production: A review. *Environmental Science and Pollution Research, 28*(2), 1292–1303. https://doi.org/10.1007/s11356-020-11218-9

Focht, D. D. (2020). Biodegradation. In *AccessScience*. McGraw Hill. https://doi.org/10.1036/1097-8542.422025. Retrieved from: (August 24, 2022).

Francia, V., Yang, K., Deville, S., Reker-Smit, C., Nelissen, I., & Salvati, A. (2019). Corona composition can affect the mechanisms cells use to internalize nanoparticles. *ACS Nano, 13*(10), 11107–11121. https://doi.org/10.1021/acsnano.9b03824

Geitner, N. K., Cooper, J. L., Avellan, A., Castellon, B. T., Perrotta, B. G., Bossa, N., Simonin, M., Anderson, S. M., Inoue, S., Hochella, M. F., Richardson, C. J., Bernhardt, E. S., Lowry, G. V., Ferguson, P. L., Matson, C. W., King, R. S., Unrine, J. M., Wiesner, M. R., & Hsu-Kim, H. (2018). Size-based differential transport, uptake, and mass distribution of ceria (CeO_2) nanoparticles in wetland mesocosms. *Environmental Science and Technology, 52*(17), 9768–9776. https://doi.org/10.1021/acs.est.8b02040

Gomes, M. H. F., Machado, B. A., Rodrigues, E. S., Montanha, G. S., Rossi, M. L., Otto, R., Linhares, F. S., & Hudson, H. W. (2019). In vivo evaluation of Zn foliar uptake and transport in soybean using X-ray absorption and fluorescence spectroscopy. *Journal of Agricultural and Food Chemistry, 67*(44), 12172–12181. https://doi.org/10.1021/acs.jafc.9b04977

Guardado-Alvarez, T. M., Chen, W., Norton, A. E., Russell, M. M., Connick, W. B., & Zink, J. I. (2016). Analyte-responsive gated hollow mesoporous silica nanoparticles exhibiting inverse functionality and an and logic response. *Nanoscale, 8*(43), 18296–18300. https://doi.org/10.1039/c6nr01640e

Ha, N., Seo, E., Kim, S., & Lee, S. J. (2021). Adsorption of nanoparticles suspended in a drop on a leaf surface of *Perilla frutescens* and their infiltration through stomatal pathway. *Scientific Reports, 11*(1). https://doi.org/10.1038/s41598-021-91073-x

Hawthorne, J., De La Torre Roche, R., Xing, B., Newman, L. A., Ma, X., Majumdar, S., Gardea-Torresdey, J., & White, J. C. (2014). Particle-size dependent accumulation and trophic transfer of cerium oxide through a terrestrial food chain. *Environmental Science and Technology, 48*(22), 13102–13109. https://doi.org/10.1021/es503792f

Hernandez-Viezcas, J. A., Castillo-Michel, H., Servin, A. D., Peralta-Videa, J. R., & Gardea-Torresdey, J. L. (2011). Spectroscopic verification of zinc absorption and distribution in the desert plant Prosopis juliflora-velutina (velvet mesquite) treated with ZnO nanoparticles. *Chemical Engineering Journal, 170*, 346–352. https://doi.org/10.1016/j.cej.2010.12.021

Huang, P. X., Wu, F., Zhu, B. L., Gao, X. P., Zhu, H. Y., Yan, T. Y., Huang, W. P., Wu, S. H., & Song, D. Y. (2005). CeO_2 nanorods and gold nanocrystals supported on CeO_2 nanorods as catalyst. *Journal of Physical Chemistry B, 109*(41), 19169–19174. https://doi.org/10.1021/jp052978u

Huang, Y., Zhao, L., & Keller, A. A. (2017). Interactions, transformations, and bioavailability of nano-copper exposed to root exudates. *Environmental Science and Technology, 51*(17), 9774–9783. https://doi.org/10.1021/acs.est.7b02523

Jagadale, T. C., Takale, S. P., Sonawane, R. S., Joshi, H. M., Patil, S. I., Kale, B. B., & Ogale, S. B. (2008). N-doped TiO_2 nanoparticle based visible light photocatalyst by modified peroxide sol-gel method. *Journal of Physical Chemistry C, 112*(37), 14595–14602. https://doi.org/10.1021/jp803567f

Kah, M., Tufenkji, N., & White, J. C. (2019). Nano-enabled strategies to enhance crop nutrition and protection. *Nature Nanotechnology, 14*(6), 532–540. https://doi.org/10.1038/s41565-019-0439-5

Kang, H., Elmer, W., Shen, Y., Zuverza-Mena, N., Ma, C., Botella, P., White, J. C., & Haynes, C. L. (2021). Silica nanoparticle dissolution rate controls the suppression of fusarium wilt of watermelon (*Citrullus lanatus*). *Environmental Science and Technology.* https://doi.org/10.1021/acs.est.0c07126

Keller, A. A., McFerran, S., Lazareva, A., & Suh, S. (2013). Global life cycle releases of engineered nanomaterials. *Journal of Nanoparticle Research, 15*(6). https://doi.org/10.1007/s11051-013-1692-4

Keller, A. A., Adeleye, A. S., Conway, J. R., Garner, K. L., Zhao, L., Cherr, G. N., Hong, J., Gardea-Torresdey, J. L., Godwin, H. A., Hanna, S., Ji, Z., Kaweeteerawat, C., Lin, S., Lenihan, H. S., Miller, R. J., Nel, A. E., Peralta-Videa, J. R., Walker, S. L., Taylor, A. A., … Zuverza-Mena, N. (2017). Comparative environmental fate and toxicity of copper nanomaterials. *NanoImpact, 7*, 28–40. https://doi.org/10.1016/j.impact.2017.05.003

Khan, M. R., Adam, V., Rizvi, T. F., Zhang, B., Ahamad, F., Jośko, I., Zhu, Y., Yang, M., & Mao, C. (2019). Nanoparticle–plant interactions: Two-way traffic. *Small, 15*(37). https://doi.org/10.1002/smll.201901794

Khan, S. B., Faisal, M., Rahman, M. M., & Jamal, A. (2011). Exploration of CeO_2 nanoparticles as a chemi-sensor and photo-catalyst for environmental applications. *Science of the Total Environment, 409*(15), 2987–2992. https://doi.org/10.1016/j.scitotenv.2011.04.019

Kranjc, E., & Drobne, D. (2019). Nanomaterials in plants: A review of hazard and applications in the agri-food sector. *Nanomaterials, 9*(8). https://doi.org/10.3390/nano9081094

Liang, Y., Sun, W., Zhu, Y. G., & Christie, P. (2007). Mechanisms of silicon-mediated alleviation of abiotic stresses in higher plants: A review. *Environmental Pollution, 147*(2), 422–428. https://doi.org/10.1016/j.envpol.2006.06.008

Lin, D., & Xing, B. (2008). Root uptake and phytotoxicity of ZnO nanoparticles. *Environmental Science and Technology, 42*(15), 5580–5585. https://doi.org/10.1021/es800422x

Lipa, J. J. (2008). Mineral nutrition and plant disease. *Journal of Plant Protection Research, 48*(2), 246. https://doi.org/10.2478/v10045-008-0034-5

Liu, J., Wolfe, K., Potter, P. M., & Cobb, G. P. (2019). Distribution and speciation of copper and arsenic in rice plants (*Oryza sativa* japonica 'Koshihikari') treated with copper oxide nanoparticles and arsenic during a life cycle. *Environmental Science and Technology, 53*(9), 4988–4996. https://doi.org/10.1021/acs.est.9b00234

Lv, J., Zhang, S., Luo, L., Zhang, J., Yang, K., & Christie, P. (2015). Accumulation, speciation and uptake pathway of ZnO nanoparticles in maize. *Environmental Science: Nano, 2*(1), 68–77. https://doi.org/10.1039/c4en00064a

Ma, X., Gurung, A., & Deng, Y. (2013). Phytotoxicity and uptake of nanoscale zero-valent iron (nZVI) by two plant species. *Science of the Total Environment, 443*, 844–849. https://doi.org/10.1016/j.scitotenv.2012.11.073

Ma, Y., Zhang, P., Zhang, Z., He, X., Li, Y., Zhang, J., Zheng, L., Chu, S., Yang, K., Zhao, Y., & Chai, Z. (2015). Origin of the different phytotoxicity and biotransformation of cerium and lanthanum oxide nanoparticles in cucumber. *Nanotoxicology, 9*(2), 262–270. https://doi.org/10.3109/17435390.2014.921344

Ma, Y., He, X., Zhang, P., Zhang, Z., Ding, Y., Zhang, J., Wang, G., Xie, C., Luo, W., Zhang, J., Zheng, L., Chai, Z., & Yang, K. (2017). Xylem and phloem based transport of CeO_2 nanoparticles in hydroponic cucumber plants. *Environmental Science and Technology, 51*(9), 5215–5221. https://doi.org/10.1021/acs.est.6b05998

Ma, Y., Yao, Y., Yang, J., He, X., Ding, Y., Zhang, P., Zhang, J., Wang, G., Xie, C., Luo, W., Zhang, J., Zheng, L., Chai, Z., Zhao, Y., & Zhang, Z. (2018). Trophic transfer and transformation of CeO_2 nanoparticles along a terrestrial food chain: Influence of exposure routes. *Environmental Science and Technology, 52*(14), 7921–7927. https://doi.org/10.1021/acs.est.8b00596

Majumdar, S., Trujillo-Reyes, J., Hernandez-Viezcas, J. A., White, J. C., Peralta-Videa, J. R., & Gardea-Torresdey, J. L. (2016). Cerium biomagnification in a terrestrial food chain: Influence of particle size and growth stage. *Environmental Science and Technology, 50*(13), 6782–6792. https://doi.org/10.1021/acs.est.5b04784

Méndez, M., Carvajal, J. J., Cesteros, Y., Marsal, L. F., Martínez-Ferrero, E., Giguere, A., Drouin, D., Salagre, P., Formentín, P., Pallarès, J., Aguiló, M., & Díaz, F. (2010). Photoluminescence and cathodoluminescence of Eu:La_2O_3 nanoparticles synthesized by several methods. In , *Vol. 8. Physics procedia* (pp. 114–120). Elsevier B.V. https://doi.org/10.1016/j.phpro.2010.10.021

Miseljic, M., & Olsen, S. I. (2014). Life-cycle assessment of engineered nanomaterials: A literature review of assessment status. *Journal of Nanoparticle Research, 16*(6). https://doi.org/10.1007/s11051-014-2427-x

Mishra, S., & Singh, H. B. (2015). Biosynthesized silver nanoparticles as a nanoweapon against phytopathogens: Exploring their scope and potential in agriculture. *Applied Microbiology and Biotechnology, 99*(3), 1097–1107. https://doi.org/10.1007/s00253-014-6296-0

Neves, V. M., Heidrich, G. M., Rodrigues, E. S., Enders, M. S. P., Muller, E. I., Nicoloso, F. T., Carvalho, H. W. P. D., & Dressler, V. L. (2019). La_2O_3 nanoparticles: Study of uptake and distribution in *Pfaffia glomerata* (Spreng.) Pedersen by LA-ICP-MS and μ-XRF. *Environmental Science and Technology, 53*(18), 10827–10834. https://doi.org/10.1021/acs.est.9b02868

Ogunkunle, C. O., Bornmann, B., Wagner, R., Fatoba, P. O., Frahm, R., & Lützenkirchen-Hecht, D. (2019). Copper uptake, tissue partitioning and biotransformation evidence by XANES in cowpea (*Vigna unguiculata* L) grown in soil amended with nano-sized copper particles. *Environmental Nanotechnology, Monitoring and Management, 12*. https://doi.org/10.1016/j.enmm.2019.100231

Part, F., Berge, N., Baran, P., Stringfellow, A., Sun, W., Bartelt-Hunt, S., Mitrano, D., Li, L., Hennebert, P., Quicker, P., Bolyard, S. C., & Huber-Humer, M. (2018). A review of the fate of engineered nanomaterials in municipal solid waste streams. *Waste Management, 75*, 427–449. https://doi.org/10.1016/j.wasman.2018.02.012

Pavlovic, J., Kostic, L., Bosnic, P., Kirkby, E. A., & Nikolic, M. (2021). Interactions of silicon with essential and beneficial elements in plants. *Frontiers in Plant Science, 12*. https://doi.org/10.3389/fpls.2021.697592

Peng, C., Duan, D., Xu, C., Chen, Y., Sun, L., Zhang, H., Yuan, X., Zheng, L., Yang, Y., Yang, J., Zhen, X., Chen, Y., & Shi, J. (2015). Translocation and biotransformation of CuO nanoparticles in rice (*Oryza sativa* L.) plants. *Environmental Pollution, 197*, 99–107. https://doi.org/10.1016/j.envpol.2014.12.008

Peng, C., Xu, C., Liu, Q., Sun, L., Luo, Y., & Shi, J. (2017). Fate and transformation of CuO nanoparticles in the soil-rice system during the life cycle of rice plants. *Environmental Science and Technology, 51*(9), 4907–4917. https://doi.org/10.1021/acs.est.6b05882

Peng, C., Tong, H., Shen, C., Sun, L., Yuan, P., He, M., & Shi, J. (2020). Bioavailability and translocation of metal oxide nanoparticles in the soil-rice plant system. *Science of The Total Environment, 713*, 136662. https://doi.org/10.1016/j.scitotenv.2020.136662

Prakash, V., Peralta-Videa, J., Tripathi, D. K., Ma, X., & Sharma, S. (2021). Recent insights into the impact, fate and transport of cerium oxide nanoparticles in the plant-soil continuum. *Ecotoxicology and Environmental Safety, 221*. https://doi.org/10.1016/j.ecoenv.2021.112403

Printz, B., Lutts, S., Hausman, J. F., & Sergeant, K. (2016). Copper trafficking in plants and its implication on cell wall dynamics. *Frontiers in Plant Science, 7*. https://doi.org/10.3389/fpls.2016.00601

Radetić, M. (2013). Functionalization of textile materials with silver nanoparticles. *Journal of Materials Science, 48*(1), 95–107. https://doi.org/10.1007/s10853-012-6677-7

Reyes-Coronado, D., Rodríguez-Gattorno, G., Espinosa-Pesqueira, M. E., Cab, C., De Coss, R., & Oskam, G. (2008). Phase-pure TiO_2 nanoparticles: Anatase, brookite and rutile. *Nanotechnology, 19*(14). https://doi.org/10.1088/0957-4484/19/14/145605

Rui, Y., Zhang, P., Zhang, Y., Ma, Y., He, X., Gui, X., Li, Y., Zhang, J., Zheng, L., Chu, S., Guo, Z., Chai, Z., Zhao, Y., & Zhang, Z. (2015). Transformation of ceria nanoparticles in cucumber plants is influenced by phosphate. *Environmental Pollution, 198*, 8–14. https://doi.org/10.1016/j.envpol.2014.12.017

Rico, C. M., Johnson, M. G., Marcus, M. A., & Andersen, C. P. (2017). Intergenerational responses of wheat (*Triticum aestivum* L.) to cerium oxide nanoparticles exposure. *Environmental Science: Nano, 4*(3), 700–711. https://doi.org/10.1039/C7EN00057J

Rico, C. M., Johnson, M. G., & Marcus, M. A. (2018). Cerium oxide nanoparticles transformation at the root-soil interface of barley (*Hordeum vulgare* L.). *Environmental Science: Nano, 5*(8), 1807–1812. https://doi.org/10.1039/c8en00316e

Sardar, R., Funston, A. M., Mulvaney, P., & Murray, R. W. (2009). Gold nanoparticles: Past, present, and future. *Langmuir, 25*(24), 13840–13851. https://doi.org/10.1021/la9019475

Schlexer, P., Widmann, D., Behm, R. J., & Pacchioni, G. (2018). CO oxidation on a Au/TiO_2 nanoparticle catalyst via the Au-assisted Mars-van Krevelen mechanism. *ACS Catalysis, 8*(7), 6513–6525. https://doi.org/10.1021/acscatal.8b01751

Schwabe, F., Tanner, S., Schulin, R., Rotzetter, A., Stark, W., von Quadt, A., & Nowack, B. (2015). Dissolved cerium contributes to uptake of Ce in the presence of differently sized CeO_2 -nanoparticles by three crop plants. *Metallomics, 7*(3), 466–477. https://doi.org/10.1039/c4mt00343h

Servin, A. D., Castillo-Michel, H., Hernandez-Viezcas, J. A., Diaz, B. C., Peralta-Videa, J. R., & Gardea-Torresdey, J. L. (2012). Synchrotron micro-XRF and micro-XANES confirmation of the uptake and translocation of TiO_2 nanoparticles in cucumber (*Cucumis sativus*) plants. *Environmental Science and Technology, 46*(14), 7637–7643. https://doi.org/10.1021/es300955b

Servin, A., Maria, M. I., Castillo-Michel, H., Hernandez-Viezcas, J., Munoz, B., Zhao, L., Nunez, J. E., Peralta-Videa, J. R., & Gardea-Torresdey, J. L. (2013). Synchrotron verification of TiO_2 accumulation in cucumber fruit: A possible pathway of TiO_2 nanoparticle transfer from soil into the food chain. *Environmental Science and Technology, 47*(20), 11592–11598. https://doi.org/10.1021/es403368j

Servin, A. D., Pagano, L., Castillo-Michel, H., De la Torre-Roche, R., Hawthorne, J., Hernandez-Viezcas, J. A., Loredo-Portales, R., Majumdar, S., Gardea-Torresday, J. L., Dhankher, O. P., & White, J. W. (2017). Weathering in soil increases nanoparticle CuO bioaccumulation within a terrestrial food chain. *Nanotoxicology, 11*(1), 98–111. https://doi.org/10.1080/17435390.2016.1277274

Sharma, P., Sharma, A., Sharma, M., Bhalla, N., Estrela, P., Jain, A., Thakur, P., & Thakur, A. (2017). Nanomaterial fungicides: In vitro and in vivo antimycotic activity of cobalt and nickel nanoferrites on phytopathogenic fungi. *Global Challenges, 1*(9), 1700041. https://doi.org/10.1002/gch2.201700041

Shi, J., Peng, C., Yang, Y., Yang, J., Zhang, H., Yuan, X., Chen, Y., & Hu, T. (2014). Phytotoxicity and accumulation of copper oxide nanoparticles to the Cu-tolerant plant *Elsholtzia splendens*. *Nanotoxicology, 8*(2), 179–188. https://doi.org/10.3109/17435390.2013.766768

Siddiqui, H., Yusuf, M., Faraz, A., Faizan, M., Sami, F., & Hayat, S. (2018). 24-Epibrassinolide supplemented with silicon enhances the photosynthetic efficiency of *Brassica juncea* under salt stress. *South African Journal of Botany, 118*, 120–128. https://doi.org/10.1016/j.sajb.2018.07.009

Silicon. (2021). https://www.britannica.com/science/silicon.

Singh, J., Dutta, T., Kim, K. H., Rawat, M., Samddar, P., & Kumar, P. (2018). "Green" synthesis of metals and their oxide nanoparticles: Applications for environmental remediation. *Journal of Nanobiotechnology, 16*(1). https://doi.org/10.1186/s12951-018-0408-4

Singh, A., Singh, N. B., Afzal, S., Singh, T., & Hussain, I. (2018). Zinc oxide nanoparticles: A review of their biological synthesis, antimicrobial activity, uptake, translocation and biotransformation in plants. *Journal of Materials Science, 53*(1), 185–201. https://doi.org/10.1007/s10853-017-1544-1

Song, B., Xu, P., Chen, M., Tang, W., Zeng, G., Gong, J., Zhang, P., & Ye, S. (2019). Using nanomaterials to facilitate the phytoremediation of contaminated soil. *Critical Reviews in Environmental Science and Technology, 49*(9), 791–824. https://doi.org/10.1080/10643389.2018.1558891

Stegemeier, J. P., Colman, B. P., Schwab, F., Wiesner, M. R., & Lowry, G. V. (2017). Uptake and distribution of silver in the aquatic plant *Landoltia punctata* (duckweed) exposed to silver and silver sulfide nanoparticles. *Environmental Science and Technology, 51*(9), 4936–4943. https://doi.org/10.1021/acs.est.6b06491

The fate of engineered nanomaterials in sewage treatment plants and agricultural applications (pp. 75–88). (2016). Organisation for Economic Co-Operation and Development (OECD). https://doi.org/10.1787/9789264249752-8-en

Thomas, P. J., Carpenter, D., Boutin, C., & Allison, J. E. (2014). Rare earth elements (REEs): Effects on germination and growth of selected crop and native plant species. *Chemosphere, 96*, 57–66. https://doi.org/10.1016/j.chemosphere.2013.07.020

Titanium dioxide: E171 no longer considered safe when used as a food additive. (2021). European Food Safety Authority.

Tolaymat, T., El Badawy, A., Genaidy, A., Abdelraheem, W., & Sequeira, R. (2017). Analysis of metallic and metal oxide nanomaterial environmental emissions. *Journal of Cleaner Production, 143*, 401–412. https://doi.org/10.1016/j.jclepro.2016.12.094

Wang, Z., Xie, X., Zhao, J., Liu, X., Feng, W., White, J. C., & Xing, B. (2012). Xylem- and phloem-based transport of CuO nanoparticles in maize (*Zea mays* L.). *Environmental Science and Technology, 46*(8), 4434–4441. https://doi.org/10.1021/es204212z

Wang, P., Menzies, N. W., Lombi, E., McKenna, B. A., Johannessen, B., Glover, C. J., Kappen, P., & Kopittke, P. M. (2013). Fate of ZnO nanoparticles in soils and cowpea (*Vigna unguiculata*). *Environmental Science and Technology, 47*(23), 13822–13830. https://doi.org/10.1021/es403466p

Wang, J., Fang, Z., Cheng, W., Yan, X., Tsang, P. E., & Zhao, D. (2016). Higher concentrations of nanoscale zero-valent iron (nZVI) in soil induced rice chlorosis due to inhibited active iron transportation. *Environmental Pollution, 210*, 338–345. https://doi.org/10.1016/j.envpol.2016.01.028

Wheeler, K. E., Chetwynd, A. J., Fahy, K. M., Hong, B. S., Tochihuitl, J. A., Foster, L. A., & Lynch, I. (2021). Environmental dimensions of the protein corona. *Nature Nanotechnology, 16*(6), 617–629. https://doi.org/10.1038/s41565-021-00924-1

Yang, Q., Shan, W., Hu, L., Zhao, Y., Hou, Y., Yin, Y., Liang, Y., Wang, F., Cai, Y., Liu, J., & Jiang, G. (2019). Uptake and transformation of silver nanoparticles and ions by rice plants revealed by dual stable isotope tracing. *Environmental Science and Technology, 53*(2), 625–633. https://doi.org/10.1021/acs.est.8b02471

Zhai, G., Walters, K. S., Peate, D. W., Alvarez, P. J. J., & Schnoor, J. L. (2014). Transport of gold nanoparticles through plasmodesmata and precipitation of gold ions in woody poplar. *Environmental Science and Technology Letters, 1*(2), 146–151. https://doi.org/10.1021/ez400202b

Zhang, P., Ma, Y., Zhang, Z., He, X., Guo, Z., Tai, R., Ding, Y., Zhao, Y., & Chai, Z. (2012). Comparative toxicity of nanoparticulate/bulk Yb_2O_3 and $YbCl_3$ to cucumber (*Cucumis sativus*). *Environmental Science and Technology, 46*(3), 1834–1841. https://doi.org/10.1021/es2027295

Zhang, T., Sun, H., Lv, Z., Cui, L., Mao, H., & Kopittke, P. M. (2018). Using synchrotron-based approaches to examine the foliar application of $ZnSO_4$ and ZnO nanoparticles for field-grown winter wheat. *Journal of Agriculture and Food Chemistry, 66*, 2572–2579.

Zhao, J., Ren, W., Dai, Y., Liu, L., Wang, Z., Yu, X., Zhang, J., Wang, X., & Xing, B. (2017). Uptake, distribution, and transformation of CuO NPs in a floating plant Eichhornia crassipes and related stomatal responses. *Environmental Science and Technology, 51*(13), 7686–7695. https://doi.org/10.1021/acs.est.7b01602

Zheng, Y., Lü, Q., Wang, J., Zhang, G., Gao, Y., & Liu, Z. (2014). Emission behaviors of Yb_2O_3 nanoparticles pumped by 980nm laser at different power densities. *Optics & Laser Technology, 63*, 39–44. https://doi.org/10.1016/j.optlastec.2014.03.012

Zhou, P., Adeel, M., Shakoor, N., Guo, M., Hao, Y., Azeem, I., Li, M., Liu, M., & Rui, Y. (2021). Application of nanoparticles alleviates heavy metals stress and promotes plant growth: An overview. *Nanomaterials, 11*(1), 1–18. https://doi.org/10.3390/nano11010026

Zhu, H., Han, J., Xiao, J. Q., & Jin, Y. (2008). Uptake, translocation, and accumulation of manufactured iron oxide nanoparticles by pumpkin plants. *Journal of Environmental Monitoring, 10*(6), 713–717. https://doi.org/10.1039/b805998e

Zuverza-Mena, N., Martínez-Fernández, D., Du, W., Hernandez-Viezcas, J. A., Bonilla-Bird, N., López-Moreno, M. L., Komárek, M., Peralta-Videa, J. R., & Gardea-Torresdey, J. L. (2017). Exposure of engineered nanomaterials to plants: Insights into the physiological and biochemical responses-a review. *Plant Physiology and Biochemistry, 110*, 236–264. https://doi.org/10.1016/j.plaphy.2016.05.037

Zuverza-Mena, N., & White, J. C. (2018). Engineered nanomaterials in terrestrial systems: Interactions with co-existing contaminants and trophic transfer. *Current Opinion in Environmental Science and Health, 6*, 60–65. https://doi.org/10.1016/j.coesh.2018.08.003

Index

Note: 'Page numbers followed by "*f*" indicate figures and "*t*" indicate tables'.

Q

R

FSC
www.fsc.org
MIX
Papier | Fördert
gute Waldnutzung
FSC® C083411

Kolibri 360 GmbH
Ferdinand-Jühlke-Straße 7
99095 Erfurt, Deutschland
produktsicherheit@kolibri360.de